NANOBIOTECHNOLOGY FOR SENSING APPLICATIONS

From Lab to Field

NANOBIOTECHNOLOGY FOR SENSING APPLICATIONS

From Lab to Field

Edited by

Ajeet Kumar Kaushik, PhD
Chandra K. Dixit, PhD

AAP | APPLE ACADEMIC PRESS

Apple Academic Press Inc.
3333 Mistwell Crescent
Oakville, ON L6L 0A2
Canada

Apple Academic Press Inc.
9 Spinnaker Way
Waretown, NJ 08758
USA

Exclusive worldwide distribution by CRC Press, a member of Taylor & Francis Group

No claim to original U.S. Government works

Printed in the United States of America on acid-free paper

International Standard Book Number-13: 978-1-77188-328-3 (Hardcover)

International Standard Book Number-13: 978-1-77188-329-0 (eBook)

Library and Archives Canada Cataloguing in Publication

Nanobiotechnology for sensing applications : from lab to field / edited by Ajeet Kumar Kaushik, PhD, Chandra K. Dixit, PhD.

Includes bibliographical references and index.
Issued in print and electronic formats.
ISBN 978-1-77188-328-3 (hardcover).--ISBN 978-1-77188-329-0 (pdf)
1. Biosensors. 2. Nanobiotechnology. I. Kaushik, Akshat Kumar,author, editor II. Dixit, Chandra K., author, editor

R857.B54N35 2016 610.28 C2016-904015-1 C2016-904016-X

Library of Congress Cataloging-in-Publication Data

Names: Kaushik, Ajeet Kumar, editor. | Dixit, Chandra K., editor.
Title: Nanobiotechnology for sensing applications : from lab to field / editors, Ajeet Kumar Kaushik, Chandra K. Dixit.
Description: Toronto ; New Jersey : Apple Academic Press, 2017. | Includes bibliographical references and index.
Identifiers: LCCN 2016026205 (print) | LCCN 2016027155 (ebook) | ISBN 9781771883283 (hardcover : alk. paper) | ISBN 9781771883290 (ebook) | ISBN 9781771883290
Subjects: | MESH: Biosensing Techniques | Nanotechnology Classification: LCC R857.N34 (print) | LCC R857.N34 (ebook) | NLM QT 36.4 | DDC 610.28--dc23
LC record available at https://lccn.loc.gov/2016026205

Apple Academic Press also publishes its books in a variety of electronic formats. Some content that appears in print may not be available in electronic format. For information about Apple Academic Press products, visit our website at **www.appleacademicpress.com** and the CRC Press website at **www.crcpress.com**

ABOUT THE EDITORS

Ajeet Kumar Kaushik, PhD

Dr. Ajeet Kaushik received his PhD in chemistry (in the area of biosensors) in 2010 after working in collaboration with the National Physical Laboratory and Jamia Milia Islamia, New Delhi, India. Presently, he is a faculty member of the Center for Personalized NanoMedicine, Department of Immunology at Florida International University (FIU), Miami, Florida, United States. He is currently exploring smart electro-active magnetic nanocarriers for on-demand site-specific delivery and controlled release of therapeutics across the blood–brain barrier to prevent human immunodeficiency virus—a new approach toward neurological complications of acquired immune deficiency syndrome (neuroAIDS). He has worked as a postdoctoral research associate in the area of fabrication of biomedical (or biological) microelectromechanical systems (BioMEMS)-based electrochemical biosensors for biomarker detection using nanostructures at the Department of Electrical and Computer Engineering of the FIU. He has also worked at the Chungnam National University, Deajeon, South Korea, as a visiting researcher, and at the Dublin City University, Dublin, Ireland, as a research assistant. To date, he has publications of more than 70 international research papers to his credit in the area of nanostructured platforms for biosensor and drug delivery applications. His main research interest is in the area of fabrication of nanomedicines and wearable sensors for personalized health care.

Chandra K. Dixit, PhD

Chandra K. Dixit has received his master's degree from Chaudhary Charan Singh University, Meerut, Uttar Pradesh, India, in 2002. He was awarded a PhD in biotechnology by Dublin City University, Dublin, Ireland, in 2012, where he worked with Professor Richard O'Kennedy, Professor Colette McDonagh, and Professor Brian MacCraith as his PhD supervisors.

Currently, he is working as a postdoctoral associate in the areas of disease diagnostics, microfluidic systems for point-of-care, nanoimprinted polymers, and electrochemical biosensors at the Department of Chemistry, University of Connecticut, Storrs, United States. Due to his previous postdoctoral research, he has acquired significant experience in the areas of biosensors, microfluidics, and disease diagnostics. His future research focus is in the field of developing cheap diagnostics for field applications. He has publications of over 20 international research papers to his credit in the areas of biosensors, microfluidics, and disease diagnostics.

CONTENTS

LIST OF CONTRIBUTORS

Robson Benjamin Alby
Department of Physics, The American College, Madurai, Tamil Nadu, India

Md Azahar Ali
Department of Electrical and Computer Engineering, Iowa State University, Ames, USA

Shekhar Bhansali
Electrical and Computer Engineering Department, Florida International University, Miami, FL 33174, USA

Karunakarn Chandran
Biomedical Research Laboratory, Department of Chemistry, VHNSN College (Autonomous), Virudhunagar, Tamil Nadu, India

Chandra K. Dixit
Department of Chemistry, University of Connecticut, Storrs, CT, USA

Ciarán Ó'Fágáin
Department of Biotechnology, Dublin City University, Glasnevin, Dublin, Ireland

Divya Goyal
Department of Physics, Panjab University, Chandigarh, India

Ahmed Hasnain Jalal
Electrical and Computer Engineering Department, Florida International University, Miami, FL 33174, USA

Vairamani Kanagavel
University Science Instrumentation Centre, Madurai Kamaraj University, Madurai, Tamil Nadu, India

Ajeet Kaushik
Centre for Personalized Medicine, Institute of NeuroImmuno Pharmacology, Department of Immunology, Herbert Wertheim College of Medicine, Florida International University, Miami, FL, USA

Richard O'Kennedy
Department of Biotechnology, Dublin City University, Glasnevin, Dublin, Ireland; Biomedical Diagnostics Institute, Glasnevin, Dublin, Ireland

Raju Khan
Analytical Chemistry Division, CSIR-North East Institute of Science and Technology, Jorhat, Assam, India

Raj Kumar
University of Brescia, Italy

Rajesh Kumar
Department of Physics, Panjab University, Chandigarh, India

Chen-Zhong Li
Bielectronics and Nanobioengineering Laboratory, Department of Biomedical Engineering, Florida International University, Miami, USA

Pandiaraj Manickam
Bio-MEMS and Microsystems Laboratory, Department of Electrical and Computer Engineering, Florida International University, Miami, USA

Jairo Nelson
Bielectronics and Nanobioengineering Laboratory, Department of Biomedical Engineering, Florida International University, Miami, USA

Chandra Mouli Pandey
Biomedical Instrumentation Section, CSIR-National Physical Laboratory, New Delhi, India

Syed Khalid Pasha
Electrical and Computer Engineering Department, Florida International University, Miami, FL 33174, USA

Manoj Kumar Patel
Department of Chemistry, College of Arts and Sciences, Oklahoma State University, Stillwater, Oklahoma, USA

K. Kamil Reza
National Physical Laboratory, New Delhi, India

Pratik Kumar Shah
Bio-MEMS and Microsystems Laboratory, Department of Electrical and Computer Engineering, Florida International University, Miami, USA

Krati Sharma
Rare Genomic Institute, St Louis, Missouri, USA

Gaganpreet K. Sidhu
Department of Physics, Panjab University, Chandigarh, India

Aparajita Singh
Electrical and Computer Engineering Department, Florida International University, Miami, FL 33174, USA

Chandan Singh
Department of Science & Technology Centre on Biomolecular Electronics, Biomedical Instrumentation Section, CSIR-National Physical Laboratory, New Delhi, India

Pratima Solanki
Special Centre for Nanosciences, Jawaharlal Nehru University, New Delhi, India

Surendra K. Yadav
University of Rome, Tor Vergata, Italy

LIST OF ABBREVIATIONS

EDC	1-ethyl-3-[3-dimethylaminopropyl] carbodiimide
DCFH	2,7-dichlorodihydrofluorescein
cGMP	3′,5′-cyclic guanosine monophosphate
APTES	3(aminopropyl) triethoxysilane
8-OHdG	8-Hydroxy-2′-deoxyguanosine
ACS	acute coronary syndrome
AMI	acute myocardial infarction
ADC	analog-to-digital converter
APA	anodic porous alumina
Abs	antibodies
Ab	antibody
Ags	antigens
Apts	aptamers
RGD	arginine–glycine–aspartic acid
AA	ascorbic acid
AST	aspartate transaminase
AFM	atomic force microscopy
AIH	autoimmune hepatitis
BAMs	bioactive molecules
BioFET	bio-field effect transistor
bio-MEMS	bio-microelectromechanical systems
BBB	blood–brain barrier
BSA	bovine serum albumin
CdO	cadmium oxide
C-dots	carbon dots
CNT	carbon nanotube
CFM	carbon-fiber microelectrode
CMWNTs	carboxyl multi-wall carbon nanotubes
CEA	carcinoembryonic Ag
CEACAM1	carcinoembryonic antigen related cell adhesion molecule-1
cTnl	cardiac troponin
cTnI	cardiac troponin I
cTnT	cardiac troponin T
CVD	cardiovascular disease

CCP	cationic-conjugated polymer
COC	cell-on-chip
CDS	cellular dielectric spectroscopy
CVD	chemical vapor deposition
CBD	chitin-binding domain
CTS-AuNPs	chitosan stabilized-gold nanoparticles
CT	cholera toxin
ChO_x	cholesterol oxidase
GCMS	chromatography-mass spectrometry
CD	coeliac disease
cDNA	complementary DNA
CMOS	complementary metal–oxide–semiconductor
CPs	conducting polymers
CE	counter electrode
CRP	C-reactive protein
CK-MB	creatine kinase-MB
CMV	cucumber mosaic virus
CV	cyclic voltammetry
Cys	cystine
cyt *c*	cytochrome *c*
DAG	dialkylglycerol
DTPA	diethylenetriamine pentaacetic acid
DAC	digital-to-analog converter
DSP	dithiobis-succinimidyl propionate
DTT	dithiotreitol
DRPs	domain-recognition proteins
DA	dopamine
dsDNA	double-stranded DNA
DOX	doxorubicin
ELPs	elastin-like polypeptides
ECG	electrocardiogram
EC	electrochemical
ECD	electrochemical detection
EBL	electron beam lithography
ET	electron transfer
ELISA	enzyme-linked immunosorbent assay
EGFR	epidermal growth factor receptor
EFO	evanescent field opto chemical nanobiosensor
ENO	exhaled nitric oxide

Fc	ferrocene
FET	field effect transistor
FBAR	film bulk acoustic resonator
FRET	fluorescence resonance energy transfer
FIB	focused ion beam
FTL	folate-targeted stealth
FDA	Food and Drug Administration
FDI	foreign direct investment
Fab	fragment Ag-binding
Fab	fragment antigen-binding
Fc	fragment crystalline
Fc	fragment crystallizable
FBI	functional biointerlayer
GCE	glassy carbon electrode
GO_x	glucose oxidase
GSH	glutathione
AuNPs	gold nanoparticles
Au NPs	gold NPs
GPCRs	G-protein-coupled receptors
GO	graphene oxide
GUI	graphical user interface
GdnHCl	guanidine hydrochloride
HA	hemagglutinin
HCV	hepatitis C virus
HCVcoreAg	hepatitis C virus core antigen
HPLC	high-performance liquid chromatography
HRP	horse radish peroxidase
hAGT	human O^6-alkylguanine transferase
HIV	human immunodeficiency virus
HID	human interface device
HLA	human leukocyte antigen
ITO	indium–tin oxide
IGFBP7	insulin-like growth factor-binding protein 7
IC	integrated circuit
ICs	integrated circuits
IDA	interdigitated array
IgG	immunoglobulin G
JEV	Japanese encephalitis virus
LabVIEW	Laboratory Virtual Instrument Engineering Workbench
LDH	lactate dehydrogenase

LB	Langmuir–Blodgett
LIF	laser-induced fluorescence
LFIA	lateral flow immunoassay
LAM	lipoarabinomannan
LSPR	local surface plasmon resonance
MZI	Mach–Zehnder interferometer
MRI	magnetic resonance imaging
MI	magneto-impedanc
MR	magneto-resistance
mRNA	messenger RNA
MNPs	metal nanoparticles
MNPs	metal NPs
MO–NPs	metal–oxide NPs
MNO_x	metal–oxides nanostructures
MOSFET	metal–oxide–semiconductor field-effect transistor
MEMS	microelectromechanical systems
miRNA	micro RNA
miniEC	miniaturized electrochemical
MBs	molecular beacons
McAbs	monoclonal Abs
MAb	monoclonal antibody
MWCNTs	multi-walled CNTs
MI	myocardial injury
NAG	N-Acetylglucosamine
NCs	nanocrystals
NEMS	nanoelectromechanical systems
NMs	nanomaterials
NPs	nanoparticles
NRs	nanorods
NMs	nanostructured material
NMO	nanostructured metal–oxide
NMEs	nanostructured microelectrodes
NW	nanowire
NNI	National Nanotechnology Initiative
NIR	near-infrared
NHS	N-Hydroxysuccinimide
NTE	neuropathy target esterase
NiO	nickel oxide
nAChRs	nicotine acetylcholine receptors

NO	nitric oxide
NTA	nitrilotriacetic acid
NA	nucleic acid
1D	one-dimensional
OTFT	organic thin film transistor
OP	organophosphorus
p-APP	*p*-aminophenyl phosphate
p-Bpa	p-benzoyl-L-phenylalanine
PPy	pancreatic polypeptide
PEDOT: PSS	(poly (3,4-ethylenedioxythiophene) polystyrene sulfonate semiconductor)
PNA	peptide nucleic acid
PBS	phosphate-buffered saline
PPFs	plasma-polymerized films
PDGF	platelet-derived growth factor
POC	point-of-care
POCT	point-of-care testing
PDMS	poly(dimethylsiloxane)
PLGA	poly(lactic-co-glycolic acid)
PMMA	poly(methyl methacrylate)
PANI	polyaniline
PDMS	polydimethylsiloxane
PEG	polyethylene glycol
PET	polyethylene terephthalate
his	polyhistidine
PCR	polymerase chain reaction
PMMA	polymethyl methacrylate
PPy	polypyrrole
pDNA	probe DNA
PSA	prostate specific antigen
PSMA	prostate-specific membrane antigen
QD	quantum dot
QCM	quartz crystal microbalance
RFID	radio-frequency identification
ROS	reactive oxygen species
rGO	reduced graphene oxide
RGO	reduced graphene oxide
RE	reference electrode
RI	refractive index
RA	rheumatoid arthritis

RNA	ribonucleic acid
SEM	scanning electron microscope
SAMs	self-assembled monolayers
STIs	sexually transmitted infections
ScFv	short-chain fragment variable
AgNPs	silver nanoparticles
SPECT	single-photon emission computed tomography
ssDNA	single-stranded DNA
SWNTs	single-wall nanotubes
SWCNTs	single-walled carbon nanotubes
SWV	square-wave voltammetry
SDF-1	stromal cell derived factor-1
SAW	surface acoustic wave
SERS	surface enhanced Raman spectroscopy
SPR	surface plasmon resonance
SVR	surface-to-volume ratio
SLE	systemic lupus erythematosus
tDNA	target DNA
3D	three-dimensional
TIR	total internal reflection
TGase	transglutaminase
TMDC	transition metal dichalcogenide
TCEP	Tris(2-carboxyethyl)-phosphine
TB	tuberculosis
2D	two-dimensional
USB	universal serial bus
UA	uric acid
VLS	vapor–liquid–solid
VEGF	vascular endothelial growth factor
VI	virtual instrument
WE	working electrode

LIST OF SYMBOLS

ε	molar absorptivity
ρ	material resistivity
Δf	the resonant peak frequency at full width at half maximum
$\Delta\sigma$	is the difference in the top and bottom surface stresses
α	absorption coefficient; excess polarizability of the particle
λ	wavelength
ρ	material resistivity
χ	susceptibility
ϑ	Poisson's ratio
σ	conductivity
ε	capacitance under strain
ω	angular frequency

PREFACE

The introduction of nanotechnology in biosensing improves everyday life-styles via specially personalized health care, diagnostics, and monitoring. Efforts are being made continuously to modify form and factors of analytical sensing devices for on-site monitoring of environment and health. This book is an attempt to describe the importance of nanobiosensors and their use in developing point-of-care (POC) systems. In this book, basic concepts are presented pertaining to the nanobiosensor fabrication, developments in the field of smart nanomaterials, nano-enabling technologies, micronano hybrid platforms, and their applications in health care.

Therefore, chapters are broadly divided into three categories: the first section expands from Chapters 1 to 5 which describe the basic concepts of nanobiotechnology, nanosurfaces, their biofunctionalization, and involved electronics. The second section (from Chapters 6 to 8) is dedicated to the designing and development of *in vitro* sensors, and in the third section (from Chapters 9 to 11) *in vivo* applications of the nanobiosensors are presented.

The fundamentals of nanotechnology for biosensor development are explained in Chapter 1 with emphasis on the overall analytical performance of the developed nanobiosensing systems, such as sensitivity, detection limit, and stability. In Chapter 2, the methods to prepare nanosurfaces and their characterization are described. Details pertaining to the tuning up of optical, electrical, and molecular properties of a nanomaterial are also discussed. Nanobiosensors, in particular those for immunodiagnostics and bioassay development, are exhaustively covered in Chapter 3. Further, in Chapter 3, grafting functionalities to the nanosurfaces and using those for biomolecule immobilization with special reference to cross-linking tools and site-specific orientation-based methods are also included.

Thin films and two-dimensional nanomaterials are revolutionizing the development of efficient biosensors. These methods are described in Chapter 4. Circuit designing and transduction techniques are very important for fabricating smart sensors. Fundamentals and recent prospects pertaining to the development of integrated electronics of analytical transducers and signal processing are presented in Chapter 5.

Several important classes of *in vitro* nanobiosensing platforms and tech-niques for the development of enzymatic, geno, and immune nanobiosensors are discussed in detail in Chapters 6–8. Prior to the use of nanomaterials for *in vivo* biosensing, it is necessary to assess their biocompatibility and nanotoxicity. These two aspects are summarized in Chapter 9. The ultimate aim of biosensor preparation related research is to promote efforts for the development of biosensor for health. As an example, the promotion of nanobiotechnology to detect cardiovascular biomarkers is described in Chapter 10. Chapter 11 is dedicated to discussing the use of previously developed (Chapters 6–8) and tested (Chapter 9) nanobiosensing platforms for *in vivo* applications, such as drug/gene delivery, cancer therapy, etc. The miniaturizing biosensors for their portability, desired sensing performance, and user friendly operations are continuously being sought with an ultimate goal of personalization in health care and management. Therefore, Chapter 12 is dedicated to discussing the development of wearable biosensors and POC sensing. Challenges and future prospects of nano-enabling sensing technology for personalized health care and monitoring are discussed in Chapter 13.

In summary, this book explores the potential of nanosystems as multidisciplinary science with the aim of designing and development of smart sensing technologies using micro/nano electrodes, novel nanosensing materials, and their integration with microelectromechanical systems, miniaturized transduction systems, novel sensing strategies, that is, field-effect transistor, complementary metal–oxide–semiconductor, system-on-a-chip, diagnostic-on-a-chip, and lab-on-a-chip, and wearable sensors performing at POC for diagnostics and personalized health-care monitoring. This can be acknowledged as sensors moving from lab-to-field.

—Ajeet Kumar Kaushik, PhD
Chandra K. Dixit, PhD

CHAPTER 1

NANOBIOTECHNOLOGY: AN ABRUPT MERGER

SYED KHALID PASHA*, AHMED HASNAIN JALAL, APARAJITA SINGH, and SHEKHAR BHANSALI

Electrical and Computer Engineering Department, Florida International University, Miami, FL 33174, USA

**E-mail: spash001@fiu.edu*

CONTENTS

This chapter touches upon the aspects of nanotechnology and how they have influenced the field of biosensing to achieve what are known as nanobiosensors. The various sensing techniques such as optical, magnetic, mechanical, and electrochemical are discussed in depth here.

1.1 NANOTECHNOLOGY AND BIOSENSING: THE MERGER

Nanotechnology and biosensing started to emerge together when National Nanotechnology Initiative (NNI) was launched by President Bill Clinton in the year 2000. This is a multiorganization effort which is working toward the vision of understanding and controlling matter at nanoscale to revolutionize technology that benefits society. Since then, the need for smaller, faster, accurate, noninvasive, point-of-care, and wearable sensing gave birth to nanobiotechnology. The first steps of miniaturization and integrating all assaying steps required for biosensing with the advances in microfluidics led to the development of lab-on-a-chip device. Further miniaturization and enhanced performance were possible through advances in nanotechnology. Superior performance such as faster and more accurate sensing requires the sensing medium to interact closely as well as capture more targets. The target and sensor interaction is analogous to "communication". Communication is well understood and faster through a good communication medium or if the "gap" between the communicators is reduced. Nanotechnology helps fill that "gap" between the biological sensing layer (antibodies, enzymes, cell receptors, nucleotides, microorganisms, or tissues) and the nanoscale biomolecular targets due to its nanoscale size. At nanoscale, these materials have relatively larger surface area for functionalizing targets and possess unique physical properties when compared to their bulk material, facilitating improved interaction between sensor and target. Due to the consistent advances in nanotechnology, nanomaterials (NMs) such as colloidal quantum dots,[1] carbon nanotubes (CNTs),[2] nanoparticles,[3] nanowires (NW),[4] nanorods,[5] and nanocomposite materials[6] have established their roles in biosensing applications. Table 1-1 highlights the various physical characteristics of these materials and the applications they enable in biosensing.

As summarized in Table 1-1, these NMs are being widely used for various types of biosensing operations utilizing their many excellent physical characteristics. They have enabled the generation of fast, ultrasensitive, and cost-effective sensors. These developments have allowed us to monitor our well-being in real-time or in short time spans and diagnose pathogens

TABLE 1-1 Various nanomaterials used to develop an efficient nanobiosensor

Nanomaterial (NM)	Properties	Type of sensing	Applications	References
Colloidal quantum dots	Broad absorption band, narrow emission band, size tunable emission (visible to near-infrared), brightness, superior fluorescent brightness, and lifetime	Optical	Immunoassay, nucleic acid detections, fluorescence resonance energy transfer sensing, and in vivo/in vitro imaging	1
Carbon nanotubes	High surface-to-volume ratio (SVR), fast electron-transfer kinetics, highest elastic modulus, good chemical compatibility	Electro-chemical	Immunosensor, enzymatic sensor, nanoscale electrode	2
Nanoparticles	High SVR, high surface energy to provide stable immobilization of large amount of biomolecules retaining their bioactivity, fast and direct electron transfer between wide range of electroactive species and electrode materials,	Electro-chemical	Immunosensor, enzymatic sensor, DNA sensor, electron-transfer mediator, electrocatalyst	3a, b
		Optical	Cellular imaging, detection of molecular binding and conformation, protein quantification	3b, c
	light-scattering properties, local surface plasmon resonance (LSPR) enhancement, mass sensitive	Magnetic	Detection of proteins, enzymes, DNA/mRNA, drugs, pathogens, and tumor cells	3d, e
		Piezoelectric	Amplification tags, DNA sensor, immunosensor	3b
Nanowires	Large SVR, tunable electrical properties	Electro-chemical	Immunosensor, DNA sensors, detection of protein–DNA interactions, protein–molecule interactions, cells, and pathogens	4a, b
		Magneto-resistive	Detection of cells	4c

TABLE 1-1 *(Continued)*

Nanomaterial (NM)	Properties	Type of sensing	Applications	References
Nanorods	Has advantages over nanoparticles: faster electron transport, higher surface area, and stronger light scattering properties (LSPR enhancement with no aggregation deficit), multiplexing capability	Optical	Immunosensor, DNA sensor, pathogen detection, multiplex biosensor, single particle biosensing	5a–d
		Electro-chemical	DNA sensor, detection of pathogen and proteins	5e–g
Organic–inorganic nanocomposites	Combination of excellent electrical and optical properties of inorganic material with beneficial inorganic properties: higher susceptibility to chemical modification, good adhesion, excellent film-forming capabilities, immobilization of biomolecules via covalent bonding	Electro-chemical and optical	Immunosensor, enzymatic sensors, DNA biosensor	6
Nanoelectromechanical systems	Single biomolecular sensing, high mass responsivity, low minimal detectable mass	Optical	Immunosensing, single molecule detection, DNA detection, pathogen detection	7

at early stages. In these applications, metallic nanoparticles and semiconducting quantum dots have been widely used for detection of biomarkers due to their high surface-to-volume ratio (SVR), fast charge transfer properties, and excellent optical properties. They are the alternatives of the fluorescent dyes which are used for labeling. Labeling is a time-consuming process and causes steric hindrance. Label-free sensing while maintaining high SVR and electrical properties has been possible due to one-dimensional (1D) structures such as NW, particularly silicon NW (Si-NW) and CNTs. They also opened up opportunities for real-time sensing applications. Si-NW sensors are generally fabricated as field effect transistor (FET) and can be mass produced due to the matured integrated circuit (IC) fabrication techniques. They have proven to be ultrasensitive sensors, where the electric field changes on the Si-NW surface when the biomolecule binds to the semiconducting Si-NW, which causes change in charge density. Similarly, CNT-based FETs have been widely utilized for sensing. CNTs have the added advantage of being chemically functionalized to almost any chemical species and have fast electron transfer kinetics for a wide range of electroactive species. Due to their outstanding mechanical property and ability to bind easily to most chemicals, CNTs are often used to form nanocomposites or matrix for sensing. Nanoparticles still dominate the optical biosensors, however have their shortcomings when it comes to monitoring biospecific interactions. It is difficult to monitor target binding since it causes minimal changes in the wavelength for detection purposes. In the alternative detection form, the intensity of the absorption can be monitored. However, the intensity change can also be due to the change in nanoparticle concentration, which often changes due to several washing steps required in an assay. Also, the aggregation of nanoparticles results in poor resolution to differentiate multiple targets or in case of specific target detection. Nanorods, which are elongated nanoparticles, overcome these challenges due to their surface plasmon bands in transverse and longitudinal directions. This gives the advantage of multiplexing with different aspect ratio of the nanorods. Nanocomposities of inorganic materials such as metals/metal oxides and organic nanostructures such as conducting/bio polymers are also widely applied in biosensing. In case of organic–inorganic nanocomposites, the inorganic counterpart provides the enhanced electrical and optical properties required for the detection and the organic counterpart immobilizes the biomolecules in an intimate contact with the sensor surface while maintaining the functionality of the biomolecule. A controlled device design with these NMs for single biomolecular sensing is a challenging process. The nanoelectromechanical system (NEMS) technology enables mass production of complex devices for controlled single biomolecular sensing and multiplexed sensing. The moving part of the NEMS sensor is functionalized on one side by the bioreceptors, and then the mechanical bending or compression induced by the target-binding event is monitored. Each section in this chapter gives a detailed insight of various detection techniques using different NMs and highlights the critical parameters required for significantly improved biosensing.

1.2 OPTICAL NANOBIOSENSOR

Comparing with other sensing methodologies, optical sensing is one of the foremost sensing mechanisms for numerous biological applications.[8] Optical sensing is a strong tool rather than the conventional analytical techniques due to its high signal to noise ratio, accuracy, fast and reversible sensing, multiplexing, immune of electromagnetic interference, cost-effective criteria, biodegradable sensing mechanism, and so on. Various nanomaterials (organic, inorganic and nanocpmposites), antibodies, antigens, opto-coupled enzymes, different microorganisms, and so on are being used in the optical biosensing systems. Absorbance, fluorescence, reflectance, dispersion and polarization are the most usual physical optical properties required for the sensing of multifarious biospecies employing the optical window (visible, infra-red and ultra-violet frequency ranges).

For the fabrication of an optical biosensor at the nanoscale, deposition of a thin film of a bioreceptor is critical. Adhesion property, thickness, and roughness of the receptors and guiding media control the sensitivity and response time of the sensors. Different thin film deposition techniques such as Langmuir–Blodgett method, electrostatic self-assembly method, and layer-by-layer technique are the mostly used for suitable layer formation with nanometer thickness for the nanobiosensing devices.[8,9]

1.2.1 TYPES OF OPTICAL NANOBIOSENSOR

Optical nanobiosensors consist of transducers in nanoscales which rely on the measurement of photons for detecting the presence of biological analytes. Based on the amount of biological analytes, the sensor produces an electromagnetic signal that is proportional to the magnitude, phase or frequency of its initial optical signal value. The working mechanism of a typical optical nanobiosensor includes the following four different steps[10]: (i) a bioreceptor is fabricated or incorporated with the optical waveguide,

(ii) an electrochemical interaction alters the optical signal, (iii) a transducer converts biochemical reaction to electromagnetic signal, and finally (iv) a signal processor converts electrical signal into a meaningful physical parameter. Most of the optical nanobiosensors are classified into two categories:

- Evanescent field opto-chemical nanobiosensors
- Surface plasmon resonance (SPR) nanobiosensors

1.2.1.1 EVANESCENT FIELD OPTO-CHEMICAL SENSOR

Optical energy is generally confined within the solid dielectric waveguide structure. A small amount of exponential loss of energy is observed between the bioreceptors and waveguide interface. This phenomenon is referred to as the evanescent wave (shown in Figure 1-1) propagation and it is the result of conservation of energy at the interface region.[11] Its intensity $I(z)$ decays exponentially with the distance z perpendicular to the interface as follows[12]:

$$I(z) = I_0 e^{\frac{-z}{d_p}} \tag{1.1}$$

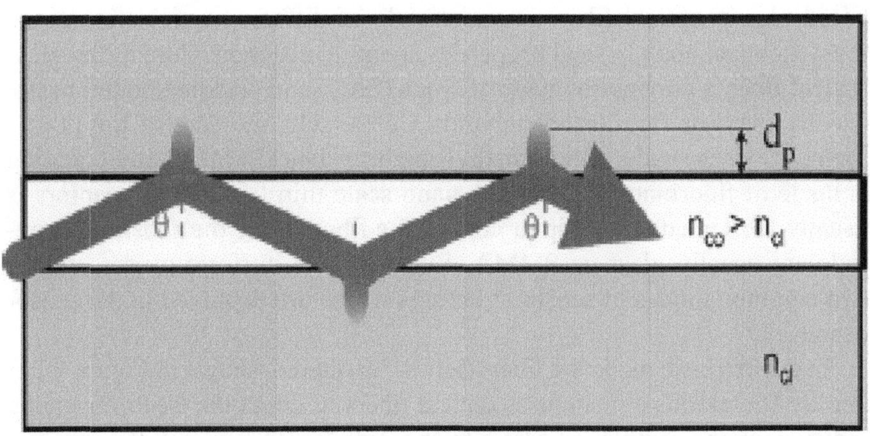

FIGURE 1-1 The cross-sectional profile of optical fiber where the evanescent field decays to $1/e$ of its value at a distance of d_p.[11]

whereas, I_0 is the magnitude of initial intensity and d_p, depth of penetration of the evanescent field is given as

$$d_{\mathrm{p}} = \frac{\lambda}{2\pi\sqrt{\{(n_{\mathrm{co}}\sin\theta)^2 - n_{\mathrm{cl}}^2\}}} \tag{1.2}$$

Here, λ is the corresponding wavelength. n_{co} and n_{cl} are the refractive indices of the core and cladding of the optical fiber respectively.

Three major types of optical nanobiosensors are common which utilize the changes in the refractive index induced by biomolecular binding, and this information is carried using evanescent field.[13] They are: (i) fiber-optic nanobiosensor, (ii) interferometric nanobiosensor, and (iii) resonant cavity-based nanobiosensor.

1.2.1.1.1 Fiber-optic nanobiosensor

Fiber-optic nanobiosensor is based on the unique ability of the nanostructured form to modify their optical response in the presence of a "recognition element", on the modified core. There are two kinds of optical fibers that are commercially available: silica-based glass optical fiber and polymer-based plastic optical fiber. The materials of different optical fibers and their chemical and physical properties are quite different. Core of the glass optical fiber is commonly made of silica (SiO_2) and cladding materials are usually glass or fluorinated polymer. Conversely, the core of the plastic optical fiber is made of polymethyl methacrylate (PMMA) and cladding is made of fluorinated polymer. A nano-scale thin film of bioreceptors is usually fabricated on the tip or core of the fiber where the substrate materials are usually glass or PMMA. Extrinsic and intrinsic modes are the two common modes of sensor structures which are deployed in the sensor design.[14]

In the extrinsic mode the fiber itself is not altered, shown in Figure 1-2 A and B. The extrinsic mode uses optical fibers to direct the electromagnetic radiation to the sample, and later to the detector. In contrast, the intrinsic mode employs the fiber itself as a transduction element. The interaction of light with the sample takes place inside the guiding region due to the evanescent field, or in the lower refractive index surrounding medium. The basic designs of intrinsic mode optical fiber sensors are shown schematically in Figure 1-2 C and D.

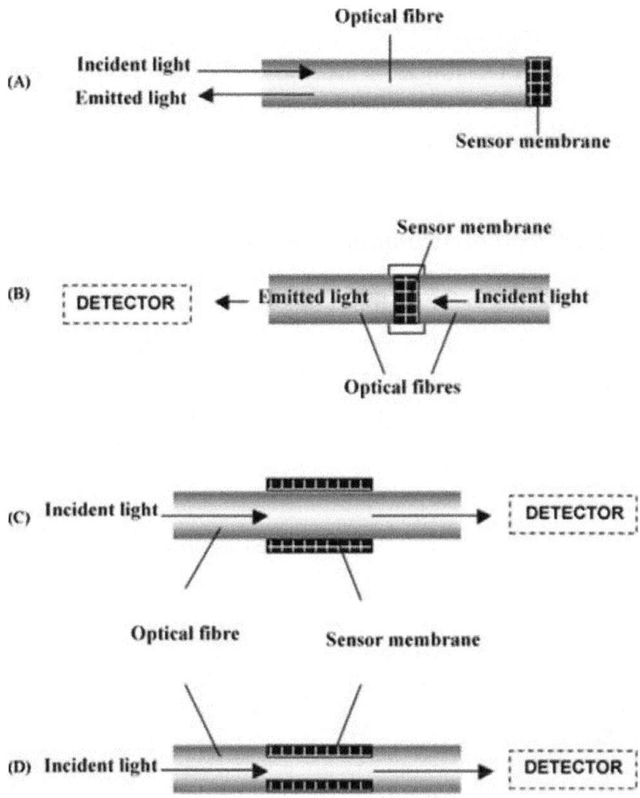

FIGURE 1-2 (A) and (B) Extrinsic sensors. (C) and (D) Intrinsic sensors.[14]

Fiber-optic nanobiosensors are usually employed for detecting biologically active living organisms or biological molecules at the nanoscale. They react with the sensing materials which results in the alteration of intensity or absorbance according to the Beer–Lambert law:

$$A = \varepsilon lc = \alpha\, l \tag{1.3}$$

where ε = molar absorptivity of the absorbing material, l = path length, c = concentration of the reacting products of absorbing material and contaminants, and α = absorption coefficient and absorbance (A) and is defined as:

$$\mathrm{A} = -\log_{10}\left(\frac{\mathrm{I}}{\mathrm{I}_0}\right) \tag{1.4}$$

where I is the intensity of light at a specified wavelength λ that has passed through a sample (transmitted light intensity) and I_0 is the intensity of the light before it enters the sample or incident light intensity. This information is carried out by the optical fiber following total internal reflection (TIR) and is traced as a modified optical signal by the photodetector.

1.2.1.1.2 *Interferometric nanobiosensor*

Interferometric sensing technique relies on measuring phase difference between two collimated light beams of two coherent light sources. The most common interferometric technique is Mach–Zehnder interferometer (MZI) for lab-in-chip nanosensor configuration for biological applications that are shown in Figure 1-3.[15]

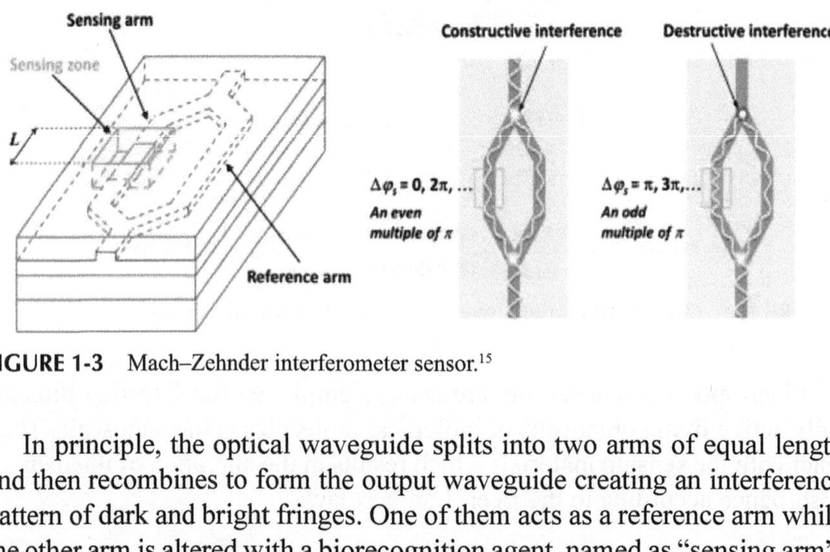

FIGURE 1-3 Mach–Zehnder interferometer sensor.[15]

In principle, the optical waveguide splits into two arms of equal length and then recombines to form the output waveguide creating an interference pattern of dark and bright fringes. One of them acts as a reference arm while the other arm is altered with a biorecognition agent, named as "sensing arm". Biochemical molecules cause biochemical reaction in the sensing region; therefore, the light that travels through this arm experiences a change in its effective refractive index and also causes destructive interference (for odd multiple of π) due to the phase shift. In contrast, constructive interference (for even multiple of π) takes place by means of light recombination in the absence of a biochemical reaction in the sensing arm. The output intensity (I) of light depends upon the difference between effective refractive index of reference (ΔN) and sensing arm and interaction length (L) and is given as[16]:

$$I \propto (1 + V \cdot \cos\Delta\phi) \tag{1.5}$$

where V is the visibility factor and $\Delta\phi = (\phi_r - \phi_s)$ is the phase shift between the guided modes of reference and sensing arms, respectively.

The phase shift ($\Delta\phi$) at the wavelength (λ) is given as

$$\Delta\phi = \frac{2\pi}{\lambda} \cdot L \cdot \Delta N \tag{1.6}$$

Young's interferometer-based sensor is also used for biosensing, where a Y-junction splits light in a parallel manner as shown in Figure 1-4.[17] Similar to the MZI, this kind of set up also has two arms (reference and sensing), and the output light is collected by a cylindrical lens and is merged on a charge-coupled device screen. Though the sensitivity of these kinds of sensor is quite high, these sensors have limitations with the long interaction length of the sensing arm, which makes the system bulky. This long length is necessary for the formation of sufficient phase shift to provide a reliable output.

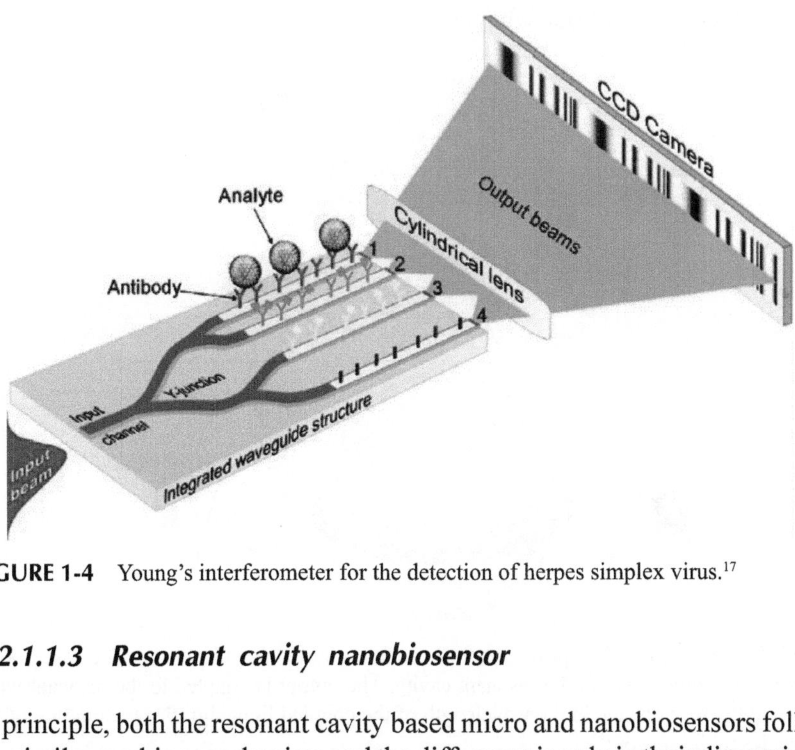

FIGURE 1-4 Young's interferometer for the detection of herpes simplex virus.[17]

1.2.1.1.3 Resonant cavity nanobiosensor

In principle, both the resonant cavity based micro and nanobiosensors follow the similar working mechanism and the difference is only in their dimensions.

In the both sensor systems, the source light is evanescently coupled with the resonator using tapered fiber and this light is confined within the resonator following the principle of TIR.[18] As the integer number of wavelengths is circulated within a closed circular optical path, the confined light forms a resonance. This resonant frequency matches with the input light amplification by stimulated emission of radiation (LASER) frequency and the resultant light is coupled with the optical waveguide. The output light intensity is followed and recorded by the photodetector via the waveguide. Any shifting of resonant frequency due to alteration of optical path length ensures the binding of biomolecules to the micro-resonator surface. For example, bovine serum albumin protein attached to the optical microsphere causes the modification of overall optical path length due to polarization and distribution of evanescent field slightly outside the microcavity. The presence of such alien molecules causes shifting of resonance wavelength (shown in Figure 1-5). The three basic components of the sensing systems are the following:

i. Biorecognition element—antibody, aptamer, or oligonucleotide
ii. Optical transducer
iii. Electrical read-out scheme

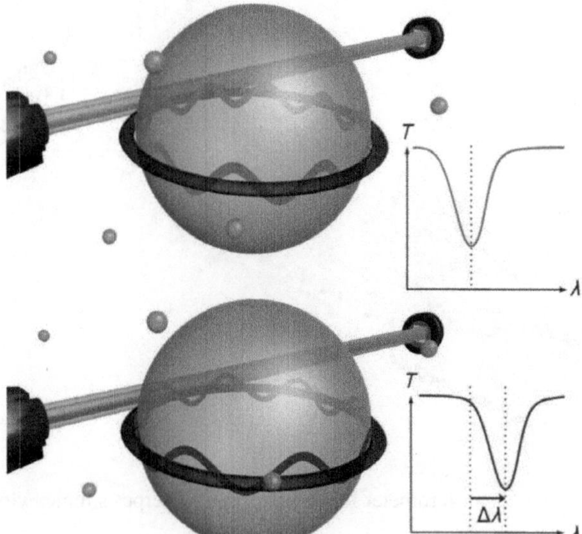

FIGURE 1-5 Biosensing principle for resonant cavity nanobiosensor. The laser beam induces a standing wave in the resonant cavity. The output is coupled to the resonant wave in the cavity. Any change in the wavelength of the coupled light signifies the binding of the molecules to the sensor.[18]

However, the normalized shifting of wavelength (or frequency, $\frac{\Delta\omega}{\omega}$) is calculated from the energy required to polarize the biomolecule and the total energy of the microcavity:

$$\frac{\Delta\lambda}{\lambda} = -\frac{\Delta\omega}{\omega} = \frac{\alpha\left|\bar{E}(\bar{r_0})^2\right|}{2\int\varepsilon\left|\bar{E'}(\bar{r})^2\right|\mathrm{d}V} \tag{1.7}$$

here, λ is the wavelength, $\Delta\lambda$ is the wavelength shifting, ω is angular frequency, $\Delta\omega$ angular frequency shifting, α is the excess polarizability of the particle, r_0 is the location of the binding, ε is the permittivity of the medium, $E(r)$ is the energy at the position of ri where the target analyte is absorbed and interact with the evanescent field and dV is the change of volume.

The benefit of these sensors is that they require less area and provide unprecedented sensitivity.[13,18]

1.2.1.2 SURFACE PLASMON RESONANCE NANOBIOSENSOR

In the surface plasmon resonance (SPR) system, optical energy propagates to the surface of the metal layer as the packets of electrons called surface plasmons. SPR sensors consist of a thin (~ 50 nm) metal layer (usually gold or silver) that is coated on a dielectric in which the light passes. During the traveling of light, p-polarized light satisfies the resonance condition where a charge density oscillates at 100 nm around the metal surface. This resonance wave is termed as surface plasmon wave. The resonant condition depends on the dielectric constant of both metal and medium, incident angle, and the wavelength. For the biosensing applications, the metal layer is incorporated with a "sensing" dielectric layer that causes shifting of resonance frequency by changing the angle of orientation of the metal layer or wavelength or reflection intensity in the presence of biochemical molecules.

SPR sensors employ three modes of detection, which are shown in Figure 1-6.[13]

1. Angular SPR nanobiosensor
2. Spectral SPR nanobiosensor
3. Local SPR nanobiosensor

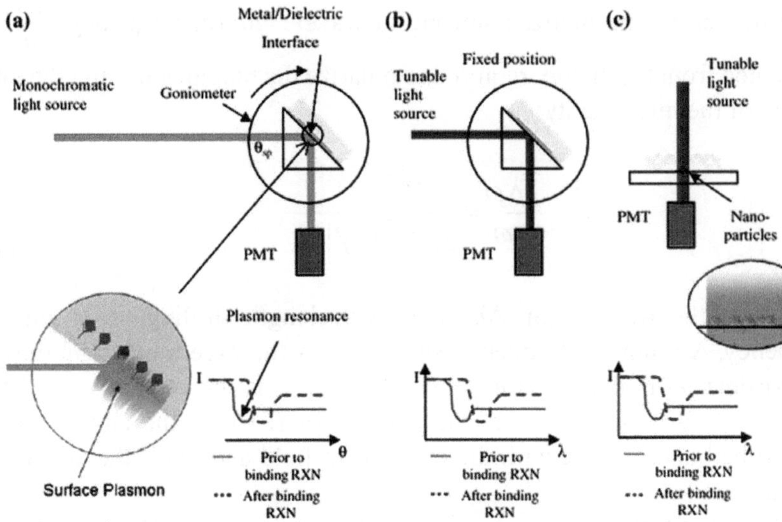

FIGURE 1-6 Different surface plasmon resonance (SPR) sensor systems.[13]

1.2.1.2.1 Angular SPR nanobiosensor

This is the most common SPR sensing technique. In this method, a thin metal layer is combined with several biorecognition agents. Therefore, dielectric constant is altered accordingly and the angle of incidence (θ_{sp}) is shifted. The quantification of target biochemical species directly depends on the shifting of angle of incidence (θ_{sp}) of plasmon excitation. The angle of excitation is defined as follows[13]:

$$\theta_{\text{sp}} = \frac{1}{n_{\text{p}}} \sqrt{\frac{\varepsilon_{\text{m}}(\lambda)\varepsilon_{\text{d}}}{\varepsilon_{\text{m}} + \varepsilon_{\text{d}}}} \tag{1.8}$$

here n_p represents the refractive index of the prism, $\varepsilon_m(\lambda)$ is the dielectric constant of the metal film which changes as a function of the excitation wavelength, λ and ε_d is the dielectric constants of the dielectric coating.

1.2.1.2.2 Spectral SPR nanobiosensor

The difference between the spectral SPR nanobiosensors comparing with the angular SPR nanobiosensors is that they have fixed incident angle and use tunable light source instead of monochromatic light source. The concentration and bioaffinity property of the target biomolecules is measured with a change in the resonant wavelength by this sensor.

This sensor is superior to angular SPR sensors because of its outstanding sensitivity. In addition, instead of measuring incidence angle spectral approach is compatible to multiplexing for high throughput screening for the image-based data collection technique. Contrarily, tunable light source enhances the cost of the system.

1.2.1.2.3 Local SPR nanobiosensor

This sensing technique is more appropriate for nanostructure or nanoparticles sensing because of its ability of coupling into plasmon mode on the surface of sub-wavelength scale. The resonant frequency strongly depends on the dielectric constant of this kind of sensor. For example, the sandwiching assay format of the antigen, amyloid-β derived diffusible ligands (ADDLs), and specific anti-ADDL antibodies in the nanostructured, this nanosensor provides quantitative binding information for both antigen and second antibody detection in sub-wavelength scale for the Alzheimer disease monitoring.[19] Generally, ion beam or electron beam lithography is applied for the fabrication of this nanostructure-based sensor. The most attractive advantages of this sensor are its simplicity of arrangement and multiplexing ability.

1.3 MAGNETIC NANOBIOSENSOR

A magnetic nanobiosensor is a compact analytical device incorporated with biologically sensitive magnetic nanoparticles to quantify biochemical molecules such as DNA, proteins, antibodies, enzymes, or any microbial cells. These sensors are featured with high accuracy and biocompatibility.[20] Most of the applications of magnetic nanobiosensors focus on biomedicine, their synthesis, and detection.

The basic principle of a magnetic nanobiosensor is based on the interaction between biomolecules at the nanoscale and magnetic field. Usually, biomolecules to be detected are immobilized on a magnetic label and passed over on-chip magnetic sensor. The sensor senses the presence of magnetic labels by the alteration of magnetic field due to their inherent properties. A schematic diagram of a typical magnetic nanobiosensor is shown in Figure 1-7.[21]

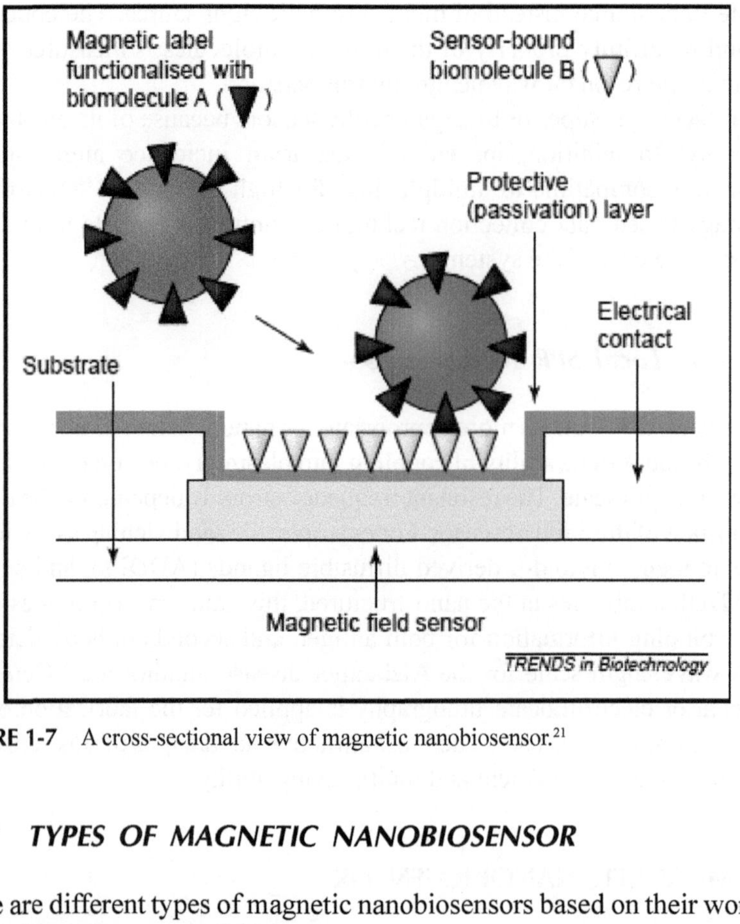

FIGURE 1-7 A cross-sectional view of magnetic nanobiosensor.[21]

1.3.1 TYPES OF MAGNETIC NANOBIOSENSOR

There are different types of magnetic nanobiosensors based on their working mechanism. They are as follows:

i. Hall effect nanobiosensor
ii. Magneto-resistance (MR)-based nanobiosensors
iii. Magneto-impedance (MI) nanobiosensors
iv. Flux gate biosensor

1.3.1.1 HALL EFFECT NANOBIOSENSOR

These kinds of sensors follow the 'Hall Effect' principle where the device has four terminals: two terminals are to pass the current due to electric field and the rest of the two are for measuring voltage perpendicularly. This voltage is zero at null magnetic field and in the presence of an orthogonal magnetic field, Lorentz field causes to amass charge carriers at one side of the conductor surface perpendicular to the current direction. Due to this fluctuation of charge carrier density across the conductor, an additional voltage called 'Hall voltage', $V_{H,}$ is created.

This Hall voltage (V_H) is expressed as

$$V_H = \frac{I_x B_z}{nqt} \tag{1.9}$$

Where I_x = current flow in the x direction, B_z = magnetic field in the z direction, q = charge of the electron = 1.6×10^{-19} C, t = thickness of the slab, and n = number of careers per unit volume.

The principle of this sensor is its output depends upon whether the film magnetization is initially parallel or antiparallel to the sensor current. Therefore, initial orientation of the film is important for the sensor performance. The sensor consists of a thin ferromagnetic film (in nanometer range) in which current passes through the x direction according to Figure 1-8 (a) and (b). When a bead containing DNA or protein is introduced in the sensor system, a corresponding signal is found as an output voltage according to Equation 1.10.[22]

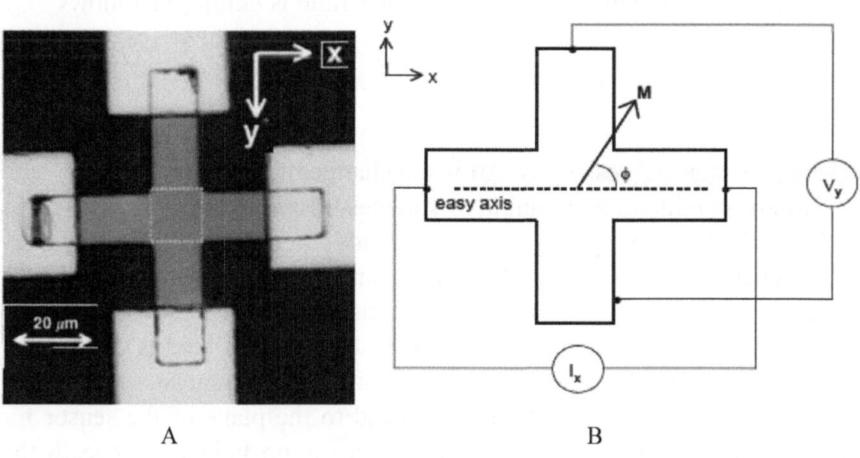

FIGURE 1-8 Hall effect nanobiosensor. (A) Micrograph image which shows the view of the magnetic layers. (B) Geometric view.[22]

Using the following value of total magnetic field in the y direction (H_y) on a bead,

$$H_y \cong H_{app} - \frac{0.38 fNH \chi V_{bead}}{4\pi z^3} \tag{1.10}$$

The output voltage of the sensor (V_y) will be

$$V_y = S_0 H_y I_x \cong V_0 \left(1 - \frac{H}{H_{app}} \frac{0.38 fN \chi V_{bead}}{4\pi z^3}\right) \tag{1.11}$$

Where H_{app} = applied external field, f = fraction of the sensor area influenced by the bead, N = independently acting beads, H = field intensity, χ = susceptibility, V_{bead} = bead volume, z = normal distance between the center of the bead and the sensor surface, and S_0 = effective sensitivity.

1.3.1.2 MAGNETO-RESISTANCE (MR) NANOBIOSENSOR

The idea of MR-based sensor was first reported by Thomson in the year 1856. The basic principle of MR sensor lies on changing resistance of the material due to application of magnetic field. The MR ratio is defined as follows

$$\text{MR} = \frac{\Delta\rho}{\rho} = \frac{R_{max} - R_{min}}{R_{min}} \tag{1.12}$$

Where ρ is material resistivity, Δρ is the change of resistivity, and R_{max} and R_{min} are the maximum and minimum voltages, respectively.

In the MR sensor system, magnetic beads are contained with nanoparticles of heterogeneous size (200–400 nm) and shape. These magnetic beads and sensors are exposed to external magnetic field in the plane or out of the plane of the sensor. The magnetic field is unidirectional and is opposite in sense to the applied field for the in-plane condition. In contrast, the resulting stray field is symmetric in shape from bead to the plane of the sensor for the out-of-plane condition. In addition, there is no field just beneath the bead itself. However, the presence and absence of the bead alters the corresponding effective field on the sensor and consequently, the resistive value as well. As a result, the voltage changes due to the change of this resistance. A setup of in-plane magneto-resistive nanobiosensor is shown in Figure 1-9. Anisotropic MR nanobiosensor, giant MR nanobiosensor, spintronics sensor, tunneling MR nanobiosensor, and so on is the major types of this kind of sensor. The main advantage of these kinds of sensors is their miniaturized structure.

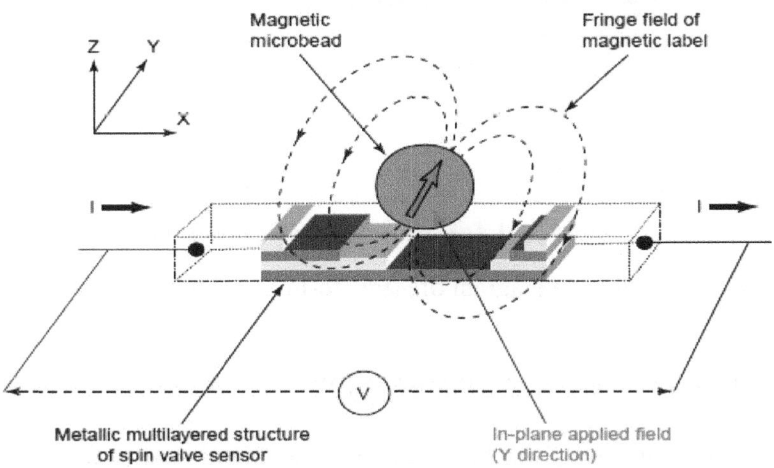

FIGURE 1-9 Magneto-resistive nanobiosensor.[21]

1.3.1.3 MAGNETO-IMPEDANCE NANOBIOSENSOR

Magneto-impedance (MI) nanobiosensor detects any disturbance created by magnetic beads by means of variation of impedance due to the stray field. Amorphous nanowires (NW) are embedded in the sensor system that notices the presence of magnetic beads. High-frequency alternating current passes through these wires that produce a magnetic field circumferentially. The presence of magnetic beads binding with the NW causes a change in magnetic field as well as the overall impedance at a given frequency. Magnetic nanoparticles embedded in ferroliquid (Figure 1-10) can be used, instead of NW, as a recognition element.[23] Higher sensitivity and their small size make this sensor attractive in the various biosensing and pharmaceutical applications.

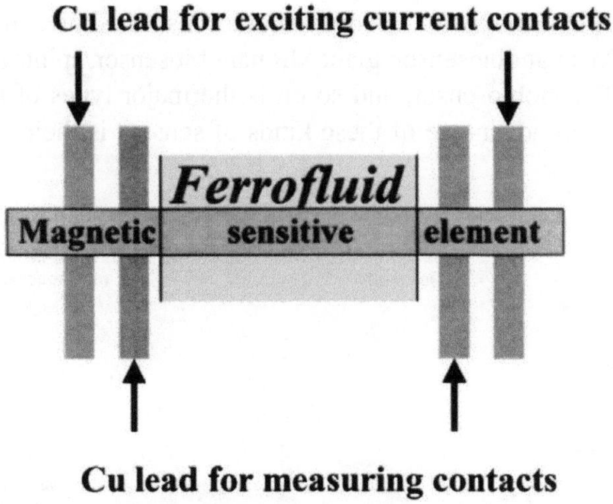

FIGURE 1-10 Principle diagram of magneto-impedance element.[23]

1.3.1.4 FLUXGATE NANOBIOSENSOR

This kind of sensors is based on the residence time difference factor. There is a fluxgate magnetometer in the sensor system, which consists of current-carrying primary and secondary coils around the ferromagnetic core. An oscillating magnetic field is produced due to the alternate current which causes a series of oscillating voltages in the inductively coupled coil. The residence time that is given by time interval between two peaks of voltage is normally equal. In the presence of perturbing magnetic beads, residence time changes. The major disadvantage of this kind of sensor is that it requires a large number of magnetic beads for the desired output signal.

1.4 MECHANICAL NANOBIOSENSORS

Another popular type of nanobiosensors is mechanical sensors, also known as nanomechanical biosensors. Mechanical biosensors have advantages over other types of nanobiosensors as they have great sensitivity and selectivity with regard to a wide variety of analytes and have been used as artificial nose to detect analytes such as DNA strands, bacteria, explosives, hazardous gases, small molecules of biological interest, and so on. The most common types of nanomechanical sensors are of the cantilever type, which have emerged from the AFM technique. The AFM cantilever can image atoms at the angstrom level and measure forces between two atoms. Such arrays of cantilever beams tuned to different frequencies have been utilized for sensing applications.

1.4.1 WORKING PRINCIPLE

A mechanical nanosensor consists mostly of a micromachined cantilever beam that is quite similar to the beams of a conventional AFM tip (Figure 1-11). It is usually made out of silicon or silicon nitride using the same micromachining methods as for the manufacture of computer chips. The transducing effect in such a system can be attributed to the bending of the cantilever due to changes in mass, stress, temperature, frequency, and so on. For example, when a force is applied to the free end of the cantilever, due to the loading of analyte molecules, there is a vertical deflection in the beam that can be measured using the optical or electrical methods. This vertical deflection can be described as

$$F = -K\Delta z \tag{1.13}$$

FIGURE 1-11 A micromachined mechanical sensor array.[26,50]

where K is the spring constant of the beam and F is the loading force that caused the vertical deflection Δz. The spring constant K defines the sensitivity and flexibility of the cantilever and is given by the relation

$$K = \text{Ewt3/4L2} \tag{1.14}$$

where E is the Young's modulus of the material used for fabrication of the beam and w, t and L represent the width thickness and length of the cantilever. Typical materials used for fabrication of the cantilevers are Si, polysilicon, silicon nitride, and metals such as gold, aluminum, and so on. With careful selection of the materials, the cantilever can be designed for chosen sensitivity and for particular frequency.[24]

Selectivity toward the target molecules is achieved by the use of target-specific molecules that include enzymes, polymers, antibodies, or other moieties that can bind to the metalized surface of the cantilevers. Usually, a thin film of gold is deposited on the surface of the silicon/ silicon nitride cantilever.

This metallic thin film also serves dual purpose:

1. It forms a reflecting layer from which a beam of laser light can be reflected to measure the change in the cantilever's bending.[25]
2. It also serves as a substrate where self-assembled monolayers can bind to it in order to facilitate the immobilization of target molecules.

In order to sense the deflection of the cantilever, the most common technique applied is that of the beam deflection or the optical lever method.[25] The cantilevers are coated with a very thin reflecting layer of metal, usually gold on Cr or Ti (to improve the adhesion of gold, the substrate). A beam of laser is reflected off from the lever into a position-sensitive detector; the change in the position of the beam spot can be easily mapped back to the cantilever deflection. Using this scheme, deflections as small as 0.1 nm can be sensed. There are several modes of operations for these sensors. Some of them are discussed in Section 1.4.2.

1.4.2 MODES OF OPERATION

1.4.2.1 DYNAMIC MODE

In this mode, frequency changes occurring due to the dampening effect of the analytes adsorbed onto the surface of the cantilever are measured. The cantilever can be assumed to be a resonator whose resonance frequency is given by

$$f = \frac{1}{2\pi}\sqrt{\frac{K_{spring}}{m}} \tag{1.15}$$

where K is the spring constant of the cantilever and m is the effective mass of the cantilever.[26] With careful design and under conditions of vacuum, it is possible to measure mass changes as less as a few atoms.[24b,26] Various vibrational modes are available and their resonant frequency relations have been listed out in the work by Timoshenko.[27]

Transverse vibration

$$f_{\mathrm{Tr}}^{\mathrm{n}} = \frac{t\lambda_{\mathrm{n}}^2}{4\pi L^2}\sqrt{\frac{E}{3\rho_{\mathrm{c}}}} \tag{1.16}$$

Lateral vibration

$$f_{\mathrm{La}}^{\mathrm{n}} = \frac{w\lambda_{\mathrm{n}}^2}{4\pi L^2}\sqrt{\frac{E}{3\rho_{\mathrm{c}}}} \tag{1.17}$$

Longitudinal vibration

$$f_{\mathrm{Lo}}^{\mathrm{n}} = \frac{(2n-1)}{4l}\sqrt{\frac{E}{\rho_{\mathrm{c}}}} \tag{1.18}$$

Torsional vibration

$$f_{\mathrm{To}}^{\mathrm{n}} = \frac{(2n-1)}{4-l}\sqrt{\frac{G\varepsilon}{\rho_{\mathrm{c}} I_{\mathrm{p}}}} \tag{1.19}$$

The w, t, L, E and ρ_c being the width, thickness, length, young's modulus and density for the beam. Also the I_p is the polar moment of inertia of the beam. Out of these modes, the transverse mode is practical for all purposes due to the comparatively lower fundamental frequency as compared to the other standing vibrational modes.

Apart from the standing vibrations, there are other factors that also influence the sensitivity of the sensors. Due to practical considerations while measuring biomolecules, the sensor has to be placed in fluids (to maintain the integrity of the biological system). This causes dampening due to the fluid and hence places limitation on the sensitivity of the sensor. The quality factor Q determines the accuracy of the measurement of the resonance frequency as it determines the slope of the amplitude and phase curves near resonance. It is given by $Q = \frac{2\pi W_s}{W_d}$, where W_s is the stored vibrational energy and W_d is the energy lost per cycle of vibration (including thermo-elastic losses, viscous damping, and acoustic losses to the support). Also, Q can be defined as $Q = \sqrt{3}\frac{f_0}{\Delta f}$, where Δf is the resonant peak frequency full width at half maximum (FWHM).

The dynamic mode of operation also takes into account the longitudinal and torsional frequencies.

1.4.2.2 STATIC MODE

This is one of the most common modes used for biosensing applications. It utilizes the interaction of adsorbed analyte molecules with the surface of the cantilever (Figure 1-12). There are changes in the environment on or around the cantilever surface that causes mechanical changes on the cantilever. Such interactions lead to either expansion or contraction of the cantilever surface. Stresses produced due to such contraction and expansion cause the cantilever to bend. Depending on whether the adsorbed target molecules produce compressive or tensile stress, the cantilever will bend upward or downward.[28] However, one should always keep in mind that the stresses measured due to the bending of the cantilever are always measured in relative and not absolute values. Stoney's formula is used most commonly to relate the cantilever deflections to the surface stresses,[26] and is given as

$$\Delta\sigma = \frac{Et^2}{3(1-v)L^2}\Delta z \tag{1.20}$$

where $\Delta\sigma$ is the difference in the top and bottom surface stresses and is related to the change in detected cantilever deflection Δz by a constant term which is a function of material properties, E is the elasticity modulus, v is the Poisson's ratio, and L and t are the effective length (from the base of the cantilever) and thickness of the cantilever, respectively. The sensitivity of the sensor to detect stress changes increases as the length of the cantilever increases. However, as the length of the cantilever increases, thermal vibrations also increase. This offsets the advantage of having a longer cantilever beam. As discussed earlier, care is to be taken while reporting the values of the surface stresses, instead of absolute deflections, to get a fair understanding of the adsorption of molecules on the surface of the sensor.

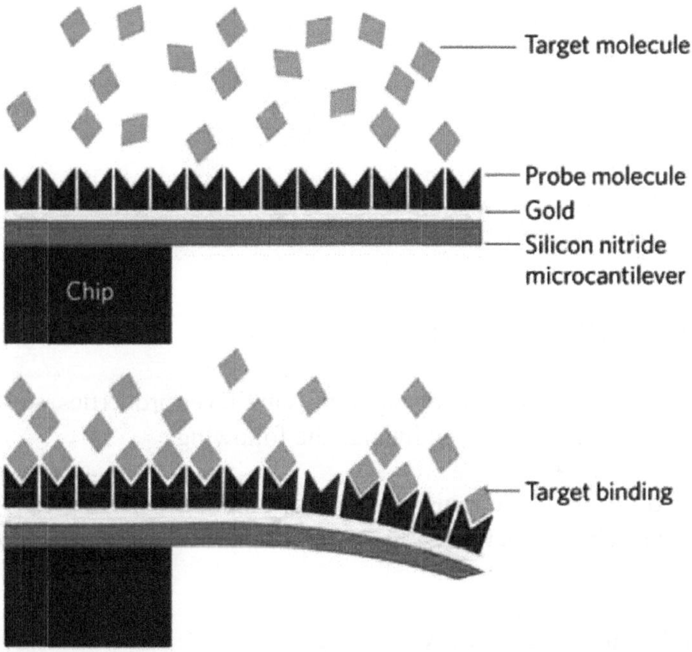

FIGURE 1-12 Illustration of static mode of operation of a cantilever sensor.[28a,51]

1.4.2.3 HEAT MODE

This mode is utilized when a cantilever is coated with a thin metal film. This structure is a composite structure with two layers of different thermal expansion coefficients (silicon and metal) and resembles a bimetallic strip (Figure 1-13). Application of heat to such a cantilever would cause it to bend due to thermal stress produced in the cantilever. Temperature changes as small as 10^{-5}K can be detected using this method and detection of calorimetric changes with high sensitivity is possible.[26]

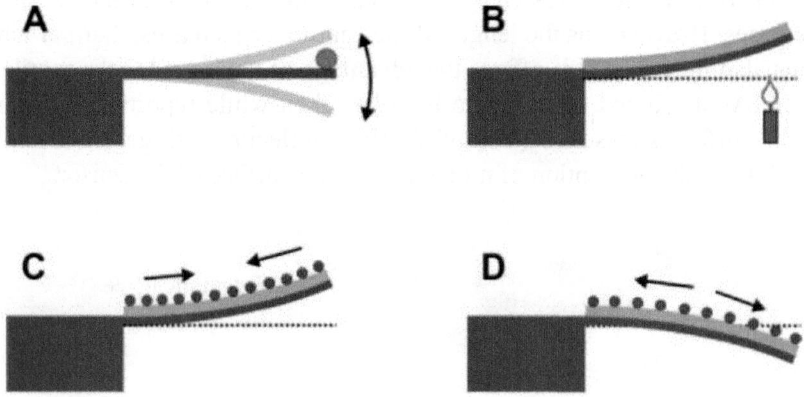

FIGURE 1-13 The figure shows (A) dynamic mode of operation of cantilever biosensor in heat mode (B) and static mode (C). (D) Operation of the given cantilever sensor.[26, 52]

1.4.3 MONITORING OF THE CANTILEVER

In order to detect the changes occurring in the cantilever properties, several methods have been adopted, which include the following:

1. Optical deflection Electronic monitoring techniques (piezoresistivity/ piezoelectricity/ capacitance)

1.4.3.1 OPTICAL DEFLECTION

Out of the two methods, optical monitoring (using laser light bounced from the surface of the cantilever) is the most commonly used. However, there are inherent limitations associated with it. The bulky electronic control units that power the laser and detector limit the application to the desktop environment. Piezoelectric-based detection has been comparatively explored as a mode of signal transduction for use in biosensors. Integration of semiconductor devices such as metal-oxide-semiconductor field-effect transistor (MOSFET) on the cantilever chip has been a challenge by itself. However, it offers the simplicity and portability to monitor the deflection of the cantilever.

1.4.3.2 ELECTRONIC MEASUREMENT

The use of embedded transistor technique can really simplify the measurements made. A MOSFET is fabricated at the base of the cantilever. The changes in the carrier mobility (resulting from the stress induced by the bending of the cantilever) cause a change in the drain current of the MOSFET. The advantage of using such a setup is the simplicity of the system, miniaturization of the associated electronics, and low power consumption.[29] Changes in capacitance due to the bending of the cantilever consequent to loading can also be measured using simple ICs. For more information on such techniques, the reader can refer to the paper reported by Hierold et al.[29]

1.4.3.2.1 Other electronic techniques

Piezoresistive and piezoelectric modes of monitoring of sensors are also popular modes of monitoring. They use the piezoelectric property of the cantilever materials coupled with integrated electronic control circuits to detect changes in the frequency of the sensors. Piezoresistive mode can be described as the change in the electrical resistance of a metal or a semiconductor due to the application of mechanical strain. In contrast, piezoelectric effect causes a change in the potential of such a material. The mechanical cantilevers that employ piezoresistive effect have an integrated piezoresistor fabricated on them. The resistance of the piezoresistor changes due to the bending of the cantilever sensor. This bending can be sensitively monitored by a simple electrical measurement. The general scheme of measurement of such a system includes placing a reference cantilever besides a sensor cantilever (Figure 1-14).[25] While the sensor cantilever is coated with a sensing element (e.g., antibodies, etc.), the reference cantilever is usually coated with a polymer like polyethylene glycol to reduce any binding on it. Both the cantilevers are connected in such a manner so as to make the two arms of a Wheatstone bridge. Any deflection due to absorption of target on the sensing cantilever is recorded.[25]

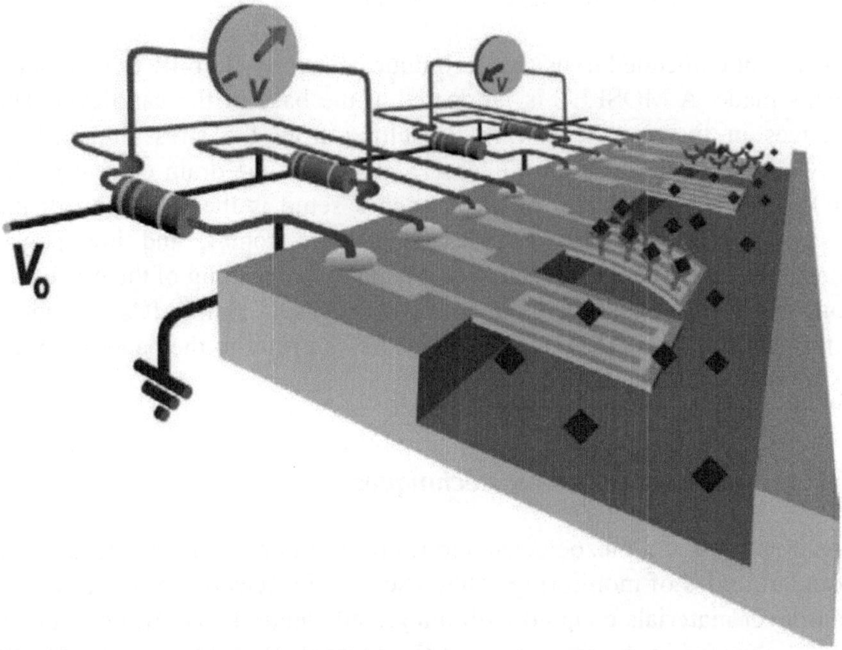

FIGURE 1-14 A schematic of the piezoresistive cantilever—the differential measurements are made with one cantilever serving as a control and the other coated with sensing molecules to detect the analytes.[25]

1.4.4 FABRICATION OF CANTILEVERS

Achieving nanometer scale precision while fabricating the cantilever nanosensors seems to be a huge task. However, it is made possible by using the established techniques of microfabrication. Silicon, silicon dioxide, and silicon nitrides are the materials of choice. They show resonance with high *q*-values and have relatively low energy dissipation, and also the ease of fabrication using the conventional microfabrication techniques is a big plus point. Apart from silicon-based materials, several other materials, such as piezoelectric, magneto-resistive, and so on have been explored to achieve special characteristics of the sensors.[24a]

We can see that due to a large number of materials available, there has been several ways in which cantilever sensors have been designed. For example, silicon-based cantilevers have been used to estimate the mass of a single cell.[30] Piezoelectric-based cantilevers were shown to measure very dilute bacterial pathogen concentrations continuously in a liquid[31] (Campbell and Mutharasan, 2007b). Metal oxide-based cantilevers were used to detect prostate cancer biomarkers.[32]

The cantilevers are fabricated by 3D etching of a thin silicon wafer. Low-pressure chemical vapor deposition technique is used to deposit the silicon nitride layers. It is easy to produce different geometries and different layouts of the cantilevers, which can be integrated with channels for liquid handling.[25] A typical microfabrication process is illustrated in Figure 1-15.[33]

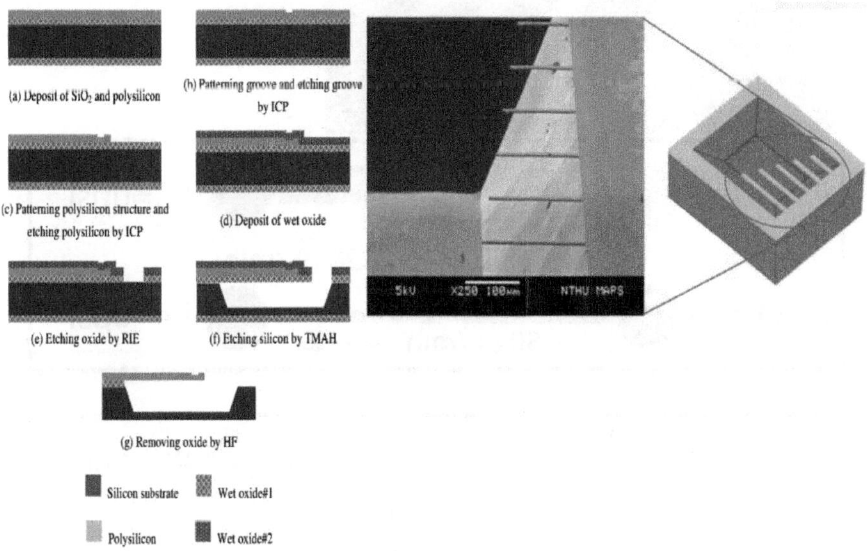

FIGURE 1-15 A schematic of the process steps involved in the fabrication of a cantilever.[33]

The first step in this process is the deposition of an oxide film on a silicon wafer (typically by wet oxidation). The next step is deposition of a polysilicon film over this wafer. This is followed by standard photolithography process of spin coating a photoresist and exposing the pattern, developing the resist, and etching to get the desired pattern onto the polysilicon film. Later on, another layer of silicon oxide is deposited. This results in the polysilicon layer being sandwiched between the two oxide layers. Another patterning is done and reactive ion etching is performed. It is later followed by wet etching in tetramethylammonium hydroxide solution to remove the bulk of silicon. The oxide layer serves as a passivation layer for the polysilicon structure during bulk etching. The resulting structure looks similar to that shown in Figure 1-16.

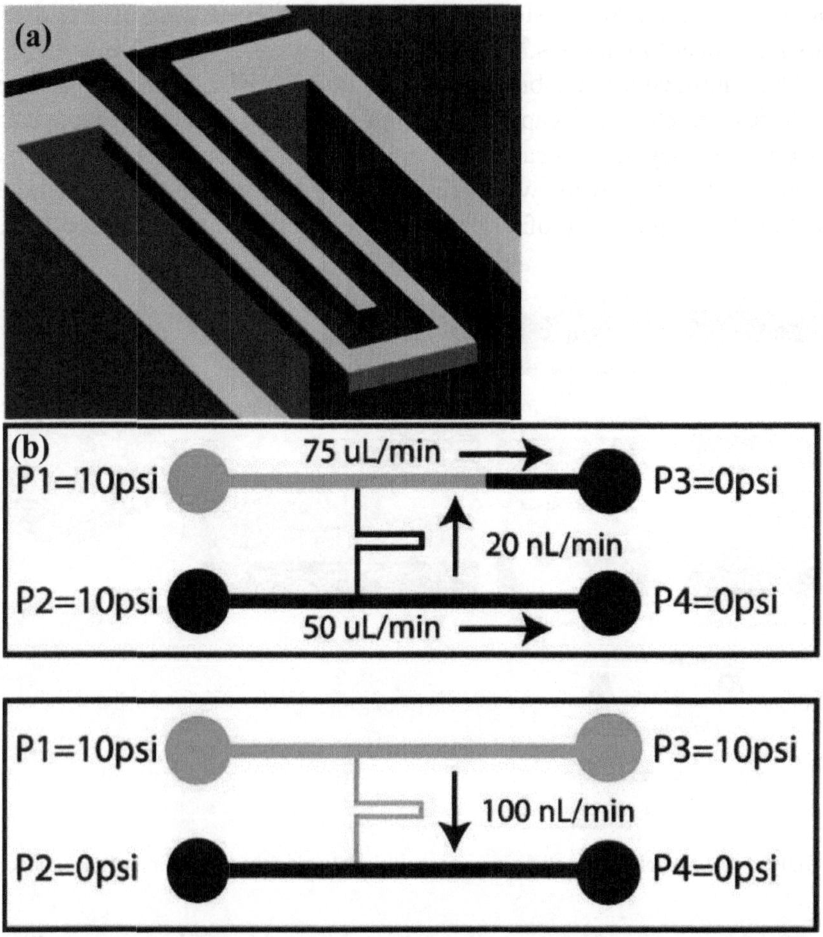

FIGURE 1-16 (A) A schematic design of fluid phase nanosensor having microchannels in the cantilever itself. (B) The schematic of the sensor fluidics used for sensing. The input and output pressures are provided for the channels.[28]

1.4.5 FUNCTIONALIZATION OF THE CANTILEVER SURFACE

The cantilever surface is the location where most of the sensing activity happens. In order to increase the sensitivity and selectively detect the analytes under study, the surface is functionalized with recognition elements such as antibodies, enzymes, nanoparticles, and so on. Some of the applications discussed here are of cantilever sensors used along with the recognition elements integrated with supporting electronics.

1.4.6 APPLICATIONS

The mechanical biosensors are versatile sensors and have been demonstrated to be useful in many applications—ranging from detection of DNA to explosives. Fritz and coworkers demonstrated the potential of nanomechanical systems for label-free biological sensing in 2000. They showed that DNA hybridization can be detected in real time by cantilever sensors.[34] Several other reports of cantilever-based sensors being used for biological sensing are available. We summarize some of these applications in the following section.

Cantilever-based nanosensors have been used for the detection of protcin biomarkers such as prostate-specific antigen in a background of human serum albumin and human plasminogen.[35]

Pathogen detection: Another interesting application was to study the mechanism of antibiotic interactions with mucopeptides (components of bacterial cell wall) with a sensitivity of 10 nM, and at clinically relevant concentrations in blood serum.[36] Such studies can offer new clues on the mechanisms that make bacterial strains increasingly resistant to antibiotics.

Nanomechanical mass spectrometry: The use of nanomechanical sensors for applications in spectrometry has also been demonstrated.[37] The use of mass-to-charge ratio in conventional spectroscopy requires ionization of the analyte molecules, which limits the sensitivity of the technique. Also, some molecules are fragile and it may not be suitable to ionize them. Nanomechanical mass spectrometer studies are advantageous in this respect as large protein molecules need not be fragmented and can be analyzed intact.

Electronic nose: Commercially available polymers have also been used to coat cantilevers for differentiating between different volatile organic compounds in air. Baller et al. developed a nanotechnology olfactory sensor to characterize and identify gaseous analytes.[24b]

1.4.7 OTHER TYPES OF MECHANICAL SENSORS

1.4.7.1 FLUID-PHASE CAPTURE AND DETECTION

While in vacuum and air, mechanical biosensors provide good resolution. However, when used in liquid samples, the system mechanics behave in a different manner. The process involves operating the device in the solution, removing it, and once the analytes have bound, desiccating them before detection. One disadvantage of liquid environment results from the errors caused by unwanted molecules binding to the surface.[28a] An alternative to place the sensors in a fluid is to constrain the liquid to micromachined channels constructed on the cantilever itself. Figure 1-16[28b] shows the design of such a biosensor.

1.4.7.2 QUARTZ CRYSTAL MICROBALANCES

Quartz crystal microbalances (QCM) are mechanical resonators that enable nanoscale mass and structural changes to be measured and offer a method for studying molecular interactions and surface phenomena. As the target molecules adsorb on the surface of the resonator, the resonant frequency downgrades due to dampening; this chenge of the resonant frequency can be accurately mapped to the addition of mass on the sensor. Such a system has showed high sensitivity—capable of detecting femto-molar concentrations of the sample—using a sandwich assay.[38] Disadvantages of using the QCM system include removal of the sample from the fluid to make correct measurements.

1.5 ELECTROCHEMICAL NANOBIOSENSORS

Electrochemical sensors are one of the most attractive forms of sensors currently in use. Commercially, the demand for electrochemical sensors in the United States is projected to increase at 4.9% annually to USD 1.9 billion in 2017. While the optical sensor market size is projected to grow faster, that of electrochemical sensor is expected to remain dominant.[39] Electrochemical form of sensing is preferred for fast, accurate, and continuous measurements for on-site monitoring of analytes, whether they are pollutants being monitored in a field, hazardous gases in confined spaces, or on-site clinical monitoring of analytes of interest in a patient such as blood glucose, and so on. Depending on the mode of operation, they can be classified as amperometric, potentiometric, or voltammetric sensors.

1.5.1 WORKING PRINCIPLE

Electrochemical sensors work by reacting with target analytes to produce an electrical signal that can be correlated to the concentration of the target analyte in the given phase. The typical electrochemical sensor setup consists of a working electrode, counter electrode, reference electrode, and the electrolyte medium. We discuss here the basic operating principle of different types of electrochemical sensors.

1.5.2 IMPORTANCE OF REFERENCE ELECTRODE

The reference electrode is important when using an external voltage to drive the reactions. It is important to have a stable potential at the sensing electrode. However, this is not the case always when there are electrochemical reactions taking place at the surface of the electrode. The reference electrode when placed in close proximity of the sensing electrode, maintains a fixed potential over the sensing electrode surface. Depending on the configuration used, there is no flow of current through the reference electrode (three-electrode setup).

1.5.3 TYPES OF ELECTROCHEMICAL SENSORS

1.5.3.1 VOLTAMMETRIC SENSORS

Voltammetry is a general term that is used to describe a collection of chemical analysis methods where the changes in the electrochemical response of the analyte are measured with respect to an applied (variable) potential. Voltammetry in its original form was introduced as polarimetry by Jaroslav Heyrovský in 1922.

The most common setup used in voltammetry is the three-electrode setup, proposed first by Hickling in 1942.[40] This involves the working electrode, auxiliary (counter) electrode, and a reference electrode. A time-varying potential is applied to the working and reference electrodes. The current responses due to changes in the analyte solution are measured across the working and the counter electrodes.

The resulting plots of current versus the variable voltage are referred to as voltammograms (Figure 1-17). Voltammograms are a reflection of the half-cell reactions of the analytes in question and can uniquely identify the species involved in the cell reactions (reduction and oxidation). The auxiliary electrode is usually a platinum wire and the reference electrodes are made up of a standard carbon electrode or a silver/silver chloride electrode. The voltammograms provide information regarding the concentration of the analytes and their respective standard state potentials.

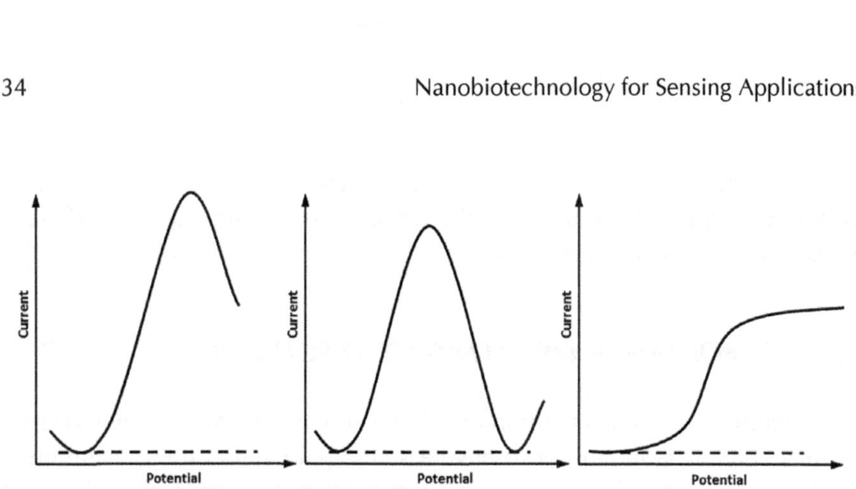

FIGURE 1-17 Different types of voltammograms.

Some of the techniques that can be classified under voltammetry are as follows:

- Linear sweep voltammetry: A linearly varying voltage is applied on the working electrode with respect to the reference electrode. The resulting current (measured between the working and the counter electrodes) provides information regarding the oxidation and reduction voltages of the species involved in the reaction.
- Staircase: A form of linear sweep voltammetry where the potential is applied in steps instead of continuum. This helps in reducing the capacitive charging currents involved in the reactions.
- Squarewave differential pulse: A form of staircase voltammetry where pulses of voltages are applied in a ramping fashion. Such pulses (normal or differential) can easily distinguish the signal against charging currents. This makes this form of voltammetry much more sensitive than cyclic voltammetry (CV).
- CV: The basis of CV is the cyclic ramping of the potential linearly with respect to time. A potential is applied between the working and the reference electrodes while the response current is measured between the working and the counter electrodes. The variations in the response current depict the oxidation and reduction potentials of the chemical species present in the analytes. These redox graphs are the signature response of different electroactive components of the analyte and can be used to determine the presence of intermediates in oxidation–reduction reactions, the reversibility of a reaction, the electron stoichiometry of a system, the diffusion coefficient of an analyte, and the formal reduction potential, used as an identification tool. Also, the concentration of an unknown solution can be determined

by generating a calibration curve of current versus concentration by using the Nernstian equations.[41]

1.5.3.2 AMPEROMETRIC SENSORS

An amperometric cell may comprise two or three electrodes. The working electrode is usually constructed from noble metals such as platinum (Pt) or gold (Au). The potential applied to the working electrode is measured and controlled as compared to the potential provided by the reference electrode. It is common to have Ag/AgCl as the reference electrode. A third electrode, the counter (or auxiliary) electrode, is sometimes included.

In industry, pH sensors are an indispensable part for process optimization and monitoring. These sensors have the potential maintained at a fixed value and measure the current in response to this potential over time. The applied potential drives the electroactive species toward electron transfer reaction (oxidation or reduction). This electron transfer at the electrode results in a current that is measured and correlated to the concentration of the analytes being monitored. Electrons are transferred from the analyte to the working electrode or vice versa. The direction of flow of electrons depends upon the properties of the analyte and can be controlled by the electric potential applied to the working electrode.

1.5.3.3 POTENTIOMETRIC SENSORS

They are based on the ion transfer between two immiscible phases. The first phase is the test solution and the other phase can be a solid or liquid supported on an inert medium called an ion-sensitive membrane.[42] Potentiometric sensors measure the potential of the reaction passively (they do not affect the reaction happening in the solution) between the reference and working electrodes. This potential is a function of ion mobility of the analyte being measured.

The reference electrode provides a constant potential and the potential at the working electrode is measured in the solution. The working electrode potential is variable and can be correlated to the concentration of a specific analyte in the solution. Such ion sensors are referred to as ion-selective electrodes.

Such sensors have been used for the determination of nonionic compounds, and by interaction of gases with electrolytes to detect a particular gas. Also, the biocatalytic activities of living cells or biological processes produce ions. These can be detected with the help of enzymes for other biocatalytic molecules. The text[42] provides a detailed theoretical explanation about the working of potentiometeric sensors. Also, one can find careful discussions on various properties of such sensors, such as different configurations, response timing, selectivity, response range, and time of potentiometric sensors.

1.5.3.4 ELECTROCHEMICAL IMPEDANCE SPECTROSCOPY

Electrochemical impedance spectroscopy (EIS)-based sensors have been an important tool in electrochemical measurements for various industrial applications. The underlying principle of impedance-based sensors is the application of sinusoidal voltage probe. The response of the system under measurement provides information about its impedance and capacitance components. Care has to be taken so that the applied AC probe is small enough to assume the system response is linear. The high-frequency response of the system under EIS study can be used to characterize the electron flow in the system and the low-frequency responses can be used to characterize the mass flow reactions happening in the system. The detection of analyte in this measurement technique is quantified by any changes in the equivalent circuit caused by the analyte itself.[43]

1.5.4 BIO/NANOFUNCTIONALIZATION AT SENSOR SURFACE

The sensing techniques described earlier are beset with certain problems such as the selectivity toward analytes when being used to measure bioanalytes. There are at times issues related to the sensitivity as electrochemical sensors need a specific sample size to detect. One of the approaches used to mitigate such limitations is functionalization of the electrochemical sensing surface with nanosized biorecognition molecules such as enzymes, antibodies, DNA, aptamers, nanoparticles, polymers, and so on. These biorecognition elements are highly specific to the bioanalytes being measured. Their integration with electrochemical sensors makes them highly sensitive. The sensitivity of the sensor is also increased due to the high surface area offered by such nanoparticles/molecules.

1.5.5 POTENTIAL APPLICATIONS OF NANOELECTROCHEMICAL SENSORS

The techniques involving nanoelectrochemical sensors have been important in the fabrication of nanosensors for the application of environmental monitoring and wearable applications. Bandodkar and Wang[44] have given a comprehensive review regarding the wearable electrochemical sensors. In order to fabricate sensors that could continuously monitor the analytes in question, electrochemistry has been a useful tool. It has been combined with other detection principles and techniques to achieve better sensing modality; for example, electrophoresis, molecular imprinting, and so on have been combined with different voltammetric techniques like CV to achieve highly selective nanosensors. Electrochemical sensors based on Graphene and carbon nanotubes have been summarized by Yuyan Shao et al.[45] These sensors have been used to detect dopamine, glucose, DNA strands, heavy metal ions, and so on. Gold nanoparticles and functionalized CNTs were reported by Pandiaraj et al.[46] to develop a CV-based nanosensor to detect cytochrome C. CV is one of the most common techniques that has been used along with NMs to fabricate sensors for the detection of biomolecules for diagnostic purposes. CV-based detection of small molecules like cortisol using immunosensing protocol was demonstrated by Kaushik et al,[47] Pasha et al,[48] and Singh et al.[49]

1.6 CONCLUSION

The nanobiosensing techniques that are briefly introduced in this chapter have changed the way sensing is done. These techniques will be discussed in the subsequent chapters of this book by other authors as well.

1.7 ACKNOWLEDGMENT

We thank Dr. Ajeet Kaushik for providing the opportunity to contribute to this book and for providing valuable guidance toward this chapter. We also thank Dr. Pandiaraj Manickam for his support and valuable inputs regarding this chapter.

KEYWORDS

- **Nanobiosensor**
- **SPR**
- **Nanomaterials**
- **Sensors**
- **Hall Effect**
- **Laser**

REFERENCES

1. Wang, Y.; Hu, R.; Lin, G.; Roy, I.; Yong, K.-T. Functionalized Quantum Dots for Biosensing and Bioimaging and Concerns on Toxicity. *ACS Appl. Mater. Interfaces* **2013,** *5* (8), 2786–2799.
2. Balasubramanian, K.; Burghard, M. Biosensors based on Carbon Nanotubes. *Anal. Bioanal. Chem.* **2006,** *385* (3), 452–468.
3. (a) Pingarrón, J. M.; Yáñez-Sedeño, P.; González-Cortés, A. Gold Nanoparticle-based Electrochemical Biosensors. *Electrochimica. Acta* **2008,** *53* (19), 5848–5866; (b) Li, Y.; Schluesener, H.; Xu, S. Gold Nanoparticle-based Biosensors. *Gold. Bull.* **2010,** *43* (1), 29–41; (c) Anker, J. N.; Hall, W. P.; Lyandres, O.; Shah, N. C.; Zhao, J.; Van Duyne, R. P.,Biosensing with Plasmonic Nanosensors. *Nat. Mate.r* **2008,** *7* (6), 442–453; (d) Ennen, I.; Albon, C.; Weddemann, A.; Auge, A.; Hedwig, P.; Wittbracht, F.; Regtmeier, A.-K.; Akemeier, D.; Dreyer, A.; Peter, M.; Jutzi, P.; Mattay, J.; Mitzel, N. W.; Mill, N.; Hütten, A. From Magnetic Nanoparticles to Magnetoresistive Biosensors. *Acta Physica. Polonica., A* **2012,** *121* (2), 420–425; (e) Koh, I.; Josephson, L., Magnetic Nanoparticle Sensors. *Sensors* **2009,** *9* (10), 8130–8145.
4. (a) Zhang, G.-J.; Ning, Y. Silicon Nanowire Biosensor and its Applications in Disease Diagnostics: A Review. *Anal. Chim. Acta.* **2012,** *749* (0), 1–15; (b) Wanekaya, A. K.; Chen, W.; Myung, N. V.; Mulchandani, A. Nanowire-Based Electrochemical Biosensors. *Electroanalysis* **2006,** *18* (6), 533–550; (c) Huang, H.-T.; Lin, Y.-H.; Ger, T.-R.; Wei, Z.-H. Detection of Magnetically Labeled Cells Using Wavelike Permalloy Nanowires. *Appl. Phys. Express* **2013,** *6* (3), 037001.
5. (a) Yu, C.; Irudayaraj, J. Multiplex Biosensor Using Gold Nanorods. *Anal. Chem.* **2007,** *79* (2), 572–579; (b) Lu, X.; Dong, X.; Zhang, K.; Han, X.; Fang, X.; Zhang, Y. A Gold Nanorods-Based Fluorescent Biosensor for the Detection of Hepatitis B Virus DNA Based on Fluorescence Resonance Energy Transfer. *Analyst* **2013,** *138* (2), 642–50; (c) Wang, X.; Li, Y.; Wang, H.; Fu, Q.; Peng, J.; Wang, Y.; Du, J.; Zhou, Y.; Zhan, L. Gold Nanorod-Based Localized Surface Plasmon Resonance Biosensor for Sensitive Detection of Hepatitis B Virus in Buffer, Blood Serum and Plasma. *Biosens. Bioelectron.* **2010,** *26* (2), 404–410; (d) Kabashin, A. V.; Evans, P.; Pastkovsky, S.; Hendren, W.; Wurtz, G. A.; Atkinson, R.; Pollard, R.; Podolskiy, V. A.; Zayats, A. V.

Plasmonic Nanorod Metamaterials for Biosensing. *Nat. Mater.* **2009,** *8* (11), 867–871; (e) Shakoori, Z.; Salimian, S.; Kharrazi, S.; Adabi, M.; Saber, R. Electrochemical DNA Biosensor Based on Gold Nanorods for Detecting Hepatitis B Virus. *Anal. Bioanal. Chem.* **2015,** *407* (2), 455–461; (f) Han, X.; Fang, X.; Shi, A.; Wang, J.; Zhang, Y. An Electrochemical DNA Biosensor Based on Gold Nanorods Decorated Graphene Oxide Sheets for Sensing Platform. *Anal. Biochem.* **2013,** *443* (2), 117–123; (g) Kim, J. S.; Park, W. I.; Lee, C.-H.; Yi, G.-C. ZnO Nanorod Biosensor for Highly Sensitive Detection of Specific Protein Binding. *J. Korean Phys. Soc.* **2006,** *49* (4), 1635–1639.

6. Kaushika, A.; Aryab, S. K.; Vasudevc, A.; Bhansalia, S. Nanocomposites Based on Chitosan-Metal/Metal Oxides Hybrids for Biosensors Applications. *J. Nanosci. Lett.* **2013,** *3*, 32.
7. (a) Salomon, S.; Leïchlé, T.; Dezest, D.; Seichepine, F.; Guillon, S.; Thibault, C.; Vieu, C.; Nicu, L. Arrays of Nanoelectromechanical Biosensors Functionalized by Microcontact Printing. *Nanotechnology* **2012,** *23* (49), 495501; (b) Kumar, R.; Somvir; Singh, S.; Kulwant. A Review on Applicatation of Nanoscience for Biosensing. *Int. J. Eng. Sci. Res.* **2014,** *3* (4), 279–283.
8. Dey, D.; Goswami, T. Optical Biosensors: A Revolution Towards Quantum Nanoscale Electronics Device Fabrication. *J. Biomed. Biotechnol.* **2011,** *2011*.
9. Consales, M.; Cutolo, A.; Penza, M.; Aversa, P.; Giordano, M.; Cusano, A. Fiber Optic Chemical Nanosensors Based on Engineered Single-Walled Carbon Nanotubes. *J. Sensor* **2008,** *2008*.
10. Sagadevan, S.; Periasamy, M. Recent Trends in Nanobiosensors and Their Applications – A Review. *Rev. Adv. Mater. Sci.* **2014,** *36*, 62–69.
11. Leung, A.; Shankar, P. M.; Mutharasan, R. A Review of Fiber-Optic Biosensors. *Sens. Actuat B, Chem.* **2007,** *125* (2), 688–703.
12. Cao, W.; Duan, Y. Optical Fiber-Based Evanescent Ammonia Sensor. *Sensor Actuat B, Chem.* **2005,** *110* (2), 252–259.
13. Erickson, D.; Mandal, S.; Yang, A. J.; Cordovez, B. Nanobiosensors: Optofluidic, Electrical and Mechanical Approaches to Biomolecular Detection at the Nanoscale. *Microfluid Nanofluid.* **2008,** *4* (1–2), 33–52.
14. Jerónimo, P. C. A.; Araújo, A. N.; Conceição B.S.M. Montenegro, M. Optical Sensors and Biosensors Based on Sol–Gel films. *Talanta* **2007,** *72* (1), 13–27.
15. Choo, S. J.; Kim, J.; Lee, K. W.; Lee, D. H.; Shin, H.-J.; Park, J. H. An Integrated Mach–Zehnder Interferometric Biosensor with a Silicon Oxynitride Waveguide by Plasma-Enhanced Chemical Vapor Deposition. *Curr. Appl. Phys.* **2014,** *14* (7), 954–959.
16. Lechuga, L. M.; Sepulveda, B.; Sanchez del Rio, J.; Blanco, F.; Calle, A.; Dominguez, C. In *Integrated micro- and nano-optical biosensor silicon devices CMOS compatible*, 2004; pp 96–110.
17. Ymeti, A.; Greve, J.; Lambeck, P. V.; Wink, T.; van, H.; Beumer; Wijn, R. R.; Heideman, R. G.; Subramaniam, V.; Kanger, J. S. Fast, Ultrasensitive Virus Detection Using a Young Interferometer Sensor. *Nano Lett.* **2007,** *7* (2), 394–397.
18. Baaske, M.; Vollmer, F. Optical Resonator Biosensors: Molecular Diagnostic and Nanoparticle Detection on an Integrated Platform. *ChemPhysChem.* **2012,** *13* (2), 427–436.
19. Haes, A. J.; Chang, L.; Klein, W. L.; Van Duyne, R. P. Detection of a Biomarker for Alzheimer's Disease from Synthetic and Clinical Samples Using a Nanoscale Optical Biosensor. *J. Am. Chem. Soc.* **2005,** *127* (7), 2264–2271.

20. Perez, J. M.; Kaittanis, C. Magnetic Nanosensors for Probing Molecular Interactions. In *Nanoparticles in Biomedical Imaging*; Bulte, J. M., Modo, M. J., Eds.; Springer: New York, 2008; Vol. 102, pp 183–197.
21. Graham, D. L.; Ferreira, H. A.; Freitas, P. P. Magnetoresistive-Based Biosensors and Biochips. *Trends Biotechnol.* **2004,** *22* (9), 455–462.
22. Ejsing, L.; Hansen, M. F.; Menon, A. K.; Ferreira, H. A.; Graham, D. L.; Freitas, P. P. Magnetic Microbead Detection Using the Planar Hall Effect. *J. MagnMagn Mater.* **2005,** *293* (1), 677–684.
23. Kurlyandskaya, G. V.; Sánchez, M. L.; Hernando, B.; Prida, V. M.; Gorria, P.; Tejedor, M. Giant-Magnetoimpedance-Based Sensitive Element as a Model for Biosensors. *Appl. Phys. Lett.* **2003,** *82* (18), 3053–3055.
24. (a) Johnson, B. N.; Mutharasan, R. Biosensing Using Dynamic-Mode Cantilever Sensors: A Review. *Biosens. Bioelectron.* **2012,** *32*, 1–18; (b) Baller, M. K.; Lang, H. P.; Fritz, J.; Gerber, C.; Gimzewski, J. K.; Drechsler, U.; Rothuizen, H.; Despont, M.; Vettiger, P.; Battiston, F. M.; Ramseyer, J. P.; Fornaro, P.; Meyer, E.; Güntherodt, H.-J. A Cantilever Array-Based Artificial Nose. *Ultramicroscopy* **2000,** *82*, 1–9; (c) Carrascosa, L. G.; Moreno, M.; Álvarez, M.; Lechuga, L. M. Nanomechanical Biosensors: A New Sensing Tool. *TrAC, Trends Anal. Chem.* **2006,** *25*, 196–206.
25. Boisen, A.; Thundat, T. Design & Fabrication of Cantilever Array Biosensors. *Mater. Today.* **2009,** *12*, 32–38.
26. Alvarez, M.; Zinoviev, K.; Moreno, M.; Lechuga, L. M. Cantilever Biosensors. *Opt. Biosen.* **2008**, 419–452.
27. (a) Timoshenko, S. *Vibration Problems in Engineering*; D.Van Nostrand Company, Inc.: New York, 1937; (b) McFarland, A. W.; Poggi, M. A.; Bottomley, L. A.; Colton, J. S. Characterization of Microcantilevers Solely by Frequency Response Acquisition. *J. Micromech. Microeng.* **2005,** *15* (4), 785.
28. (a) Arlett, J. L.; Myers, E. B.; Roukes, M. L. Comparative Advantages of Mechanical Biosensors. *Nat. Nanotechnol.* **2011,** *6*, 203–215; (b) von Muhlen, M. G.; Brault, N. D.; Knudsen, S. M.; Jiang, S.; Manalis, S. R., Label-Free Biomarker Sensing in Undiluted Serum with Suspended Microchannel Resonators. *Anal. Chem.* **2010,** *82* (5), 1905–1910.
29. Hierold, C. From Micro- to Nanosystems: Mechanical Sensors Go Nano. *J Micromech Microeng* **2004,** *14*, S1–S11.
30. (a) Ilic, B.; Czaplewski, D.; Craighead, H. G.; Neuzil, P.; Campagnolo, C.; Batt, C., Mechanical Resonant Immunospecific Biological Detector. *Appl. Phys. Lett.* **2000,** *77* (3), 450–452; (b) Ilic, B.; Czaplewski, D.; Zalalutdinov, M.; Craighead, H. G.; Neuzil, P.; Campagnolo, C.; Batt, C. Single Cell Detection with Micromechanical Oscillators. *J. Vac. Sci. Technol., B* **2001,** *19* (6), 2825–2828.
31. Campbell, G. A.; Mutharasan, R. Detection of Pathogen Escherichia coli O157:H7 Using Self-Excited PZT-Glass Microcantilevers. *Biosen. Bioelectron.* **2005,** *21* (3), 462–473.
32. Vancura, C. Liquid-Phase Chemical and Biochemical Detection Using Fully Integrated Magnetically Actuated Complementary Metal Oxide Semiconductor Resonant Cantilever Sensor Systems. *Anal. Chem.* **2007,** *79* (4),1646–1654.
33. Hung, J.-N.; Hocheng, H. Frequency Effects and Life Prediction of Polysilicon Microcantilever Beams in Bending Fatigue. *MOEMS* **2012,** *11* (2), 021206-1-021206-6.
34. Fritz, J.; Baller, M. K.; Lang, H. P.; Rothuizen, H.; Vettiger, P.; Meyer, E.; Güntherodt, H.; Gerber, C.; Gimzewski, J. K. Translating Biomolecular Recognition into Nanomechanics. *Science (New York, N.Y.)* **2000,** *288*, 316–318.

35. Wu, G.; Datar, R. H.; Hansen, K. M.; Thundat, T.; Cote, R. J.; Majumdar, A. Bioassay of Prostate-Specific Antigen (PSA) Using Microcantilevers. *Nat. Biotechnol.* **2001,** *19*, 856–860.
36. Tamayo, J.; Kosaka, P. M.; Ruz, J. J.; San Paulo, Á.; Calleja, M. Biosensors Based on Nanomechanical Systems. *Chem. Soc. Rev.* **2013,** *42* (3), 1287–1311.
37. Bantscheff, M.; Schirle, M.; Sweetman, G.; Rick, J.; Kuster, B. Quantitative Mass Spectrometry in Proteomics: A Critical Review. *Anal. Bioanal. Chem.* **2007,** *389* (4), 1017–1031.
38. (a) Kurosawa, S.; Nakamura, M.; Park, J.-W.; Aizawa, H.; Yamada, K.; Hirata, M. Evaluation of a High-Affinity QCM Immunosensor Using Antibody Fragmentation and 2-methacryloyloxyethyl Phosphorylcholine (MPC) Polymer. *Biosens. Bioelectron.* **2004,** *20*, 1134–1139; (b) Ko, S.; Park, T. J.; Kim, H.-S.; Kim, J.-H.; Cho, Y.-J. Directed Self-assembly of Gold Binding Polypeptide-Protein A Fusion Proteins for Development of Gold Nanoparticle-Based SPR Immunosensors. *Biosens. Bioelectron.* **2009,** *24* (8), 2592–2597.
39. *Chemical Sensors—Demand and Sales Forecasts, Market Share, Market Size, Market Leaders*; Study no. 3058, The Freedonia Group, 2013; p 284. http://www.freedonia-group.com/Chemical-Sensors.html
40. Hickling, A. Studies in Electrode Polarisation. Part IV.-The Automatic Control of the Potential of a Working Electrode. *T. Faraday Soc.* **1942,** *38*, 27–33.
41. (a) Skoog, D.; Holler, F.; Crouch, S. *Principles of Instrumental Analysis*; Thomson Brooks/Cole: Belmont, 2007; (b) Kissinger, H. Cyclic Voltammetry. *J. Chem. Educ.* **1983,** *60*, 702–706; (c) DuVall, S. H.; McCreery, R. L., Control of Catechol and Hydroquinone Electron-Transfer Kinetics on Native and Modified Glassy Carbon Electrodes. *Anal. Chem.* **1999,** *71*, 4594–4602; (d) Nicholson, R. S. Theory and Application of Cyclic Voltammetry f m Measurement of Electrode Reaction Kinetics. *Anal. Chem.* **1965,** *37*, 1351–1355.
42. Sensors, P., Potentiometric Sensors 10.1. **2012.**
43. Suni, I. I. Impedance Methods for Electrochemical Sensors Using Nanomaterials. *TrAC, Trends Anal. Chem.* **2008,** *27*, 604–611.
44. Bandodkar, A. J.; Wang, J. Non-invasive Wearable Electrochemical Sensors: A Review. *Trends Biotechnol.* **2014,** *32*, 363–371.
45. Shao, Y.; Wang, J.; Wu, H.; Liu, J.; Aksay, I. a.; Lin, Y. Graphene Based Electrochemical Sensors and Biosensors: A Review. *Electroanalysis* **2010,** *22*, 1027–1036.
46. Pandiaraj, M.; Madasamy, T.; Gollavilli, P. N.; Balamurugan, M.; Kotamraju, S.; Rao, V. K.; Bhargava, K.; Karunakaran, C. Nanomaterial-Based Electrochemical Biosensors for Cytochrome c Using Cytochrome c Reductase. *Bioelectrochemistry* **2013,** *91*, 1–7.
47. Kaushik, A.; Vasudev, A.; Arya, S. K.; Pasha, S. K.; Bhansali, S. Recent Advances in Cortisol Sensing Technologies for Point-of-Care Application. *Biosens. Bioelectron.* **2014,** *53* (0), 499–512.
48. Pasha, S. K.; Kaushik, A.; Vasudev, A.; Snipes, S. A.; Bhansali, S. Electrochemical Immunosensing of Saliva Cortisol. *J. Electrochem. Soc.* **2014,** *161* (2), B3077–B3082.
49. Singh, A.; Kaushik, A.; Kumar, R.; Nair, M.; Bhansali, S. Electrochemical Sensing of Cortisol: A Recent Update. *Appl. Biochem. Biotechnol.* **2014,** *174* (3), 1115–1126.
50. Shu, W.; Liu, D.; Watari, M.; Riener, C. K.; Strunz, T.; Welland, M. E.; Balasubramanian, S.; McKendry, R. A. DNA Molecular Motor Driven Micromechanical Cantilever Arrays. *J. Am. Chem. Soc.* **2005,** *127* (48), 17054–17060.

51. Wu, G.; Datar, R. H.; Hansen, K. M.; Thundat, T.; Cote, R. J.; Majumdar, A. Bioassay of Prostate-Specific Antigen (PSA) Using Microcantilevers. *Nat. Biotech.* **2001,** *19* (9), 856–860.
52. Fritz, J. Cantilever Biosensors. *Analyst* **2008,** *133* (7), 855–863.

CHAPTER 2

NANOSURFACE PREPARATION AND BIOFUNCTIONALIZATION: TYPES AND METHODS

SURENDRA K. YADAV[1,*], RAJ KUMAR[2], and K. KAMIL REZA[3]

[1]*University of Rome, Tor Vergata, Italy*

[2]*University of Brescia, Brescia, Italy*

[3]*National Physical Laboratory, New Delhi, India*

**E-mail: surendraky@gmail.com*

CONTENTS

In recent years, nanotechnology has ushered a new era of fabrication techniques for nanomaterials and modifications of the nanosurface in a very subtle way. The progression in the field of nanofabrication has improved the quality of platform available for biofunctionalization. In this chapter, we have discussed the types and methods of nanosurface preparation and simultaneously focused on the biofunctionalization of the nanosurface. The objective of this chapter is to provide a quick glance to an assortment of major techniques available for nanosurface preparation. Also, we have discussed about few well known techniques and methods for the modification of nanosurface with different biomolecules, nanostructured thin film fabrication techniques, and biofunctionalization of nanosurface.

2.1 INTRODUCTION

Necessity is the mother of our invention
Plato (c. 370 B.C.)

The quote by Plato clearly depicts the human nature in exploration of science and technology. Today in this highly competitive world, every research institute/industry is striving to achieve more and more sophisticated know-how through nanotechnology as it ushers a new era of world. Nanotechnology intends to imitate nature by taking advantage of the unique properties of nanoscale matter to come up with more efficient ways of controlling and manipulating molecules at the atomic scale. The "nanoscale" is typically measured in nanometers or billionths of a meter (nanos, the Greek word for "dwarf". being the source of the prefix), and materials built at this scale often exhibit distinctive physical and chemical properties due to quantum mechanical effects.[1] Nanomaterials are used for achieving unique properties or improving up on established functional materials. Nanotechnology is a truly multidisciplinary area of research and development, bringing together the scientific disciplines such as physics, chemistry, biology, engineering, and medicine. The recent interest in nanostructures results from their numerous potential applications such as in material development, biomedical sciences, electronics, optics, magnetism, energy harvesting, energy storage, and electrochemistry.[2,3]

The progress of electronic industry heavily depends on the parameters of interest in the device. The relentless increase in device density in the microelectronics industry provides a very prominent example of the relative importance of surfaces to device fabrication and performance, as the feature size is currently a few tens of nanometers. If the surface is arbitrarily considered to encompass the typical escape depth of an excited electron (1–10 nm), then the progression of microelectronic feature sizes from 65 to diffusion length of electron and below is rapidly approaching the point where the device itself is "all surface". In addition to occupying a greater fraction of device volume for smaller devices, surfaces also provide the opportunity to create unusual measures of atoms and molecules. The bulk structure of a given material is tailored at its surface, and it is often possible to orient and/or bond materials with quite different properties at an interface between two different materials.

The fabrication techniques for the synthesis of nanosurface have evolved as well. The micro- and nanofabrication techniques have been extensively utilized in the industries and laboratories for the preparation of thin film-based platforms. There are varieties of techniques available for the fabrication of nanostructured based thin films. Mostly, nanostructures such as noble metals like gold, silver; carbon allotropes (carbon nanotube, graphene); semiconductors (metal–oxides); quantum dots, and so forth are being used for the production of thin films. Nanofabrication involves a wide range of surface modification procedures that are important in many different areas of nanotechnology. In this chapter, we emphasize on physical and chemical methods for surface modification at the mono- and multilayer levels with particular attention to the interactions between the bulk substrate material and the surface modifier occurring at their interface. In this regard, we focus on the methods for orienting and bonding surface layers on solid substrates such as silicon, metals, carbon, and oxides. Such interactions are often classified into two categories: "physisorption" involving relatively weak attractions between the substrate and surface modifier (e.g., electrostatic attraction) and "chemisorption" involving strong surface bonds that are often covalent. Of particular interest to the reader will be the orientation and structure of the interfacial region resulting from various surface modification procedures. There are several thin film preparation techniques currently in use and we now discuss some of them briefly describe.

Thin films are the preferred form of nanosurface because of the convenient way of fabrication and ease of functionalization with biomolecules. Moreover, these two-dimensional (2D) platforms can also be used as electrode surface for a wide range of application in engineering and medical sectors. Nanostructured thin film requires very precise and selective surface modification in order to be used as a bioelectrode or biocompatible surface. The biofunctionalization method of the thin film surface is a widely explored field in the diagnostic sector to be utilized as the smart bioelectrode surface. Generally, thin film surfaces are biofunctionalized by two ways: physical and chemical methods by various bioconjugation groups. These biofunctionalized electrode surfaces have led to various well-established technologies like biosensor, solar cell, fuel cell, diode, and so forth. In the following sections, we have discussed few well known techniques for the biofunctionalization of nanosurface.

2.2 NANOSTRUCTURED THIN FILM FABRICATION TECHNIQUES

2.2.1 SPUTTERING

Different deposition techniques are available to deposit the nanostructured metal–oxide thin films for different application like biosensors, chemical gas sensors, light-emitting diode, solar cell, and so forth. Thin films can be deposited by pulsed laser, sputtering, spray pyrolysis, spin coating, and atomic layer deposition. Among all the techniques, sputtering is one of the best techniques to deposit nanostructured thin film at the research and industrial levels. Thin films can be deposited by radio frequency (RF), direct current (DC), and pulsed DC magnetron sputtering in inert and reactive atmosphere at different deposition temperature with variation in pressure and power.[4,5]

Nanoparticles growth can be achieved either by bottom-up or top-down approaches. There are various fabrication methods that can be applied for bottom-up approach as well as for top-down approach. Bottom-up approach consists of assembling molecular building blocks to synthesize a nanostructure. We list here some of the methods that are currently in use:

1. Chemical vapor deposition (CVD)
2. Oxide-assisted growth
3. Low temperature
4. Laser ablation catalytic growth
5. Solution–liquid–solid
6. Fluid–liquid–solid
7. Vapor–liquid–solid (VLS)

Top-down approach allows fabrication of nanosized objects starting from bulk material and are based on standard microfabrication methods with deposition, etching, and ion beam milling on planar substrates in order to reduce the lateral dimensions of the films to the nanometer size. We now discuss these methods briefly.

1. Focused ion beam (FIB)
2. Photolithography
3. Electrochemical etching
4. Electron beam lithography (EBL)
5. Thermal oxidation
6. Physical vapor deposition

2.2.2 VAPOR PHASE GROWTH MECHANISM

Wagner and Ellis[6] proposed the idea of growing the nanostructured materials by vapor–liquid growth in a tubular furnace. Similarly, nanostructures' morphology can be developed by the VLS and vapor–solid mechanism, depending on source material and destination growth sites. The deposition technique mainly consists of a tubular furnace, able to reach up to 1500 °C, connected to a rotary vacuum pump, in order to control the pressure inside the alumina furnace's tube.[7] Two mass flow controllers were used to flow the transport gases (argon or oxygen) inside the system. It is possible to tune the morphology of the fabricated nanostructures by changing the condensation temperature, pressure inside the alumina tube, carrier gas flow and composition, deposition time, and catalyst on the target substrates. Further, ZnO,[8] NiO,[9] SnO_2,[10] In_2O_3,[11] CuO,[12] and other metal–oxide nanowires (MOX NWs) were successfully deposited on different kinds of substrates, depending on the applications involved, examples in this regard include silicon for morphological and structural characterization, alumina and silicon membrane for functional measurements as chemical sensors, glass, GaN for solar cells and other optical devices, and Kapton and other polymers as proof of concept for flexible electronics. Pt and Au catalysts can be used to induce the deposition of NWs onto alumina silicon, fused silica, and substrates.

2.2.2.1 TOP-DOWN FABRICATION METHOD

FIB: FIB technique was applied especially in the semiconductor industry, materials science and approaching in the Nanomedicine, biological sciences at nano level for specific applications. The principle of the FIB technique is similar to that of a scanning electron microscope (SEM). The mechanism of the FIB system consists of focused beam of ions which can be operated at high beam currents for site-specific sputtering or low beam currents for imaging. As Figure 2-1 shows, the gallium primary ion beam (Ga^+) hits the sample surface and sputters a small amount of material, which leaves the surface as either secondary ions (i+ or i−) or neutral atoms (n0). The primary beam also produces secondary electrons (e^-). As the primary beam rasters on the sample surface, the signal from the sputtered ions or secondary electrons is collected to form an image.

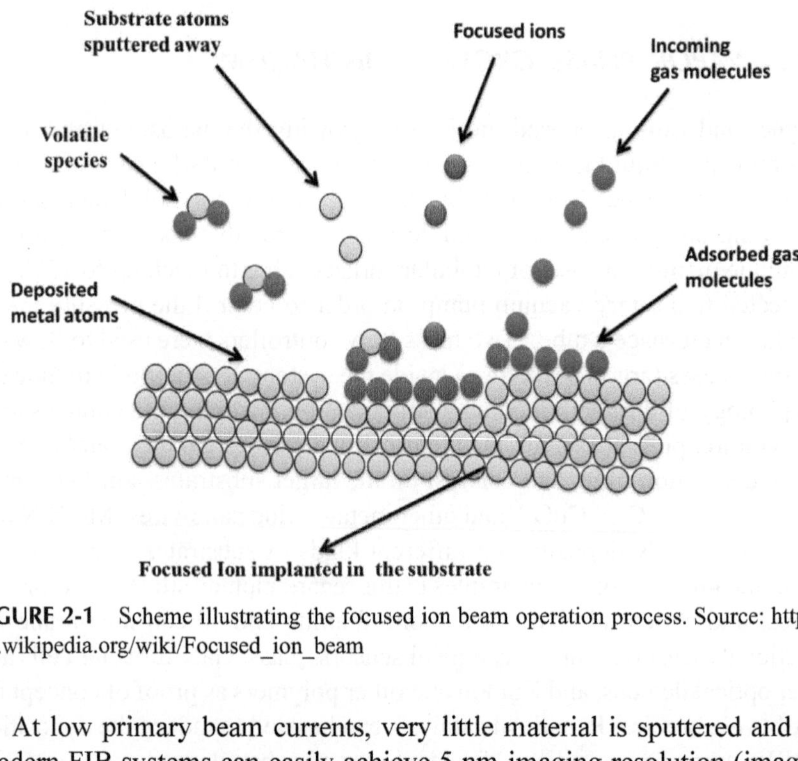

FIGURE 2-1 Scheme illustrating the focused ion beam operation process. Source: https://en.wikipedia.org/wiki/Focused_ion_beam

At low primary beam currents, very little material is sputtered and the modern FIB systems can easily achieve 5 nm imaging resolution (imaging resolution with Ga ions is limited to ~5 nm by sputtering and detector efficiency). At higher primary currents, a great deal of material can be removed by sputtering, allowing precision milling of the specimen down to a submicrometer or even a nanometer scale. If the sample is nonconductive, a low energy electron flood gun can be used to provide charge neutralization. In this manner, by imaging with positive secondary ions using the positive primary ion beam, even highly insulating samples may be imaged and milled without a conducting surface coating, as would be required in a SEM.

Photolithography: Photolithography—also called as ultraviolet lithography or optical lithography—is a method applied in microfabrication to pattern a thin film or the bulk of a substrate. Different steps are included in this process such as masking, cleaning, photoresist, exposure and development, etching, deposition of thin film, and photoresist removal. A beam of light is applied to grow the geometric pattern from a photomask to a light-sensitive chemical "photoresist", or simply "resist", on the substrate. Then, a series of chemical treatments either engraves the exposure pattern into or enables deposition of a new material in the desired pattern upon the material underneath the photoresist. All the main steps involved in a photolithography technique are explained in this section. This procedure removes organic or inorganic contaminations that are present on the wafer surface, either by wet chemical treatment or dry approaches, mostly using plasma treatment. The wafer is covered with photoresist by spin coating. A viscous, liquid solution of photoresist is dispensed onto the wafer, and the wafer is spun rapidly to produce a uniformly thick layer of few micrometers. The photoresist-coated wafer is then prebaked to drive off excess photoresist solvent, typically at 90–100 °C for 30–60 s on a hot plate. After prebaking, the photoresist is exposed to a pattern of intense light. The exposure to light causes a chemical change that allows some of the photoresist to be removed by a special solution, called "developer" by analogy with photographic developer. Positive photoresist—the most common type—becomes soluble in the developer when exposed; however, with negative photoresist unexposed regions are soluble in the developer. A post exposure bake is performed before developing, typically to help reduce standing wave phenomena caused by the destructive and constructive interference patterns of the incident light. The resulting wafer is then "hard-baked" to solidify the remaining photoresist, to make a more durable protecting layer in future ion implantation, wet chemical etching, or plasma etching.

In etching, a liquid ("wet") or plasma ("dry") chemical agent removes the uppermost layer of the substrate in the areas that are not protected by the photoresist. In semiconductor fabrication, dry etching techniques are generally used, as they can be made anisotropic to avoid significant undercutting of the photoresist pattern. This is essential when the width of the features to be defined is similar to or less than the thickness of the material being etched. In other cases, target material (usually a thin metal layer) is deposited on the whole surface of the wafer. This layer covers the remaining resist as well as parts of the wafer that were cleaned of the resist in the previous developing step. After a photoresist is no longer needed, it must be removed from the substrate. The photoresist is washed out together with parts of the target material covering it, and only the material that was in the "holes" remains—this is the material that is in direct contact with the underlying layer. This usually requires a liquid "resist stripper", which chemically alters the resist so that it no longer adheres to the substrate. Up until this point, the photolithography process is comparable to a high precision version of the method used to make printed circuit boards. Subsequent stages in the process have more in common with etching than with lithographic printing. The photolithography process is used because it can create extremely small patterns (down to a few tens of nanometers in size), affords exact control over the shape and size of the objects it creates, and also because it can create patterns over an entire surface in a cost-effective manner. The main disadvantages are that it requires a flat substrate to start with, it is not very effective at creating shapes that are not flat, and it can require extremely clean operating conditions.

EBL: EBL refers to a lithographic process that uses a focused beam of electrons to form the circuit patterns needed for material deposition on (or removal from) the wafer. This is in contrast to optical lithography, which uses light for the same purpose. Further, EBL offers higher patterning resolution than optical lithography because of the shorter wavelength possessed by the 10–50 keV electrons that it employs. Given the availability of technology that allows a small diameter focused beam of electrons to be scanned over a surface, an EBL system no longer needs masks to perform its task (unlike optical lithography, which uses photomasks to project the patterns). An EBL system simply "draws" the pattern over the resist wafer using the electron beam as its drawing pen (Figure 2-2). Thus, EBL systems produce the resist pattern in a serial manner, causing it to be slower than optical systems.

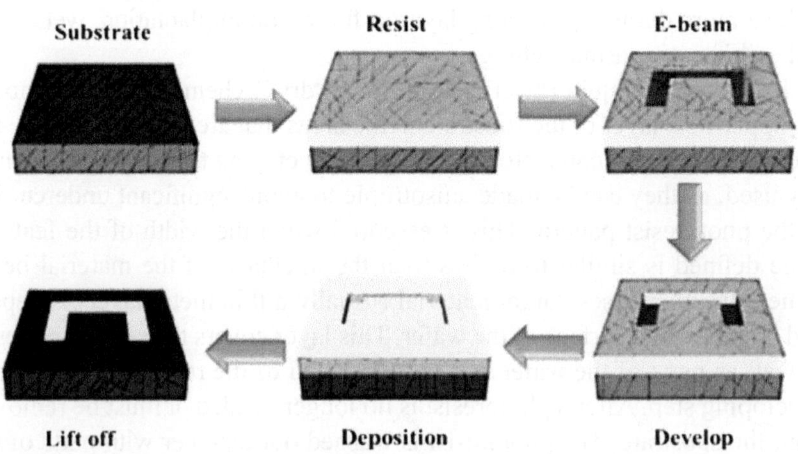

FIGURE 2-2 Illustration of the different steps involved in the electron beam lithography process. Image courtesy of LNBD, Technion.

A typical EBL system consists of the following parts: (a) an electron gun or electron source that supplies the electrons, (b) an electron column that shapes and focuses the electron beam, (c) a mechanical stage that positions the wafer under the electron beam, (d) a wafer handling system that automatically feeds wafers to the system and unloads them after processing, and (e) a computer system that controls the equipment. The resolution of optical lithography is limited by diffraction, but this is not a problem in electron lithography. This is due to the short wavelengths (0.2–0.5 Å) exhibited by the electrons in the energy range used by the EBL systems. However, the resolution of an electron lithography system may be constrained by other factors, such as electron scattering in the resist and by various aberrations in electron optics. Just like optical lithography, electron lithography also uses positive and negative resists. The resolution achievable with any resist is limited by two major factors: the tendency of the resist to swell in the developer solution and electron scattering within the resist. The primary advantage of the EBL process is that it is one of the ways to overcome the diffraction limit of light and make features in the nanometer regime. This form of maskless lithography has found wide usage in photomask making used in photolithography, low-volume production of semiconductor components, and research and development. The key limitation of the EBL process is the throughput—the very long time it takes to expose an entire silicon wafer or glass substrate.

A long exposure time leaves the user vulnerable to beam drift or instability that may occur during the exposure. Also, the turn-around time for reworking or re-design is lengthened unnecessarily if the pattern is not being changed the second time. Figure 2-3a,b demonstrate production level high

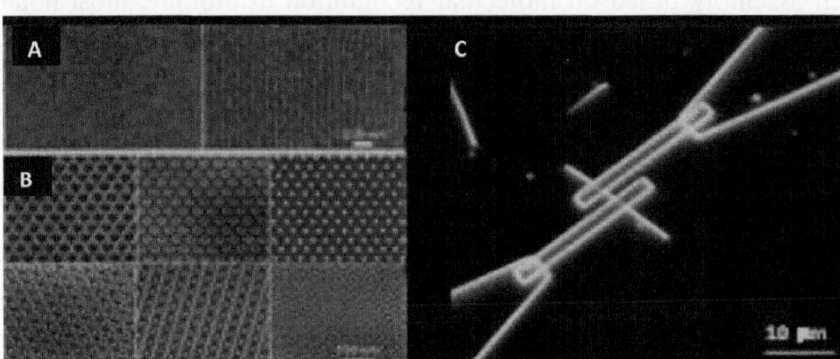

FIGURE 2-3 Nanoscale patterns fabricated on a silicon substrate. (A) Dot and line array patterns from a silicon mask, (B) various patterns obtained from a β-Si3N4mask, and (C) silicon nanowire contacted with two electrodes. Courtesy of SPIE, "A novel technique for projection-type electron-beam lithography" by Ki-Bum Kim, SPIE Newsroom, (2008).

resolution EBL of lines and dots formed in the resist. In this example, 20 nm sized features were delivered and have the pattern transfer capability for a wide range of materials. Nanometer level-to-level alignment accuracy is made possible by in-house software and marker design. Figure 2-3c shows single silicon NW contacted by two adjacent electrodes to enable the measurement of the silicon NW electrical properties.

2.2.2.2 BOTTOM-UP FABRICATION

New bottom-up techniques are being explored as a compliment to traditional top-down methods. In contrast to the removal of excess materials to obtain nanoscale dimensions, bottom-up techniques simply construct the desired features from fundamental building blocks, usually spontaneously through self-assembly without the need for patterning. In the bottom-up approach, materials and devices are built from molecular components which assemble themselves chemically by principles of molecular recognition. In molecular recognition, molecules can be designed so that a specific configuration or arrangement is favored due to noncovalent intermolecular forces. Thus, two or more components can be designed to be complementary and mutually attractive so that they make it more complex and useful complete single molecule. Bottom-up approaches should be capable of producing devices in parallel and be much cheaper than top-down methods, but could potentially be overwhelmed as the size and complexity of the desired assembly increases. Most useful structures require complex and thermodynamically unlikely arrangements of atoms. Nevertheless, there are many examples of self-assembly based on molecular recognition in biology, most notably Watson–Crick base pairing and enzyme–substrate interactions. The challenge for nanotechnology is whether these principles can be used to engineer new functional material to constructs structures in well-designed coordination, fabricating robust structures, and so forth.

Self-assembly: A key approach in the bottom-up fabrication technique is the self-assembly. Self-assembly in the classical sense can be defined as the spontaneous and reversible organization of molecular units into ordered structures by noncovalent interactions. The first property of a self-assembled system that this definition suggests is the spontaneity of the self-assembly process—the interactions responsible for the formation of the self-assembled system act on a strictly local level, or in other words, the nanostructure builds itself. There are at least three distinctive features such as disorder to ordered structure, interaction and building blocks that make self-assembly a distinct concept: the self-assembled structure must have a higher order than the isolated components, be it a shape or a particular task that the self-assembled entity may perform.

Interactions: Self-assembled structures rely on slack interactions (e.g., van der Waals, capillary, hydrogen bonds, etc.) with respect to more "traditional" covalent, ionic, or metallic bonds. Although typically less energetic by a factor of 10, these weak interactions play an important role in materials' synthesis. It can be instructive to note how slack interactions hold a prominent place in materials, but especially in biological systems, although they are often considered marginal when compared to strong (i.e., covalent, etc.) interactions.

Building blocks: The building blocks are not only atoms and molecules, but span a wide range of nano-and mesocopic structures with different chemical compositions, shapes, and functionalities. These nanoscale building blocks can in turn be synthesized through conventional chemical routes or by other self-assembly strategies. There are two types of self-assembly: intra-molecular and inter-molecular. Intra-molecular self-assembling molecules are often complex polymers with the ability to assemble from the random coil conformation into a well defined stable structure (secondary and tertiary structures). An example of intra-molecular self-assembly is protein folding. Inter-molecular self-assembly is the ability of molecules to form supramolecular assemblies (such as quaternary structure). A simple example is the formation of a micelle by surfactant molecules in a solution.

Self-assembly in biology: Self-assembly processes can occur spontaneously in nature, such as in cells and other biological systems, as well as in human-engineered systems. A self-assembly process usually results in an increase in internal organization of the system. Many biological systems use self-assembly to assemble various molecules and structures. Imitating these strategies and creating novel molecules with the ability to self-assemble into supramolecular assemblies is an important technique in nanotechnology. In self-assembly, the final (desired) structure is "encoded" in the shape and properties of the molecules that are used, in contrast to traditional techniques, such as lithography, where the desired final structure must be carved out from a larger block of matter. One example of a biology based self-assembly process is the successful assembly of 3D multicomponent nanoscale structures by scientists at the Brookhaven National Laboratory of the U.S. Department of Energy (DOE). These structures incorporated light-absorbing and emitting particles that allowed tunable optical properties. In this work, illustrated in Figure 2-4, the scientists used DNA linkers with three binding sites (black "strings") to connect gold nanoparticles (orange and red spheres)

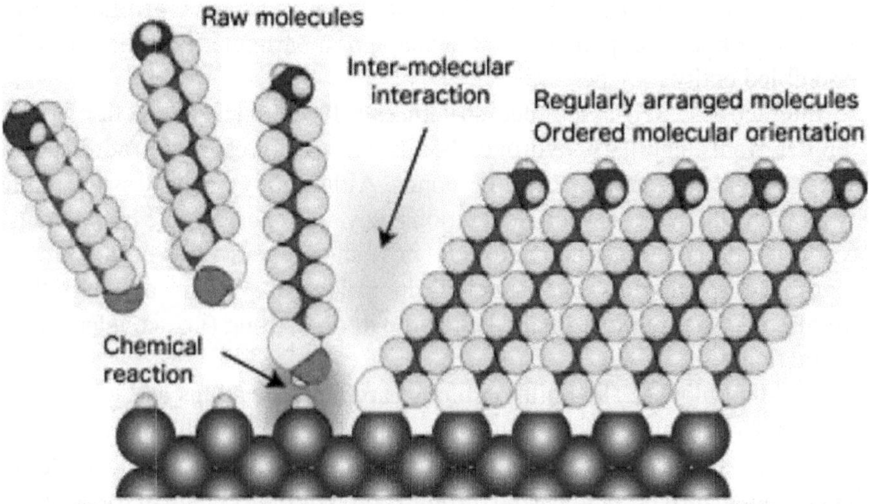

FIGURE 2-4 Self-assembly of monolayers. The head groups are anchored onto the substrate while the tail groups assemble together far from the substrate. Image courtesy of Dr. Hiroyuki Sugimura, Nanoscopic Surface Architecture Laboratory, Department of Materials Science and Engineering Kyoto University, Japan. http://www.mtl.kyoto-u.ac.jp/english/laboratory/nanoscopic/nanoscopic.htm

and fluorescent dye molecules (blue spheres) tagged with complementary DNA sequences. The DNA linker molecules had three binding sites. The two ends of the strands were designed to bind to complementary strands on plasmonic gold nanoparticles—particles in which a particular wavelength of light induces a collective oscillation of the conductive electrons, leading to strong absorption of light at that wavelength. The internal part of each DNA linker was coded to recognize a complementary strand chemically bound to a fluorescent dye molecule. This set up resulted in the self-assembly of 3D body centered cubic crystalline structures with gold nanoparticles located at each corner of the cube and in the center, with dye molecules at defined positions in between.

Self-assembly of monolayers: Self-assembled monolayers (SAMs) of organic molecules are molecular assemblies formed spontaneously on surfaces by adsorption and that are organized into relatively large ordered domains. In some cases, the molecules that form the monolayer do not interact strongly with the substrate. In other cases, the molecules possess a functional group that has a strong affinity to the substrate that anchors the molecule. Such a SAM consisting of a head group, tail, and functional end group is depicted in Figure 2-4. Common head groups include thiols, silanes, and phosphonates. Further, SAMs are formed by the chemisorption of head groups onto a substrate from a vapor or liquid phase, followed by the slow organization of tail groups. Initially, at small molecular density, adsorbate molecules form either a disordered mass of molecules or an ordered 2D "lying down phase". At higher molecular coverage, over a period of minutes to hours, the molecules begin to form 3D crystalline or semi crystalline structures on the substrate surface. The head groups assemble together on the substrate, whereas the tail groups assemble far from the substrate. Areas of close-packed molecules nucleate and grow until the surface of the substrate is covered in a single monolayer. Thin film SAMs can be placed on nanostructures, functionalizing the nanostructure. This is advantageous because the nanostructure can now selectively attach itself to other molecules or SAMs. This technique is useful when designing biosensors or other devices that need to separate one type of molecule from its environment.

2.3 BIOFUNCTIONALIZATION

To quote Richard P. Feynman:

> *cells are very tiny, but they are very active; they manufacture various substances; they walk around; they wiggle; and they do all kinds of marvelous things—all on a very small scale. Also, they store information. Consider the possibility that we too can make a thing very small which does what we want—that we can manufacture an object that manoeuvres at that level!.......Encyclopaedia Britannica, Nanotechnology.*

Generally immobilization methods for bioactive molecules (BAMs) must be evaluated with respect to several properties that are, in part, contradictory. In all cases, immobilized BAMs must display their bioactive domain(s) to the cells of the host tissue in a native conformation and must be accessible to the cells, that is, they need to be at a certain distance from the surface. The integrity of the BAMs to be immobilized must not be affected and no harmful substances involved in the immobilization process should remain at the surface to be accidently released in vivo. Some molecules (e.g., RGD peptides) must be immobilized irreversibly; others (e.g., growth factors or antibiotics) have to be released in a specific concentration–time profile to be effective. Realizing a defined release behavior of surface-bound molecules is probably the most critical issue in this field. In the past BAMs have been immobilized at titanium surfaces adsorptively and covalently via electrochemical techniques or using self-assembled layers. As we discuss in this regard shortly, all methods have their own advantages and limits with respect to their feasibility, their impact on the integrity and activity of the BAMs as well as on binding stability and release characteristics of the immobilized molecules.[13]

Surface functionalization is one of the important segments that have been researched extensively. This holds the key for successful interactions between the matrix and biomolecules. The most important segment is to decide the surface to be functionalized with what kind of functional group/groups. It is a crucial step which requires a deliberate and delicate approach. However, it is a tiny step in experiments with longer duration but often plays a significant role in determining the outcome of the experiment. The success of the whole experiment requires every step to be optimized and carefully selected parameters and boundary conditions. Surface functionalization at the nanoscale provides the ideal platform available for a reaction to occur with ease in a stable way. There are numerous approaches to implement the surface modification phenomenon by applying various functional groups at the nanoscale platform.[14]

Biofunctionalization is a process which is based on ligand exchange and chemical functionalization at the nanoscale surface of the biomolecules. The biomolecules generally include DNA, proteins, vitamins, peptides, enzymes, and so forth. The biofunctionalization at the nanoscale surface leads to hybrid materials with organic and inorganic features. The bonding process might be facilitated by physical and chemical methods depending upon the biomolecules and matrix interaction at the nanosurface. There are four classes of surface modifications by the biomolecules: (a) ligand attachment by the specific group on to the inorganic surface, (b) covalent bonding is responsible for strongest bonding between the biomolecule and matrix surface, (c) the electrostatic interaction due to presence of opposite charges in the biomolecule and surface charges of the nanomaterial, and (d) the affinity-based interactions lead to a noncovalent system[14] (Figure 2-5).

Heterobifunctional groups are one of the special kinds of coupling reagents having two different functional sides. These kinds of functional groups act as a bridge to connect two different types of molecules. The affinities of the reactive side have specific affinity toward a particular molecule based on its natural structure. Aminothiol is one such coupling reagent which has on one side the -NH_2 group (amine group) and on another side the -SH (thiol group). Cysteamine in one such simple coupling agent commonly used for bioconjugations. It has a structure which helps to combine thiols

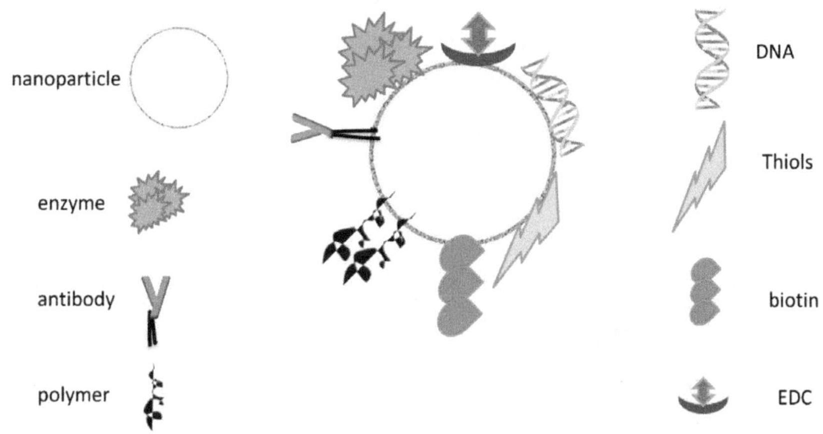

FIGURE 2-5 Biofunctionalization of nanosurface by various agents using physical and chemical methods.

and amine groups via two different carbon–hydrogen chains. The structure we are referring to here is $HS\text{-}CH_2CH_2\text{-}NH_2$.

We now discuss few biofunctionalization methods on nanosurface.

2.3.1 1-ETHYL-3-[3-DIMETHYLAMINOPROPYL] CARBODIIMIDE (LIGAND ATTACHMENT)

1-Ethyl-3-[3-dimethylaminopropyl] carbodiimide (EDC) is a cross-linking group for a strong bond formation. This is the most commonly used cross-linking group for nanoparticle's surface conjugation. Further, EDC is water soluble, making it more conducive for most biochemical reactions to occur in the presence of aqueous buffer solutions. Cross-linking occurs through an amide bond or phosphoramidate linkage by coupling of carboxylate groups and primary amines.[15] EDC are called zero-length reagents. *N*-substituted carbodiimides can react with carboxylic acids to form an intermediate state which is highly reactive which eventually reacts with a primary amine to form an amide bond. In addition, EDC is also used for nucleic acid labeling through 5′ phosphate groups, amide bond formation in proteins, and so forth. Also, *N*-Hydroxysuccinimide (NHS) or *N*-Hydroxysulfoxuccinimide is coupled with EDC as a stable bioconjugation to achieve a better and higher efficiency. In addition, EDC/NHS coupling is suitable for biosensor application with high sensitivity and stability.[16,17] Ali et al.[16] used EDC/NHS coupling for detection of low-density lipoprotein using amine functionalized reduced graphene oxide. Reza et al.[17] successfully carried out biofunctionalization of multiwalled carbon nanotube in the presence of EDC/NHS cross-linker for urea sensor (Figure 2-6a).

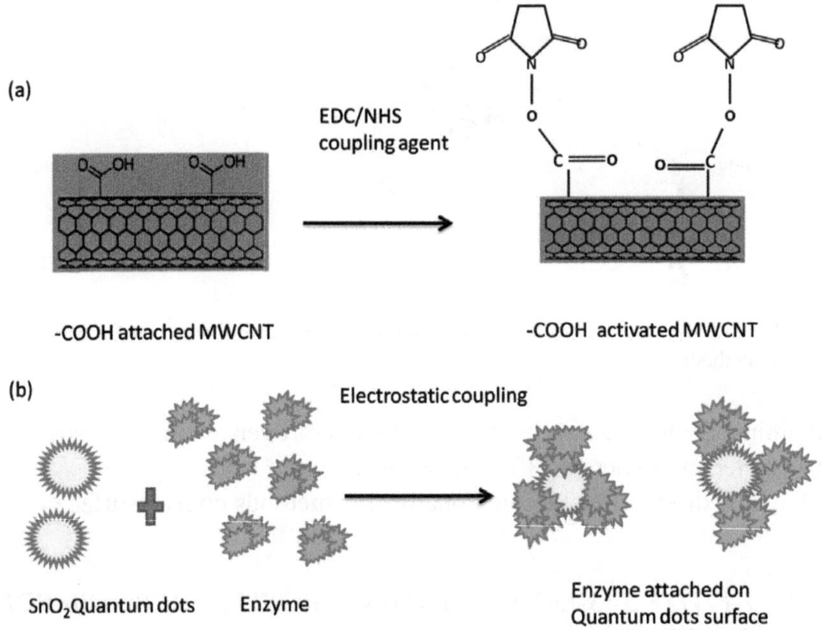

FIGURE 2-6 (A) 1-Ethyl-3-[3-dimethylaminopropyl] carbodiimide/*N*-Hydroxysuccinimide and (B) electrostatic coupling.

2.3.2 ELECTROSTATIC INTERACTION

The synthesis mechanisms play a crucial role in deciding the surface charge nature of the nanomaterials. The type of charge on nanomaterials decides what kind of functional groups can be attached to the nanosurface. If the surface of a nanomaterial is negatively charged, then an oppositely charged functional group like an amine group might be attached electrostatically for a stable cross-linking. Further, identical charge on the nanomaterials prevents them from aggregation. The presence of a metal atom as a cofactor and vitamins as a coenzyme help an enzyme to interact and bind with its substrate. Highly pure tin oxide quantum dots were electrostatically attached with the enzyme xanthine oxidase,[18] and few quantum dot probes are currently being used for fluorescent tagging of biomolecules based on electrostatic interactions[19] (Figure 2-6b).

The electrostatic interactions appear universally if charges (either positive or negative) become separated by a finite distance due to ionization or attachment of ionic species. In vacuum or dielectric (ion-free) media, for example, air and organic nonpolar liquids, the electrostatic interactions are governed by the Coulomb's law. Thus, there appear repulsive forces between like charges and attractive forces between opposite charges. Electrostatic interactions are between and among cations and anions, which are species with formal charges of ...−2, −1, +1, +2, ... Electrostatic interactions can be either attractive or repulsive, depending on the signs of the charges. Favorable electrostatic interactions cause the vapor pressure of sodium chloride and other salts to be very low. If you leave table salt (NaCl) out on a table, how long does it take before it sublimes? A very long time indeed. The electrostatic interactions within a sodium chloride crystal are called ionic bonds. However, when a single cation and a single anion are close together, within a protein, or within a folded RNA, those interactions are considered to be noncovalent electrostatic interactions. Electrostatic interactions can be very strong, and fall off gradually with distance ($1/r$). As discussed later in this chapter, electrostatic interactions are highly attenuated (dampened) by water.

2.3.3 BIOTIN–AVIDIN SYSTEM (AFFINITY-BASED SYSTEM)

The biotin–avidin molecular system is a naturally occurring coupling reagent system for bioconjugations.[14] Biotin is a small vitamin molecule having a carboxylic group, whereas avidin is a protein molecule. Each (strept) avidin molecule can bind with maximum four biotin sites making it a stronger noncovalent arrangement. Various functional groups like carbohydrates, carboxylates, amines, and so forth can be attached to any nanoscale surface using the biotin–avidin coupling system. The chemical stability and acute specificity of this coupled system has been explored in different types of conjugation chemistry. The isoelectric point of proteins depends on pH values. As a result of that, proteins can have either positive or negative state of charge which enables them to interact electrostatically with the inorganic materials for bond formation. Moreover, the functional groups present in the proteins can be bounded with the carboxylic groups or amine groups of various nanomaterials.

2.3.4 ASYMMETRIC GROUP

Generally symmetric nanoparticles (spherical, triangular, rod-shaped, tetrapods, etc.) are preferred for their geometric regularity for ease of functionalization. However, in few cases the non-regular-shaped nanoparticles also have shown greater affinity toward the biomolecule attachment. Asymmetric biofunctionalization of a nanosurface can be seen with single-stranded DNA (ssDNA) and gold nanostructure for a nonperiodic arrangement of nanocrytals,[20] whereas Cd/Te tetrapod nanostructures were asymmetrically functionalized via a site selective modification method to prepare the desired complex structures.[21]

2.3.5 POLYMERS IN FUNCTIONALIZATION

The functionalization of a nanosurface can also be accomplished by certain long chain molecules like polymers because of their biocompatible, nontoxic, and stable nature in organic and aqueous solvents. The most common polymer in this regard is polyethylene glycol (PEG) which is a molecule with chemically stability stable in many organic and inorganic solvents as well as in water in certain states. This biocompatible polymer is additive in nature and its nontoxic behavior makes it a suitable bioconjugate for the desired applications in biomedical sectors.[22] Further, PEG modified nanoparticle's surfaces are stable and water soluble in biological surroundings. Moreover, PEG molecules can specifically bind to other functionalized molecules like amino acid containing groups (e.g., cysteine, lysine, etc.) for nanoparticle's functionalization.[23] Consequently, these functionalized nanoparticles would be available for a variety of applications. The entrapment of nanoparticles by other polymers like chitosan, polyaniline, and so forth has been prolific for biocompatible stable film formation. Chitosan modified metal–oxide surfaces have been utilized for their large, active surface area along with natural additive functional amine groups for biomolecule attachments.[24,25]

Another noncovalent method for immobilizing biomolecules on nanoparticles is to entrap them in biocompatible films such as Nafion and DNA. The coating films not only prevent the aggregation of nanoparticles but also provide abundant positions for functionalization of nanoparticles with the second biomolecule. An electroactive polymer can generate a large number of electrons during electrochemical oxidation to amplify the electrochemical signal and therefore enhance the detection sensitivity. It has been shown that ssDNA can wrap around single-wall nanotubes (SWNTs) via an aromatic interaction to form a soluble DNA–SWNTs complex, which is promising for construction of a highly sensitive biosensor for target biomolecules.[26]

2.3.6 MALEIMIDE COUPLING

Maleimides are maleic acid derivatives. They are a combination of maleic acid and imides group. They are part of organic synthesis as well as bioconjugation process for specific reactions. Thiols have specific affinity toward maleimide for a stronger bond formation. Maleimides being electrophilic in nature are specific toward thiols group. However, the reaction between thiol and maleimide is pH dependent. The reaction is favorable above pH 7.0 in most of the cases for addition of thiols and imides. Maleimides also work as a coupling agent for biomolecule labeling in many heterobifunctional cross-linking agents.[27]

2.3.7 CLICK CHEMISTRY

Generally, click chemistry is a rapid synthesis process for heteroatom's formation in a very stable and natural way pioneered by the Sharpless group.[28] However, there are some methods in click chemistry that are used for the conjugation purpose as well. The conjugation method based on click chemistry has been utilized to form heterocyclic rings. The formation of 1, 3-dipolar cycloaddition between azides and alkynes in the presence of the Cu(I) catalyst has been used for the synthesis of conjugated polymers.[29] Click chemistry is a high temperature synthesis process which is why it is not suitable for cross-linking process. However, the low-temperature synthesis of a triazole ring of 1, 3-dipolar cycloaddition leads to bioconjugation purpose as well as good product yield. Many applications based on cross-linking chemistry are now a reality using carbon and carbon multiple bonds, epoxidation, dihydroxylation, and so forth. There are reports pertaining to nanoparticle's surface modifications based on cross-linking chemistry for a stable bond formation in drug delivery.[30] This methodology has good selectivity and specificity even in long-chain nanomaterials.

2.4 CONCLUSIONS

In this chapter, we have discussed about the various thin film nanofabrication techniques. The fabrication methods such top -down and bottom-up approaches are extensively used in industries as well as in research laboratories. The most common methods for thin-film fabrication such as the

CVD system, sputtering system, photolithography, self-assembly, and so forth have been discussed. These innovative methods have improved the quality of a nanosurface and provided a better platform for the functionalization of biomolecules. We have also discussed the different biofunctionalization methods based on the type of affinity between the nanosurface and the biomolecules. We have also discussed quite a few approaches such as EDC, avidin–biotin system, click chemistry, and so forth which are generally utilized for bioconjugation processes. The process of bioconjugation between nanosurface and biomolecules has been widely explored and utilized in biosensors, solar cells, drug delivery system, and so forth.

ACKNOWLEDGMENTS

We would like to express our sincere gratitude to all the scientists and researchers throughout the world for their immense contributions in this field of nanosurface preparation and biofunctionalization. Particularly, we sincerely acknowledge the authors whose names appear in the reference list. We have included in this chapter various techniques and concepts (developed by myriad authors) that lead to application of biofunctionalization process in many fields. We are also very grateful to our family, friends, and teachers for their continuous encouragement and support, without which this chapter would not have made it to its present form.

KEYWORDS

- **Nanosurface**
- **Biofunctionalization**
- **SEM**
- **Nanoscale**
- **Asymmetric**
- **Chitosan**

REFERENCES

1. S.Tom Picraux, Encyclopaedia Britannica: Nanotechnology: http://www.britannica.com/EBchecked/topic/962484/nanotechnology.
2. Official website of the United States National Nanotechnology Initiative, http://www.nano.gov/nanotech-101/what.
3. Feynman, R. P. There's Plenty of Room at the Bottom. *Eng. Sci.* **1960,** *23* (5), 22–36.
4. Safi, I. Recent Aspects Concerning DC Reactive Magnetron Sputtering of Thin Films: A Review. *Sur. Coat. Tech.* **2000,** *127* (2), 203–218.
5. Wasa, K.; Kitabatake, M.; Adachi, H. *Thin Film Materials Technology: Sputtering of Control Compound Materials*. Springer Science & Business Media, **2004**.
6. Wagner, R. S.; Ellis, W. C. Transactions of the Metallurgical Society of Aime, June, **1965**, vol. 233, 1053.
7. Zappa, D. Metal Oxide Nanostructures for Sensing Applications. Ph.D. Thesis, http://www.ing.unibs.it/~dario.zappa/,**2013**.
8. Park, W. I.; Kim, D. H.; Jung, S. W.; Yi, G. C. Metalorganic Vapor-Phase Epitaxial Growth of Vertically Well-Aligned ZnO nanorods. *Appl. Phys. Lett.* **2002,** *80*, 4232.
9. Oka, K.; Yanagida, T.; Nagashima, K.; Tanaka, H.; Kawai, T. Nonvolatile bipolar resistive memory switching in single crystalline NiO heterostructured nanowires. *J. Am. Chem. Soc.* **2009,** *131* (10), 3434–3435.
10. Choi, Y. J.; Hwang, I. S.; Park, J. G.; Choi, K. J.; Park, J. H.; Lee, J. H. Novel Fabrication of an SnO2 Nanowire Gas Sensor with High Sensitivity. *Nanotechnology* **2008,** *19* (9), 095508.
11. Zhang, D.; Liu, Z.; Li, C.; Tang, T.; Liu, X.; Han, S.; Zhou, C. Detection of NO_2 down to ppb levels using individual and multiple In2O3 nanowire devices. *Nano. Lett.* **2004,** *4* (10), 1919–1924.
12. Liao, L.; Zhang, Z.; Yan, B.; Zheng, Z.; Bao, Q. L.; Wu, T.,; Yu, T. Multifunctional CuO Nanowire Devices: p-type Field Effect Transistors and CO Gas Sensors. *Nanotechnology* **2009,** *20* (8), 085203.
13. Rene, B.; Michael, J.; Schwenzer, B.; Scharnweber, D. Biological Nano-Functionalization of Titanium-Based Biomaterial Surfaces: A Flexible Toolbox. *J. R. Soc. Interface* **2010,** *7*, S93–S105.
14. Sperling, R.; Parak, W. Surface Modification, Functionalization and Bioconjugation of Colloidal Inorganic Nanoparticles. *Philos. T. Roy. Soc. A: Math. Phys. Eng. Sci.* **2010,** *368* (1915), 1333–1383.
15. Conde, J.; Dias, J. T.; Grazú, V.; Moros, M.; Baptista, P. V.; De La Fuente, J. M. Revisiting 30 Years of Biofunctionalization and Surface Chemistry of Inorganic Nanoparticles for Nanomedicine. *Front. Chem.* **2014,** *2*, 48.
16. Ali, M. A.; Kamil Reza, K.; Srivastava, S.; Agrawal, V. V.; John, R.; Malhotra, B. D. Lipid–Lipid Interactions in Aminated Reduced Graphene Oxide Interface for Biosensing Application. *Langmuir* **2014,** *30* (14), 4192–4201.
17. Reza, K. K.; Srivastava, S.; Yadav, S. K.; Biradar, A. Biofunctionalized Carbon Nanotubes Platform for Biomedical Applications. *Mater. Lett.* **2014,** *126*, 126–130.
18. Reza, K. K.; Singh, M. K.; Yadav, S. K.; Singh, J.; Agrawal, V. V.; Malhotra, B. Quantum Dots Based Platform for Application to Fish Freshness Biosensor. *Sens. Actuators B.* **2013,** *177*, 627–633.
19. Lei, J.; Ju, H. Signal Amplification Using Functional Nanomaterials for Biosensing. *Chem. Soc. Rev.* **2012,** *41* (6), 2122–2134.

20. Loweth, C. J.; Caldwell, W. B.; Peng, X.; Alivisatos, A. P.; Schultz, P. G. DNA-Based Assembly of Gold Nanocrystals. *Angew. Chem. Intern Ed.* **1999,** *38* (12), 1808–1812.
21. Liu, H.; Alivisatos, A. P. Preparation of Asymmetric Nanostructures Through Site Selective Modification of Tetrapods. *Nano. Lett.* **2004,** *4* (12), 2397–2401.
22. Van Vlerken, L. E.; Vyas, T. K.; Amiji, M. M. Poly (ethylene glycol)-modified Nanocarriers for Tumor-Targeted and Intracellular Delivery. *Pharma. Res.* **2007,** *24* (8), 1405–1414.
23. Roberts, M.; Bentley, M.; Harris, J. Chemistry for Peptide and Protein PEGylation. *Adv. drug del. Rev.* **2012,** *64*, 116–127.
24. Reza, K. K.; Singh, N.; Yadav, S. K.; Singh, M. K.; Biradar, A. Pearl Shaped Highly Sensitive Mn_3O_4 Nanocomposite Interface for Biosensor Applications. *Biosen. Bioelectron.* **2014,** *62*, 47–51.
25. Kaushik, A.; Solanki, P. R.; Ansari, A. A.; Sumana, G.; Ahmad, S.; Malhotra, B. D. Iron Oxide-Chitosan Nanobiocomposite for Urea Sensor. *Sens. Actuator B.* **2009,** *138* (2), 572–580.
26. Lei, J.; Ju, H. Signal Amplification Using Functional Nanomaterials for Biosensing. *Chem. Soc. Rev.* **2012,** *41* (6), 2122–2134.
27. Hermanson, G. T. *Bioconjugate Techniques*; Academic Press: Amsterdam, **2013**.
28. Kolb, H. C.; Finn, M.; Sharpless, K. B. Click Chemistry: Diverse Chemical Function from a Few Good Reactions. *Angew. Chem. Intern Ed.* **2001,** *40* (11), 2004–2021.
29. Lutz, J. F. 1, 3-Dipolar Cycloadditions of Azides and Alkynes: A Universal Ligation Tool in Polymer and Materials Science. *Angew. Chem. Intern Ed.* **2007,** *46* (7), 1018–1025.
30. Archanaá Krovi, S. Clickable Polymer Nanoparticles: A Modular Scaffold for Surface Functionalization. *Chem. Comm.* **2010,** *46* (29), 5277–5279.

CHAPTER 3

ANTIBODY IMMOBILIZATION CHEMISTRIES FOR NANOSURFACES

CHANDRA K DIXIT[1*], CIARÁN Ó'FÁGÁIN[2], and RICHARD O'KENNEDY[2,3]

[1]*Department of Chemistry, University of Connecticut, Storrs, USA*

[2]*Department of Biotechnology, Dublin City University, Glasnevin, Dublin, Ireland*

[3]*Biomedical Diagnostics Institute, Dublin City University, Glasnevin, Dublin, Ireland*

[*]*E-mail: chandra.kumar_dixit@uconn.edu*

CONTENTS

In Chapter 2, we had a summary of few biofunctionalization approaches for nanosurfaces. In this chapter, we describe approaches for biomolecule immobilization in the perspective of antibodies. Our emphasis in this chapter is on the site-directed and oriented capture methods. We discuss the impact of immobilization strategy on the functionality of captured antibodies and their packaging densities. We also outline how specific immobilization strategies could be optimally employed to minimize variations in any diagnostics or in vivo targeting applications.

3.1 INTRODUCTION

Mostly, biomolecules—such as proteins and nucleic acids—possess functional groups. These functional groups are usually manipulated to capture these biomolecules. Enzymes and antibodies constitute the majority of the sensing components in biosensors. In this chapter, we mainly focus on the antibody and discuss biomolecule immobilization with the perspective of the functional groups of an antibody, such as carboxyls, amines, sulfhydryls, hydroxyls, and carbohydrates.

Chemically, an antibody, for example, immunoglobulin G (IgG) is a complex glycoprotein [1] with many pendant functionalities, such as primary and secondary amines and carboxyl and hydroxyl groups. IgG has a molecular weight of approximately 150 KDa and is composed of two heavy and two light chains as shown in Figure 3-1. The heavy chains are linked together by disulfide bonds and a disulfide bond also links each heavy chain to its corresponding light chain. However, the structural and molecular composition changes with each class/isotype (e.g., IgA, IgD, IgE, IgG, and IgM) and subclass (e.g., IgG_1, $IgG_{2a\ and\ 2b}$, and IgG_4) of antibodies. There are two antigen binding sites on each basic antibody unit. The binding functionality of an antibody is controlled by the amino acid sequence in this binding region and specifically in the hypervariable regions at the N-terminal location. Mutations in this hypervariable region are responsible for antigen specificity.

Antibodies, due to their high antigen specificity and affinity [2, 3], can make excellent probes and, therefore, have multiple applications in fields such as biosensors and diagnostics. The development of these bioanalytical detection systems requires antibody immobilization on various solid supports. Some of the most important prerequisites for designing an efficient immobilization strategy are (a) the antibody must retain its functional conformation, and thus its activity, after immobilization, (b) the surface for use in attachment should either possess or be amenable to the grafting of desired chemical functionalities, and (c) attachment must occur easily with maximum efficiency and minimum loss in antibody-binding capacity. In this chapter, we discuss antibody immobilization strategies and hence critically analyze those strategies that provide (i) site-specific capture guided through various tags and functional groups on antibodies and (ii) orientation achieved through biomolecules [4–8].

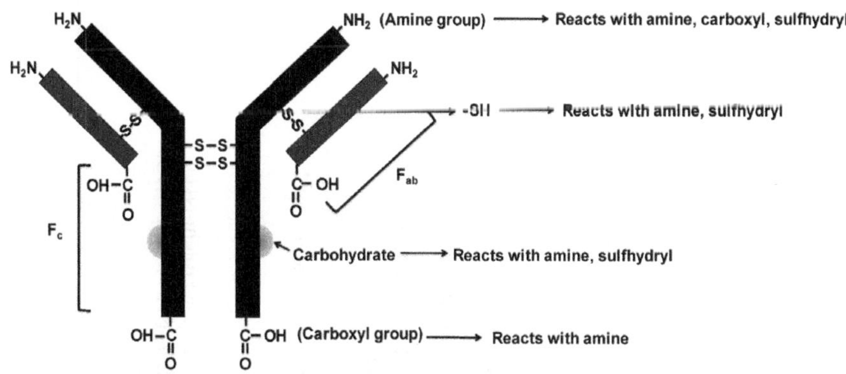

FIGURE 3-1 Model of a typical Immunoglobulin G with the available functional groups. An antibody consists of two heavy (black bars) and two light (blue bars) chains. Both of the heavy chains are linked together with two disulfide bridges at the hinge region. Similarly, each light chain is linked to its respective heavy chain through a disulfide bridge. There are numerous reactive functional groups on the heavy and light chains of an antibody. Amine and carboxyl groups are the main reactive functional groups present in a native-state antibody. Sulfhydryl groups can be generated by reducing the disulfide bonds of an antibody with Dithiotreitol or Tris(2-carboxyethyl)-phosphine. Carbohydrates in the fragment crystalline section can be activated by periodate treatment [9, 10]. The functional groups mentioned can be treated with appropriate cross-linking reagents in order to immobilize an antibody on a prefunctionalized surface, or they can be involved in "reagent-free" antibody capture by modulating the charges on these groups (e.g., via ionic interactions).

Therefore, a precise knowledge of an antibody's chemical and functional properties is important because sometimes the antibody must be pretreated (Table 3-1) in order to perform a particular immobilization scheme. There are several strategies to ensure the criteria specified in points (a) to (c) discussed earlier and site-specific orientated presentation of antibodies on the surface is the most important [8, 11]. However, scarcity of information has limited understanding and development of controlled immobilization. In addition, the antibody is prone to loss of specificity and functional conformation (due to steric hindrance [2, 12] or molecular flattening and spreading [5, 6, 13–16]); however, these can compromise the sensitivities of developed immunoassays.

TABLE 3-1 Commonly employed antibody modification strategies for immobilization

Strategy for antibody modification	Target functional group/point of modification	Reactions	Results
Activation of carbohydrates by oxidation with sodium ($NaIO_4$)/potassium periodate (KIO_4)	Hydroxyls of carbons at position 3 and 4 in the sugar ring of carbohydrate, present in the fragment crystalline region of antibodies	C(OH)–C(OH) → 2(-CHO)	Generates highly reactive aldehyde groups
Alkylation by reduction	(a) of disulfide bonds (-S-S-)	(a) Disulfide bonds (Step 1): -S-S- (DTT/TCEP) → -SH + -SH (Step 2): -SH + alkylating agent → -S-alkane	Generates reactive thiol groups and reactive alkyl derivative
	(b) of amines of lysine ($-NH_2$)	(b) lysine: $-NH_2$ + alkylating agent → -NH-alkane	Generates reactive amino alkanes
Acetylation	α-NH_2 of the protein located at the N-terminal	CH_3COOCH_3 + α-NH_2-protein→ CH_3CO-NH-protein	
(a) Chlorination/tosylation (Tresyl and Tosyl chloride derivatives [Figure 3.4])	of hydroxyl group (-OH)	$CH_3C_6H_4SO_2Cl$ + Protein-OH → $CH_3C_6H_4SO_2$-O-Protein + HCl	Allows hydroxyl group activation
(b) Sulfonamidation (Tresyl and Tosyl chloride derivatives [Figure 3.4])	of amine groups ($-NH_2$ 1°, 2°)	$CH_3C_6H_4SO_2Cl$ + Protein-NH_2 → $CH_3C_6H_4SO_2$-NH-Protein + HCl	Generates reactive sulfonamide derivative using amino groups

Adsorption is a very commonly employed approach for antibody immobilization. It can be used with any type of biosensor surface, that is, metallic- or plastic-based surface. It is mainly categorized as physico-/chemisorption, based on the nature of interactions involved. The complexity of the adsorption phenomenon is further increased with the complexity of the biomolecule and is governed by numerous factors such as the shape, structural, and chemical properties of antibodies. There are, however, many drawbacks associated with adsorption based antibody systems (some are listed in Table 3-2) which make it less useful for biosensors that require highly orientated layers of antibody.

TABLE 3-2 Drawbacks encountered with antibody adsorption on surfaces

Drawbacks	Rational	References
Conformation associated activity loss	Hydrophobic interactions cause molecular flattening, resulting in spreading over the surface.	[13–18]
Antibody leaching	Hydrophobic interactions mainly depend upon water of hydration, which is strictly governed by solvent polarity. A slight change in solvent polarity due to the presence of other proteins causes antibody to leach and/or exchange with other protein(s).	[19–23]
Uncontrolled packing and	For example, the number of antibodies interacting with the surface is not controlled, causing the creation of pockets containing excess or less antibody.	[5, 24–28]
random orientation	The binding site of the antibody may adsorb to the surface, preventing interaction with the requisite antigen.	
Lack of stability on adsorbed surface	Adsorbed antibody may not be stable, leading to inactivation.	

However, the problem of random orientation of antibody can be effectively overcome by employing an adsorbed sublayer of precaptured proteins, such as protein A (prA), G, and kappa-light (κ_1)-specific protein L [29–31] that leads to controlled and systematic capture of antibody molecules. Hence, a homogeneous molecular packaging of antibody over the surface can be achieved. Anti-tag precapture methods also employ anti-tag antibodies, such as anti-polyhistidine (His)-tagged antibody [32] and others, which are listed in Table 3-3. In addition, hapten-conjugated whole antibody (or antibody fragments) could be immobilized on anti-hapten antibody-adsorbed surfaces [9].

Biosensor-based applications demand precisely designed strategies for immobilization of antibodies onto the appropriate surfaces. Since adsorption

TABLE 3-3 Summary of all the antibody immobilization methods with their respective advantages and disadvantages

Strategy	Overview	Advantages (+) Disadvantages (–)	References
Physisorption (Section 2.1)	Antibody (Ab) immobilization with physical interactions between Ab and surface. Involves hydrophobic and electrostatic interactions with hydrophobic and charged surfaces, respectively.	(+) i. Could be used for any type of Ab ii. "Label-free" and "reagent-free" (no Ab modification or surface modification required) (–) i. Highly susceptible to protein exchange and leaching ii. Conformational instability of the adsorbed Ab (denaturation) due to the underlying processes such as molecular spreading or flattening iii. Random molecular distribution (nonhomogenous density of Ab on surface and no control over functional orientation of the molecules)	[5, 6, 13–16, 19–31]
Covalent immobilization (Section 2.2)			
Direct chemistries			
Epoxide-mediated (Figure 2)	Epoxide-functionalized surface Epoxide ring opens up following the reaction of oxygen species with amine or sulfhydryl	(+) i. Leach-proof immobilization ii. No Ab modification required iii. "Reagent-free", "label-free" single-step immobilization iv. Commercial availability (–) i. An epoxide functionalization of the surface is required ii. Randomly orientated Ab	[45–48]

TABLE 3-3 *(Continued)*

Strategy	Overview	Advantages (+) Disadvantages (–)	References
Aldehyde-mediated	Aldehyde-functionalized surface Aldehyde reacts with amino group that results in an imine group between surface and Ab. Reduction of imine to amine should be performed in order to stabilize the bond	(+) i. No Ab modification required (native Ab) ii. Single-step strategy iii. "Label free" and "reagent free" (–) i. Reversible imine-bond formation (require additional step of reduction for stabilization of the imine bond) ii. No site specificity iii. Uncontrolled orientation	[38–44]
Diels–Alder cycloaddition (Figure 3)	Polyethylene glycol (PEG)-functionalized with ligand such as biotin or domain-specific protein A at ω-carbon and cyclopentadiene at α-carbon. PEG grafted through the cyclo-pentadiene on the N-maleimide-functionalized surface	(+) i. Covalent attachment with higher steric freedom (PEG chain length can control the steric freedom of the captured Ab) ii. Orientated capture of Ab on protein A or biotin iii. Site specific iv. "Label free" and "reagent free" v. Single-step procedure to generate functionalized surface	[49–53]
Photoactive chemistry (Figure 4)	*β*-cyclodextrin-functionalized surfaces genetically incorporated *p*-benzoyl-L-phenylalanine (*p*-Bpa) Ab	(+) i. Covalently captured Ab ii. Site specific iii. "Label free" and "reagent free" single step procedure iv. Commercially available (–) i. Incorporation of *p*-Bpa may introduce conformational changes in Ab fragments such as short-chain fragment variable and fragment antigen binding	[54–57]

TABLE 3-3 *(Continued)*

Strategy	Overview	Advantages (+) Disadvantages (–)	References
Linker-mediated chemistries		Linker length plays an important role in minimizing the coupling related conformational stress.	
Sulfhydryl-targeting linkers	Primarily contains maleimide or vinyl sulfone as reactive centers which target the sulfhydryl group. These could be either homo- (HS-SH) or hetero-bifunctional ($HS-NH_2$)	(+) i. Site-specific immobilization in case of cysteine-tagged Ab ii. Commercially available (–) i. Generation of thiol groups is required on antibody either by converting pendant amino groups of Ab into sulfhydryl (this may also change the amines of the Ab's antigen recognition domain, causing compromised Ab functionality) or by reducing disulfide bonds to generate free sulfhydryl groups with chemical agents such as DTT, TCEP, etc.	[58–67]
Site-specific immobilization (Section 2.3)			
Affinity tags			
Polyhistidine (His-tag)	Nitrilo (NTA) or iminodiacetic acid-functionalized surface Imidazole of histidine forms a metal coordination complex	(+) i. Single-step methodology ii. Noncovalent coordination complex formation iii. Site specific iv. Reusable v. Commercial availability	[71–80]

TABLE 3-3 *(Continued)*

Strategy	Overview	Advantages (+) Disadvantages (–)	References
Polyhistidine (His-tag)	Nitrilo (NTA) or iminodiacetic acid-functionalized surface Imidazole of histidine forms a metal coordination complex	(–) i. Poor ligand affinity (histidine repeats could be increased from 6 to 10 in number to obtain higher affinity) ii. Leaching (can be minimized with multimeric histidine tags) iii. Low selectivity (metal-functionalized surface does not discriminate tag from the endogenous histidine-rich proteins)	[71–80]
Biotin–avidin system	Avidin/streptavidin-functionalized surface Chemical conjugation or genetic incorporation of biotin to the Ab	(+) i. Strong specificity and selectivity (–) ii. Conformational stress could be introduced due to Ab congestion on avidin surface (tetrameric molecule)	[81–93]
Peptide nucleic acid (PNA) tags (Figure 5)	Surface-functionalized with complementary nucleic acid sequence PNA could be incorporated to Ab in a selective way either by its carboxyl end or amine end using routine EDC chemistry	(+) i. High stability over a thermal and pH range ii. Single step, "reagent-free" and "label-free" method iii. High degree of steric freedom by varying length of the backbone	[94–102]
Elastin-like polypeptides (ELPs) Thermally responsive protein	ELP-functionalized surface ELP-fused Ab or fragments	(+) i. Reversible temperature-dependent immobilization ii. Highly selective (–) iii. Immobilized system must be stored in a temperature range of 2–37 °C (below which the ELP depolymerizes and above which it precipitates)	

TABLE 3-3 *(Continued)*

Strategy	Overview	Advantages (+) Disadvantages (–)	References
Enzyme-mediated			
O^6-alkylguanine transferase (AGT) SNAP-tag (Figure 7)	Benzylguanine-functionalized surface AGT-fused Ab	(+) i. Covalent immobilization ii. Site specific iii. Commercially available	[114–120]
Transglutaminase (TGase) (Figure 8)	T26 tagging on N- or C-terminal of Ab with a flank of $GGGS_3$ Amine-functionalized surface	(+) i. Site-specific and covalent immobilization ii. “Reagent-free” and “label-free” single step immobilization	[119,120]
Site-specific orientation-based (Section 2.4)			
Enzyme domains			
Chitin-binding domain (CBD) (Figure 9)	Chitin-functionalized surface CBD-fused Ab	(+) i. High selectivity and site specificity ii. Covalently captured Ab (enzyme domain interact covalently with substrate) (–) i. Enzyme domain could be susceptible to conformational change in different physiological conditions	[123–130]
Polyhydroxyalkanoate (PHA) depolymerase substrate-binding domain (PSBD)	PHA-functionalized surface PSBD-fused Ab	----ditto----	[131]

TABLE 3-3 *(Continued)*

Strategy	Overview	Advantages (+) Disadvantages (–)	References
Proteins specific to Ab domains			
Fc-specific			
protein A (prA), protein G (prG), and recombinant protein AG (prAG)	Immobilization of Fc-specific protein on surface	(+) i. Strong orientation ii. Site specific and high selectivity iii. Ab captured in its native state iv. Strong binding (–) i. Could only be used on Ab with an Fc domain	
k_l-specific			
Protein L (prL)	prL immobilized on surface	(+) i. Strong orientation ii. Site specific and high selectivity iii. Ab captured in its native state iv. Strong binding (–) i. Could only be used on Ab with k_l domain	

is a complex process, the design of a highly repeatable adsorption based immobilization strategy is laborious, complicated, and quite a challenging task. Some of the problems associated with adsorption based immobilization strategies may be overcome by covalent methods, which we now critically assess.

Covalent immobilization of antibodies is performed with surfaces that possess pendant reactive chemical functionalities. In covalent immobilization, the antibody molecules react with the functional groups of the surface via their free reactive groups such as amine or carboxyl groups, leading to the formation of covalent bonds. These covalent methods are readily available and relatively easy to perform. For example, one can use variable length linkers to control the distance of the immobilized antibodies from the surface. This may have a significant impact on antibody functionality as it reduces immobilization associated steric stress on the antigen-binding site [33]. In addition, the importance of covalently immobilized antibody systems for achieving high-sensitivity assays has frequently been demonstrated with various diagnostic platforms [34–37]. Leaching and exchange of antibody are greatly reduced following covalent bond formation. Improved analyte sensitivity may be attributed to better retention of protein (i.e., decreased antibody leaching and antibody exchange). Improved protein retention should maintain antibody surface coverage and homogeneity [34, 35]. This, in turn, increases the surface density of the immobilized antibody; consequently, this results in greater analyte sensitivity. Covalent approaches can be broadly classified as either direct or linker-mediated. However, with a few exceptions, covalent immobilization approaches do not actually solve the problem of site-specific immobilization, although they do have certain advantages over adsorption-based strategies (Table 3-3). One of the major drawbacks of covalent methods is the random immobilization of the antibody, which may drastically affect antibody functionality.

We now discuss several new methods of site-specific antibody capture. However, these may help resolve of some of the inherent immobilization issues discussed so far.

3.2 IMMOBILIZATION CHEMISTRIES

Site-directed immobilization can be performed either by direct chemical linkage, where a reactive species is generated on a specific part of an antibody that binds to the surface, or it can be facilitated in a linker-mediated fashion.

3.2.1 *DIRECTED CHEMISTRIES*

Linkage strategies mainly based on aldehyde or epoxide chemistries are important because these allow single-pot antibody immobilization on surfaces in a controlled site-directed fashion. Carbohydrates in the fragment crystalline (Fc) region of an antibody can be treated with meta-periodates of highly reactive alkali metals such as sodium ($NaIO_4$) or potassium (KIO_4), thereby activating the hydroxyls of sugar moieties. The resulting activated diol groups can be conjugated efficiently with amine-functionalized surfaces [7, 38]. In addition, the oxidized antibody molecules could be conjugated to labels such as biotin [39] and subsequently immobilized on the relevant surfaces. However, the major drawback associated with oxidizing an antibody is that the chemicals used are highly reactive and, apart from activating carbohydrates, they may oxidize amino acids such as methionine, tryptophan, or histidine at different parts of the antibody [40–42]. Therefore, modifications to this approach have been reported where the native antibody was captured on polymeric, metallic, and nanoparticle-based surfaces functionalized with aldehydes [43, 44].

Similarly, epoxide-grafted solid supports are also widely used for antibody immobilization. An epoxide is a cyclic ether that contains one oxygen in the ring and is a strained structure which makes it highly reactive in comparison to other ether molecules. An epoxide is highly reactive toward secondary amines of amino acids, particularly histidine, and also reacts with sulfhydryls of an antibody/fragment (Figure 3-2) [10, 45, 46]. The strong reactivity of an epoxide toward amine groups allows the immobilization of an antibody via its lysine-rich regions or His tags. In addition, either a full-length antibody (or a fragment, such as fragment antigen binding [Fab], or genetic variants,

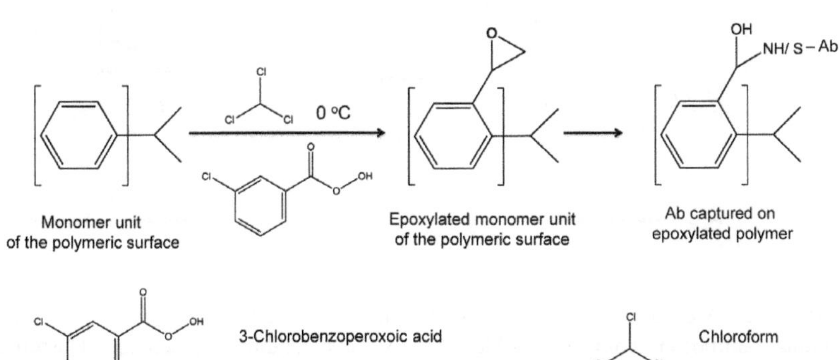

FIGURE 3-2 Polystyrene surface is functionalized with 3-chlorobenzoperoxoic acid which oxidizes the surface to generate epoxide groups. Furthermore, the antibody is immobilized through secondary amines and/or sulfhydryls.

such as short-chain fragment variable [scFv]) with exposed sulfhydryl groups could also be captured on epoxide-functionalized surfaces [10, 47, 48].

Cycloaddition is another important approach with potential for use in antibody immobilization. Immobilization of various protein molecules onto N-maleimide-functionalized surfaces was demonstrated with Diels–Alder reactions and azide–alkyne [3+2] cycloadditions [49]. The core of this reaction is addition of cyclopentadiene to N-maleimide which is grafted to the surface (Figure 3-3) [50, 51]. Peptide-modified antibodies may be conjugated to cyclopentadiene; however, these can subsequently be captured on maleimide-grafted surfaces. This approach was successfully demonstrated for cell adhesion to the extracellular matrix via arginylglycylaspartic acid oligopeptides using click chemistries [52]. In addition, a high degree of orientation can also be introduced to the system, along with site specificity, by introducing either Fc-binding proteins (such as protein A or G) or biotin. A cyclopentadiene-fused polyethylene glycol (PEG) chain with protein A at the "ω" carbon of the PEG chain was reportedly employed for capturing antibodies at the ω-end via its protein A [53]. The advantage of this strategy is that the presence of biotin or an Fc-binding protein obviates the need for

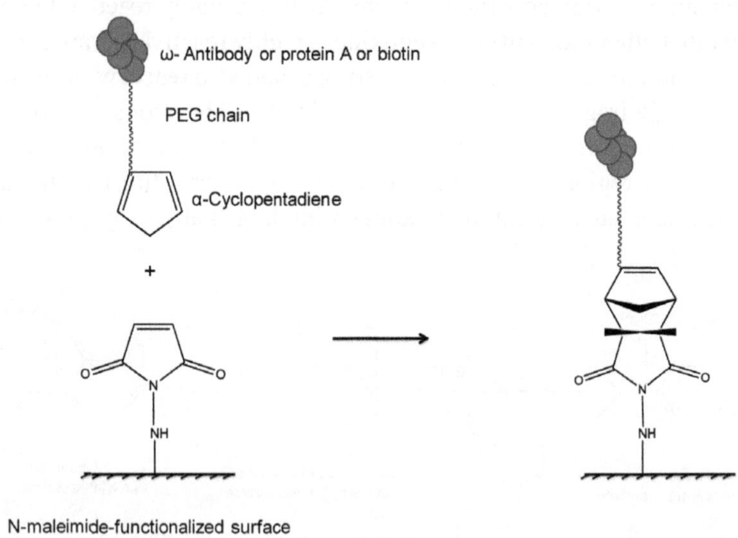

FIGURE 3-3 Mechanism of covalent immobilization of Fc binding and other proteins (including antibody) with Diels–Alder [2+3] cycloaddition reactions. Customized polyethylene glycol (PEG) was synthesized with cyclopentadiene at the 'α' carbon and biotin or protein A at the 'ω' carbon, which was immobilized on N-maleimide-functionalized surfaces. Similarly, an antibody molecule could also be bound directly to the 'ω' position of PEG, replacing biotin or protein A.

either antibody modification or direct antibody conjugation to the surface and hence confers a high degree of orientation. The surface chemistries deployed in this type of method usually generate a homogeneous surface and result in regularly distributed antibodies on the surface. However, this strategy is very time consuming and requires long preparatory phases to conjugate PEG with protein and cyclopentadiene, thus making it less attractive.

Photoactivation based antibody immobilization is another potentially efficient approach. Recently, several groups [54, 55] reported the successful immobilization of photoactive variants of amino acids such as a p-benzoyl-L-phenylalanine (p-Bpa)-modified antibody on *β*-cyclodextrin-modified surfaces by irradiating the desired surface with ultraviolet light (340–360 nm). The mechanism of antibody immobilization using this strategy is explained in Figure 3-4. However, unlike most other direct chemistries, this is highly selective in immobilizing antibodies on to surface because the reaction will take place only at the position where p-Bpa is present on the antibody. In addition, p-Bpa can be genetically introduced to the antibody at desired positions [56, 57].

In addition to direct chemistries, there is another important category for the covalent immobilization of an antibody where linkers are preferably used. These linker types provide a range of choice for immobilization and may be easier to use than other covalent immobilization strategies.

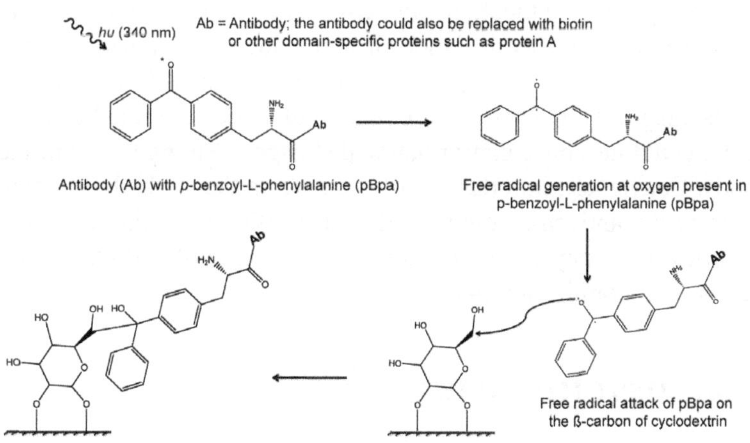

FIGURE 3-4 Photoactivable cross-linking of an antibody to a cyclodextrin-modified surface. Antibody, antibody fragment, or an antibody-capturing protein is conjugated with p-benzoyl-L-phenylalanine (p-Bpa). The p-Bpa is photoactivated. This generates an oxygen radical on the benzoyl moiety. The oxygen radical attacks the hydroxyl-containing carbon center of cyclodextrin with the subsequent formation of a covalent bond.

3.2.2 LINKER-MEDIATED CHEMISTRIES

Linkers are widely employed for a range of applications; however, their use is crucial for antibody immobilization. Linkers provide versatility in antibody immobilization due to their availability in several formats, which includes amine-, carboxyl-, and sulfhydryl-specific linkers. However, amine- and carboxyl-specific linkers for antibody immobilization are not the suitable candidates due to the abundance of these groups on an antibody and, further, they are fairly evenly distributed over the antibody surface. Sulfhydryl-specific linkers can be considered only where sulfhydryl molecules are present at specific sites on an antibody (Figure 3-1). Maleimide, haloacetyl, or vinyl sulfone-containing linkers are used for targeting sulfhydryl groups of the antibody or the surface [58, 59]. These linkers are commercially available in various combinations and can cross-link the sulfhydryl functional group to amine, carbohydrate, or other sulfhydryls. The utility of sulfhydryl targeting hetero-bifunctional cross-linkers in protein coupling is already proven. However, their use in immobilization is mainly restricted to either amine-functionalized or gold-coated surfaces [60, 61]. The major drawback of this approach is that either the disulfide bonds of an antibody have to be cleaved in order to generate free sulfhydryls, or a sulfhydryl-containing molecule must be added externally (by chemical modification) or internally (genetically introduced cysteine). However, in case of such antibody modification, whether external or internal, the possibility of losing functionality exists. Cysteine-based antibody/fragment immobilization using chemical modifications, such as cyanogen bromide activation, has also been reported [62]. This robust approach affords minimal antibody leaching [63]. However, introduction of cysteine in the basic antibody structure could introduce conformational changes, leading to a nonfunctional antibody. Conversely, introduction of cysteine on the antibody/fragment well away from the antigen-recognition domains, using reduction and alkylation chemistries [64], may enable the more efficient use of sulfhydryl linkers in solid phase immobilization [49, 65–67].

3.3 SITE-DIRECTED CAPTURE

Site-directed immobilization is where an antibody can be captured specifically through a desired region. There are many reports claiming efficient site-directed antibody immobilization. These strategies include either the use of affinity biomolecules such as biotin–avidin, enzyme, or domain-specific proteins (e.g., protein A) for antibody capture. A high degree of orientation can be introduced to the immobilized antibodies and the antigen-binding region of the antibody is correctly positioned for antigen recognition.

3.3.1 AFFINITY TAGS

Affinity is the strength of interaction between two molecules that involves a single binding site on both of them such that the resultant binding is the outcome of averaged attractive and repulsive forces. If a tag has a strong affinity toward a certain molecule then it binds to that molecule with significantly high specificity and is referred to as an "affinity tag". These tags are successfully used for affinity based protein immobilization [5, 68–70]. Tags can also be incorporated into the antibody either genetically or chemically [5, 6]. There is an array of commercially available affinity tags but only those used for antibody immobilization are described here.

In a His tag [71] there is a long chain repeated sequence of histidine residues with varying numbers (ranging from six to ten). This tag has a strong affinity for divalent transition metal ions including divalent nickel (Ni^{+2}) and copper (Cu^{+2}) [72–74]. The transition metal ions are chelated on the support layer, which consists of nitriloacetic acid or iminodiacetic acid. They provide a matrix for the selective immobilization of tagged antibodies where this tag is genetically incorporated at either the N- or C-terminal of the antibody using various recombination tools [75]. However, there are many drawbacks associated with this strategy. Firstly, nonhomogeneous antibody distribution on the nitrilotriacetic acid (NTA)-grafted surface is the foremost [76, 77]. This may be attributed to the immobilization of a mixture of functional and nonfunctional antibody on the NTA matrix. A second important disadvantage is the possible dissociation of the His-tagged protein under conditions of repeated washing or continuous flow. Thirdly, the selectivity of NTA to distinguish the His-tagged antibody from other histidine-rich naturally occurring cellular proteins is very poor [78]. Competition may occur between tagged antibody and histidine-rich cellular proteins (e.g., histidine-rich glycoprotein and histidine-rich calcium-binding protein [79–81]) for binding to NTA. Such competition may possibly lead to heterogeneous surface, along with some undesired noise in immunoassays; however, there have been no such reports to date in this regard. Several reports describe the lack of specificity associated with this system as a result of histidine's relatively low affinity toward metal ions (K_d in the order of 1–10 μM). However, a higher/better affinity may be achieved either by increasing the number of histidine residues in the tag or by employing two tags in tandem [79, 80].

Biotin (water-soluble vitamin B_7) plays a critical role in gluconeogenesis. It is a fused ring structure composed of tetrahydroimidazolone and tetrahydrothiophene with valeric acid ($C_5H_{10}O_2$) substitution on the thiophene ring. . Biotin acts as an affinity tag if it is linked to antibodies, either genetically or chemically. These biotinylated antibodies then can easily be captured on avidin-, streptavidin-, or neutravidin-functionalized polymeric, metallic [81, 82], silicate [83], and amino-polypyrrole-coated surfaces [84]. Initially, this tag was used for histochemical diagnostics [85] but its commercial relevance in immobilization and purification was realized later. In addition, this approach was successfully employed to develop microarrays [86] and other optical sensors based on surface plasmon resonance (SPR) [87]. Since avidin is multivalent (each molecule can accommodate four biotin molecules), high densities of the immobilized antibody molecules may be obtained; however, this may lead to a higher steric hindrance [88]. In addition, cellular extracts will need to be purified in order to remove any endogenous biotin that may also compete for the surface-bound avidin or streptavidin, thus reducing the overall binding efficiency of the immobilization procedure [89]. The strong affinity ($K_d \approx 10^{-15}$ M) of the biotin for avidin is a major advantage; indeed, the interaction was thought to be essentially irreversible until a temperature-dependent regeneration method was developed using a hydro-thermostat [90]. However, this strong affinity may introduce steric hindrance in the immobilized antibodies, as an antibody bound through biotin loses its freedom of rotation [91, 92]. Reducing the affinity of biotin, using genetically engineered variants of streptavidin or biomolecules mimicking the biotin-binding event with streptavidin, could minimize steric stress on the antibody and improve the regenerative capacity of the bound antibody [93].

Use of complementary peptide nucleic acids (PNAs) is another approach with considerable potential for antibody immobilization. The concept of PNA-mediated immobilization is similar to that of oligonucleotide-based hybridization-mediated antibody capture. Here we discuss PNA alone because of its novelty in providing enhanced stability against chemical and thermal denaturation that could potentially lead to antibody leaching. It is well known that PNAs are chemically synthesized single-stranded nucleic acid analogues. The backbone repeat unit of the PNA consists of an amino acid derivative of alkylamide and is shown in Figure 3-5. The nucleotide is attached to the PNA backbone repeat unit at the 4-oxoethyl position via an aminoethyl-glycine linkage. Following this, PNA is then synthesized by repeated addition of these nucleotide-functionalized amino acids via peptide bonds replacing the normal phosphodiester backbone [94, 95]. Therefore, PNAs can interact with other nucleic acids in a highly specific manner forming PNA–PNA, PNA–RNA, and PNA–DNA hybridization constructs similar to the hybridization of complementary DNA strands. We know that PNAs are nonionic and achiral molecules that resist enzymatic hydrolysis. They are also stable against chemical and thermal degradation due to the involvement of rigid peptide bonds. These properties of PNA could be harnessed in developing efficient immobilization strategies such that the PNA-conjugated antibody could be immobilized on surfaces grafted with the antisense oligonucleotide or PNA by hybridization (Figure 3-5) [96, 97]. Subsequently, the distance between surface and antibody could be controlled by varying the PNA monomers in the backbone [98–101]. Although this strategy is not commonly used for antibody immobilization, it could be an efficient single-step approach that could provide higher steric freedom to the captured antibodies, with strong site specificity and orientation. There are reports of the use of complementary oligonucleotide tags [102], which are simpler to make than PNAs. The hybridized oligonucleotide tags can be separated from their complementary strands grafted on the surface by varying either pH or divalent ion strength.

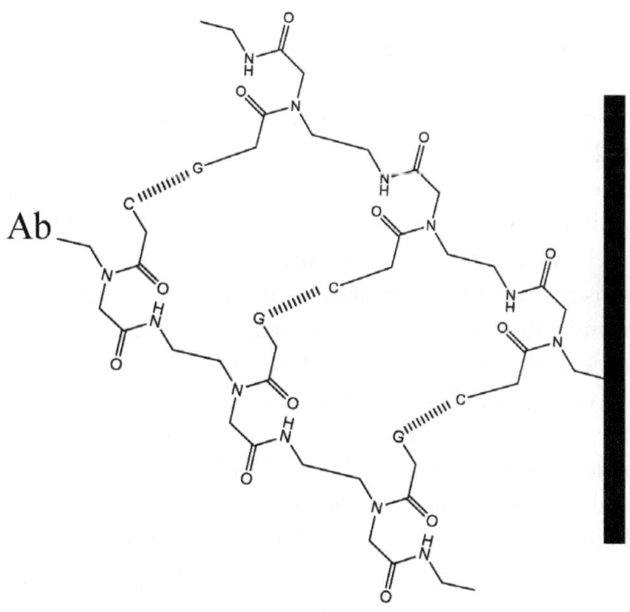

FIGURE 3-5 Immobilization of an antibody functionalized with a peptide nucleic acid (PNA) on a surface grafted with a complementary nucleotide base sequence. Antibody could be covalently linked to a PNA sequence. This PNA-functionalized antibody can be immobilized on a second PNA sequence that contains complementary nucleotide sequences, which easily hybridize directly to it.

"Elastin-like" polypeptides (ELPs) are a relatively novel category of small thermally responsive proteins [103, 104]. They can easily be tagged on to various antibodies, antibody fragments, or other proteins [105]. Further, ELPs possess "biopolymer-like" properties such that they polymerize in a temperature range of 4–37 °C. At temperatures above 37 °C these protein polymers precipitate out of the solution and below 4 °C they remain in solution phase. Therefore, ELP-tagged antibodies could efficiently be captured as a monolayer on a specifically tailored surface [106]. Moreover, ELPs have been successfully used for antibody immobilization in microarrays and related applications [107].

FLAG® proteins, a commercially available product, may also be employed for immobilization. These are octa-peptide sequences (NH_2-DYKDDDDK-COOH) that can be genetically incorporated into antibodies and antibody fragments (Fabs and scFvs) [108]. Antibodies tagged with FLAG can be captured on surfaces immobilized with monoclonal anti-FLAG antibody. Use of the FLAG tag introduces specificity and orientation for the immobilization process [109–111].

3.3.2 ENZYME–SUBSTRATE

The use of enzymes for catalyzing antibody immobilization has been reported [112]. However, a critically important factor is that the enzyme-mediated reaction must be a "two substrate component" system, where an enzyme facilitates the reaction of one substrate moiety with the other. An antibody can be conjugated to one of the two substrates of the system. This antibody–substrate complex can then be immobilized on the surface grafted with the second substrate via an enzyme-catalyzed reaction.

Antibody could also be conjugated to the active domain of an enzyme that is involved in catalyzing some reaction. These conjugates can be subsequently captured on the surfaces either grafted with the substrate or their analogues because the active domain of the enzyme has a tendency to form covalent bonds with its substrate. The immobilization of cutinase-conjugated calmodulin was one of the first demonstrations of this strategy. Cutinase-conjugated calmodulin was captured on a phosphonate inhibitor-functionalized thiolated self-assembled monolayer on a gold surface [113]. Although this report was not for antibody immobilization specifically, we discuss its potential implications.

Human O^6-alkylguanine transferase (hAGT) plays an important role in DNA repair. It binds to a DNA molecule at a specific error site and alkylates it by transferring an alkyl group from its substrate (Figure 3-6). Methyltransferase is one of the most important enzymes of this category. Immobilization of proteins tagged with this enzyme has already been demonstrated [114–117]. Similarly, the hAGT-fused antibody can be covalently captured on surfaces functionalized with benzylguanine. Fused hAGT enzyme replaces the guanine by covalently binding to a benzyl moiety. Immobilization of hAGT-fused glutathione S-transferase and tandem repeats of "immunoglobulin-type" domains of a muscle protein, called titin, was demonstrated [118].

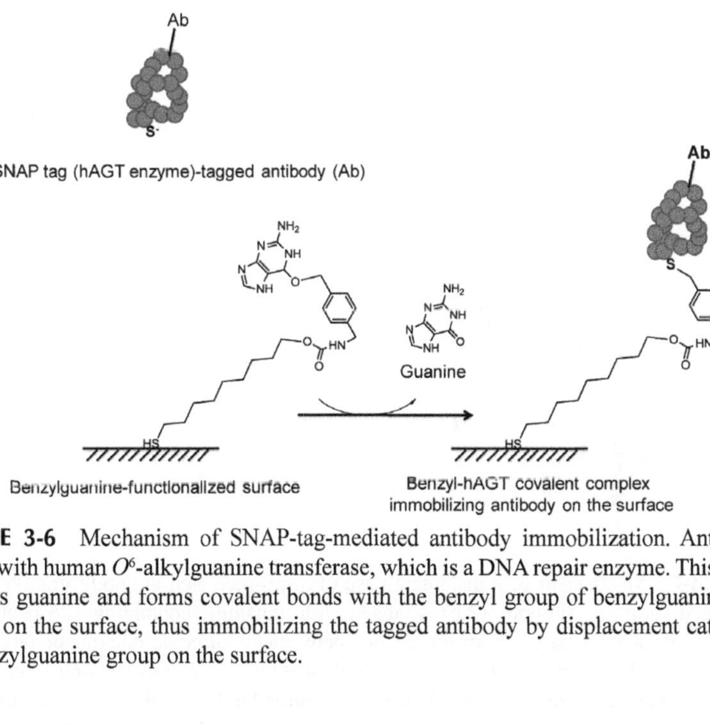

FIGURE 3-6 Mechanism of SNAP-tag-mediated antibody immobilization. Antibody is tagged with human O^6-alkylguanine transferase, which is a DNA repair enzyme. This enzyme removes guanine and forms covalent bonds with the benzyl group of benzylguanine that is grafted on the surface, thus immobilizing the tagged antibody by displacement catalysis of the benzylguanine group on the surface.

This technology is commercially available under the name SNAP-tag® from New England Biolabs. In addition, if the enzyme's substrate is changed from benzylguanine to benzylcytosine then it is known as a CLIP-tag™. These tags can easily be genetically incorporated into recombinant antibodies, or chemically fused to full-length antibodies [119]. There are many applications of these tags for specific and covalent immobilization of antibodies. Antibody micropatterning is also feasible using these strategies [120].

Transglutaminase (TGase) is another enzyme which specifically catalyzes the formation of a covalent cross-linkage; however, in this case, of the peptide-bound glutamine with a primary amine. This enzyme is specific to a 12-mer peptide sequence that is known as a "glutamine-donor substrate" sequence and is designated as T26 (HQSYVDPWMLDH). Molecular tagging of a T26 sequence onto the N- or C-terminal of the antibody fragment with a flank of diglycyl serine makes it available to use. This strategy was demonstrated for the immobilization of tagged anti-bovine serum albumin antibody on aminated surfaces by inducing the TGase-mediated amine-T26 reaction [121]. The underlying mechanism of immobilization is shown in Figure 3-7.

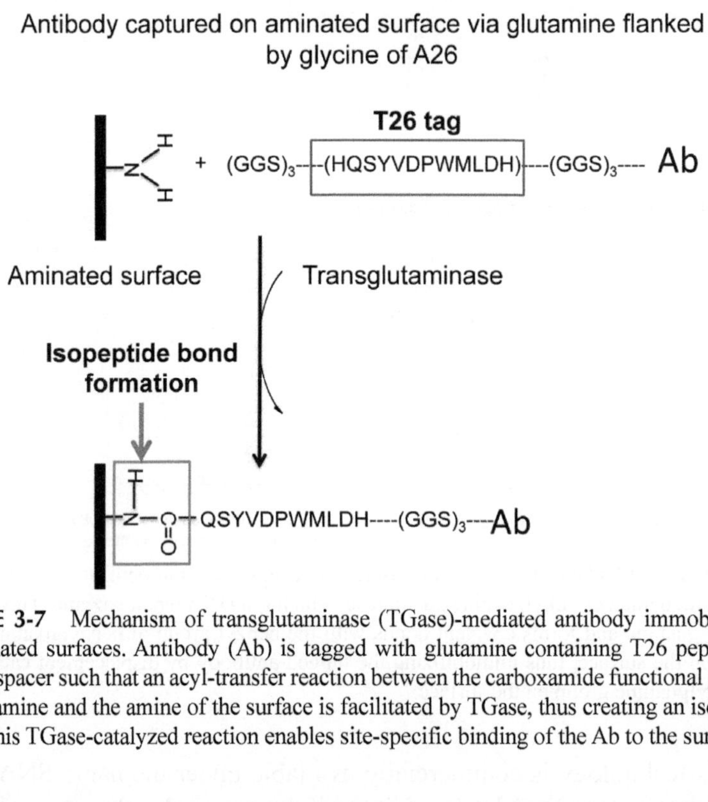

FIGURE 3-7 Mechanism of transglutaminase (TGase)-mediated antibody immobilization on aminated surfaces. Antibody (Ab) is tagged with glutamine containing T26 peptide and glycine spacer such that an acyl-transfer reaction between the carboxamide functional group of the glutamine and the amine of the surface is facilitated by TGase, thus creating an isopeptide bond. This TGase-catalyzed reaction enables site-specific binding of the Ab to the surface.

Unlike covalent chemical immobilization, this technique avoids chemical modification of the antibody or fragment, thus increasing the probability of their retaining functionality [122]. The use of enzymes on poly-L-lysine and PEG-grafted surfaces had also been demonstrated to capture cells and peptides.

Chitin-binding domain (CBD), is a part of the hydrolase enzyme chitinase, which catalyzes the hydrolysis of chitin polymers. These domains can be incorporated genetically into recombinant antibodies and can be chemically conjugated to full-length antibodies. An early use of CBD in immobilization was reported by Kuranda and co-workers [123]. Later, this strategy was employed for whole cell-based assay development, wherein whole cells were immobilized on chitin-grafted surfaces via their surface-expressed CBD [124]. Applications of this strategy extend from chromatography [125, 126] to immunoassays. Many CBD-antibody constructs have been developed [127, 128] and their utility demonstrated for assay developments on the chitin-coated surfaces (Figure 3-8) [125, 129, 130]. Similarly, use of the polyhydroxyalkanoate depolymerase substrate-binding domain [131] in antibody immobilization-related applications has also been reported.

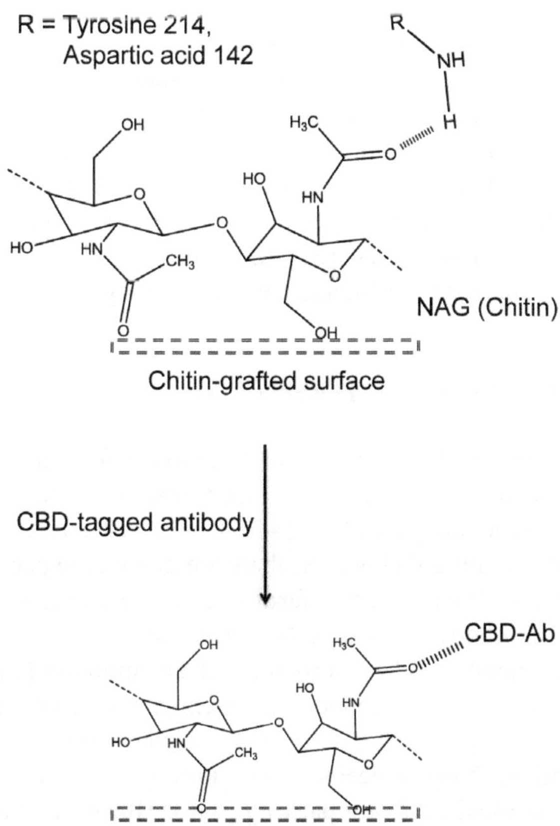

FIGURE 3-8 Schematic representation of immobilization of an antibody conjugated to the substrate-binding domain of an enzyme on surfaces grafted with the substrate of the respective enzyme. Antibody tagged with chitin-binding domain (CBD) can interact with chitin via hydrogen bonds formed between the hydrogen of the aspartic acid (142) of the CBD and oxygen of the amide of the chitin, which is grafted onto the surface. Hydrophobic interactions between the tyrosine (214) of the CBD and N-acetylglucosamine of chitin also play a crucial role in this immobilization.

3.4 SITE-SPECIFIC

Site specificity can be easily introduced using the tag-based strategies and enzyme-mediated methods. However, the orientation of immobilized molecules is of major importance for immunoassay development. Strategies based on domain-recognition proteins (DRPs), such as protein A, are widely used for immobilizing antibody molecules. These DRPs have high avidity toward specific domains on the antibody and bind strongly to them, introducing strong directed orientation to the immobilized antibodies. In addition, the covalent immobilization of these DRPs on a surface could certainly achieve a homogenous distribution of antibody over the surface [132, 133], minimizing leaching and assay variability associated with protein exchange [134]. However, DRPs must be correctly orientated on the surface so as to provide, in turn, proper orientation to the antibody molecules. Use of DRPs with multiple antibody-binding domains, such as protein A with five antibody-binding domains, may minimize the issues associated with the orientation of these DRPs. The important applications of these orientated antibodies include enzyme immunoassays [134], SPR, and microdevices such as quartz crystal microbalance [135, 136] and chromatography [137].

3.5 CONCLUSIONS AND PERSPECTIVE

Adsorption is the most commonly used approach for antibody immobilization and applications extend from immunoassays for disease diagnostics to chemical residue analyses [138–142]. It is a cost-effective method and little expertise is required. However, there are many drawbacks in adsorbed antibody formats. These is no control over (a) the amount and packing density of the antibody, (b) antibody distribution on the surface, (c) site-specificity, orientation, and conformation of the antibody [13, 14, 17, 18], and (d) susceptibility of the antibody to leaching and protein exchange [25, 32]. Introduction of precapture stages that include a sublayer of Fc-binding proteins or anti-antibody molecules may introduce a good degree of orientation with controlled antibody packing [143]. However, this will significantly increase the time and cost involved for immunoassay development. Tag-assisted strategies could be employed to address some of these issues such as site specificity and improved control over the active conformation of the immobilized antibody. However, stable interaction of the tag with its affinity ligand strictly depends on various physical and chemical parameters such as temperature, solvent polarity, pH, and salt concentration. Therefore, a slight change in the conditions may introduce leaching [144, 145] and/or activity loss. In addition, the homogeneous distribution of the tagged antibody is dependent on the distribution of the affinity ligand on the surface. Hence, a strategy was required that could address most of these drawbacks whilst maintaining the cost-effectiveness of the immunoassay development. Covalent strategies have the greatest potential to overcome most of the previously mentioned drawbacks. Covalently captured antibodies have least susceptibility for leaching and protein exchange. In addition, a high degree of site specificity and orientation could be introduced to the antibody by employing enzyme-mediated covalent strategies. Covalent immobilization methods have a variety of applications, for example, in biosensor development [20, 146], and in the preparation of antibody-immobilized matrices for affinity chromatography [147–150]. (The latter is beyond the scope of this chapter. The interested reader should consult Refs. [149, 150].) In addition, these approaches are readily employed for the designing and development of antibody-based microarrays [151–153]. Recently, covalent immobilization strategies have also been employed for the development of high-sensitivity enzyme-linked immunoassays [36, 154].

Similarly, site-directed and orientation-based strategies can also be employed for capturing antibodies for applications that require uniform distribution of directed and conformationally active antibodies [143].

Theoretically, covalent strategies offer greater potential in tag-assisted strategies in terms of both the cost and time involved. Introduction of a tag is performed by using either recombinant molecular biological approaches or by chemical methods. There are few prerequisites for using covalent strategies, except for enzyme-mediated approaches. Therefore, covalent immobilization-based strategies offer major possibilities for developing immunoassays. Equally, recombinant engineering of antibodies has huge scope for incorporation of tags resulting in optimal immobilization of antibodies with the required orientation.

ACKNOWLEDGMENTS

This work was supported partially by Science Foundation Ireland under Grant No. 10/CE3/B1821.

KEYWORDS

- **Nanosurface**
- **Biomolecules**
- **Antibody**
- **Sulfhydryls**
- **ELPs**
- **Chitinase**

REFERENCES

1. Wang, W.; Singh, S.; Zeng, D. L.; King, K.; Nema, S. Antibody Structure, Instability, and Formulation. *J. Pharm. Sci.* **2007,** *96* (1), 1–26.
2. Borgan, K. L.; Schoenfisch, M. H. Influence of Antibody Immobilization Strategy on Molecular Recognition force Microscopy Measurements. *Langmuir* **2005,** *21* (7), 3054–3060.
3. Saerens, D.; Ghassabeh, G. H.; Muyldermans, S. Antibody Technology in Proteomics. *Brief Funct. Genomic. Proteomic.* **2008,** *7* (4), 275–282.
4. Jung, Y.; Jeong, J. Y.; Chung, B. H. Recent Advances in Immobilization Methods of Antibodies on Solid Supports. *Analyst* **2008,** *133* (6), 697–701.
5. Holthues, H.; Pfeifer-Fukumura, U.; Hartmann, I.; Baumann, W. An Immunoassay for Terbutryn Using Direct Hapten Linkage to a Glutaraldehyde Network on the Polystyrene Surface of Standard Microtiter Plates. *Fresenius. J. Anal. Chem.* **2001,** *371* (7), 897–902.
6. Seurynck-Servoss, S. L.; Baird, C. L.; Miller, K. D.; Pefaur, N. B.; Gonzalez, R. M.; Apiyo, D. O.; Engelmann, H. E.; Srivastava, S.; Kagan, J.; Rodland, K. D.; Zangar, R. C. Immobilization Strategies for Single-chain Antibody Microarrays. *Proteomics* **2008,** *8* (11), 2199–2210.
7. Nogi, T.; Sangawa, T.; Tabata, S.; Nagae, M.; Tamura-Kawakami, K.; Beppu, A.; Hattori, M.; Yasui, N.; Takagi, J. Novel Affinity Tag System Using Structurally Defined Antibody-Tag Interaction: Application to Single-Step Protein Purification. *Protein. Sci.* **2008,** *17* (12), 2120–2126.
8. Kaur, J.; Singh, K. V.; Raje, M.; Varshney, G. C.; Suri, C. R. Strategies for Direct Attachment of Hapten to a Polystyrene Support for Applications in Enzyme-Linked Immunosorbent Assay. *Anal. Chim. Acta.* **2004,** *506*, 133–135.
9. Wang, Z.; Jin, G. Covalent Immobilization of Proteins for the Biosensor Based on Imaging Ellipsometry. *J. Immunol. Methods.* **2004,** *285* (2), 237–243.
10. de Boer, A.; Hokke, C.; Deelder, A.; Wuhrer, M. Serum Antibody Screening by Surface Plasmon Resonance Using a Natural Glycan Microarray. *Glycoconj. J.* **2008,** *25* (1), 75–84.

11. Arnau, J.; Lauritzen, C.; Petersen, G. E.; Pedersen, J. Current Strategies for the Use of Affinity Tags and Tag Removal for the Purification of Recombinant Proteins. *Protein Expr. Purif.* **2006,** *48* (1), 1–13.
12. Danczyk, R.; Krieder, B.; North, A.; Webster, T.; HogenEsch, H.; Rundell, A. Comparison of Antibody Functionality Using Different Immobilization Methods. *Biotechnol. Bioeng.* **2003,** *84* (2), 215–223.
13. Guo A, Zhu XY, In *Functional Protein microarrays in Drug Discovery.* 1st edition. Edited by Predki PF. Boca Raton, FL: CRC Press; **2007**, 53–71.
14. Liu, J.; Gustafsson, A.; Breimer, M. E.; Kussak, A.; Holgersson, J. Anti-pig Antibody Adsorption Efficacy of α-Gal Carrying Recombinant P-selectin Glycoprotein ligand-1/immunoglobulin Chimeras Increases with Core 2β1, 6-N-acetylglucosaminyltransferase Expression. *Glycobiology* **2005**, *15* (6), 571–583.
15. Doran, P. M. Loss of Secreted Antibody from Transgenic Plant Tissue Cultures Due to Surface Adsorption. *J. Biotechnol.* **2006,** *122* (1), 39–54.
16. Wegner, G. J.; Wark, A. W.; Lee, H. J.; Codner, E.; Saeki, T.; Fang, S.; Corn, R. M. Real-Time Surface Plasmon Resonance Imaging Measurement for the Multiplexed Determination of Protein Adsorption/Desorption Kinetics and Surface Enzymatic Reactions on Peptide Microarrays. *Anal. Chem.* **2004,** *76* (19), 5677–5684.
17. (a) Karajanagi, S. S.; Vertgel, A. A.; Kane, R. S.; Dordick, J. S. Structure and Function of Enzymes Adsorbed onto Single-Walled Carbon Nanotubes. *Langmuir* **2004,** *20* (26), 11594–11599; (b) Haynes, C. A., Norde, W. Structures and Stabilities of Adsorbed Proteins. *J. Colloid. Interf. Sci.* **1995,** *169* (2), 313–328.
18. Miller, R.; Aksenenko, E. V.; Fainerman, V. B.; Pison, U. Kinetics of Adsorption of Globular Proteins at Liquid/Fluid Interfaces. *Colloids. Surf. A.: Physicochem. Eng. Aspects.* **2001,** *183–185*, 381–390.
19. Wang, X.; Wang, Y.; Xu, H.; Shan, H.; Lu, J. R. Dynamic Adsorption of Monoclonal Antibody Layers on Hydrophilic Silica Surface: A Combined Study By Spectroscopic Ellipsometry and AFM. *J. Colloid. Interf. Sci.* **2008,** *323* (1), 18–25.
20. Dixit, C. K.; Vashist, S. K.; MacCraith, B. D.; O'Kennedy, R. Evaluation of Apparent Non-Specific Protein Loss Due to Adsorption on Sample Tube Surfaces and/or Altered Immunogenicity. *Analyst* **2011,** *136* (7), 1406–1411.
21. Peluso, P.; Wilson, D. S.; Do, D.; Tran, H.; Venkatasubbaiah, M.; Quincy, D.; Heidecker, B.; Poindexter, K.; Tolani, N.; Phelan, M.; Witte, K.; Jung, L. S.; Wagner, P.; Nock, S. Optimizing Antibody Immobilization Strategies for the Construction of Protein Microarrays. *Anal. Biochem.* **2003,** *312* (2), 113–124.
22. Xu, H.; Lu, J. R.; Williams, D. E. Effect of Surface Packing Density of Interfacially Adsorbed Monoclonal Antibody on the Binding of Hormonal Antigen Human Chorionic Gonadotrophin. *J. Phys. Chem. B.* **2006,** *110* (4), 1907–1914.
23. Huetz, P.; Ball, V.; Voegel, J-C.; Schaaf, P. Exchange Kinetics for a Heterogeneous Protein System on a Solid Surface. *Langmuir* **1995,** *11* (8), 3145–3152.
24. Ghatnekar-Nilsson, S.; Dexlin, L.; Wingren, C.; Montelius, L.; Borrebaeck, C. A. Design of Atto-Vial Based Recombinant Antibody Arrays Combined with a Planar Wave-Guide Detection System. *Proteomics* **2007,** *7* (4), 540–547.
25. Hu, X.; Spada, S.; White, S.; Hudson, S.; Magner, E.; Wall, J. G. Adsorption and Activity of a Domoic Acid Binding Antibody Fragment on Mesoporous Silicates. *J. Phys. Chem .B.* **2006,** *110* (37), 18703–18709.
26. Bee, J. S.; Chiu, D.; Sawicki, S.; Stevenson, J. L.; Chatterjee, K.; Freund, E.; Carpenter, J. F.; Randolph, T. W. Monoclonal Antibody Interactions with Micro- and Nanoparticles:

Adsorption, Aggregation, and Accelerated Stress Studies. *J. Pharm. Sci.* **2009,** *98* (9), 3218–3238.
27. Kusnezow, W.; Jacob, A.; Walijew, A.; Diehl, F.; Hoheisel, J. D. Antibody Microarrays: An Evaluation of Production Parameters. *Proteomics* **2003,** *3* (3), 254–264.
28. Kusnezow, W.; Hoheisel, J. D. Solid Supports for Microarray Immunoassays. *J. Mol. Recognit.* **2003,** *16* (4), 165–176.
29. Trilling, A. K.; Beekwilder, J.; Zuilhof, H. Antibody Orientation on Biosensor Surfaces: A Minireview. *Analyst* **2013,** *138*, 1619–1627.
30. Jung, Y.; Lee, J. M.; Jung, H.; Chung, B. H. Self-directed and Self-Oriented Immobilization of Antibody by Protein G-DNA Conjugate. *Anal. Chem.* **2007,** *79* (17), 6534–6541.
31. Berruex, L.G.; Freetag, R.; Tennikova, T. B. Comparison of Antibody Binding to Immobilized Group Specific Affinity Ligands in Monolith Affinity Chromatography. *J. Pharm. Biomed. Anal.* **2000,** *24*, 95–104.
32. Sasakura, Y.; Kanda, K.; Yoshimura-Suzuki, T.; Matsui, T.; Fukuzono, S.; Han, M. H.; Shimizu, T. Protein Microarray System for Detecting Protein-Protein Interactions Using an Anti-His-Tag Antibody and Fluorescence Scanning: Effects of the Heme Redox State on Protein-Protein Interactions of Heme-Regulated Phosphodiesterase from *Escherichia coli. Anal. Chem.* **2004,** *76* (22), 6521–6527.
33. Wong, S. S. Reactive Groups of Proteins and Their Modifying Agents. In *Chemistry of Protein Conjugation and Cross-Linking*; Wong, S. S., Ed.; CRC Press: Boca Raton, 1991; 1st edn, pp 7–48.
34. Corso, C. D.; Dickherber, A.; Hunt, W. D. An Investigation of Antibody Immobilization Methods Employing Organosilanes on Planar ZnO Surfaces for Biosensor Applications. *Biosens. Bioelectron.* **2008,** *24* (4), 811–817.
35. Das, R. D.; Maji, S.; Das, S.; RoyChaudhuri, C. Optimization of Covalent Antibody Immobilization on Macroporous Silicon Solid Supports. *Appl. Surf. Sci.* **2010,** *256* (20), 5867.
36. Dixit, C. K.; Vashist, S. K.; O'Neill, F. T.; O'Reilly, B.; MacCraith, B. D.; O'Kennedy, R. Development of a High Sensitivity Rapid Sandwich ELISA Procedure and Its Comparison with the Conventional Approach. *Anal. Chem.* **2010,** *82* (16), 7049–7052.
37. Darain, F.; Gan, K.; Tjin, S. Antibody Immobilization on to Polystyrene Substrate On-Chip Immunoassay for Horse IgG Based on Fluorescence. *Biomedical Microdevices* **2009,** *11* (3), 653–661.
38. Shmanai, V.; Nikolayeva, T.; Vinokurova, L.; Litoshka, A. Oriented Antibody Immobilization to Polystyrene Macrocarriers for Immunoassay Modified with Hydrazide Derivatives of poly(meth)acrylic Acid. *BMC Biotech.* **2001,** *1* (1), 4.
39. Abuknesha, R. A.; Jeganathan, F.; Wu, J.; Baalawy, Z. Labeling of Biotin Antibodies with Horseradish Peroxidase Using Cyanuric Chloride. *Nat. Protocol.* **2009,** *4* (4), 452–460.
40. McIntyre, J. A. Method of Altering the Binding Specificities of Monoclonal Antibodies by Oxidation-Reduction. U.S. Patent 7, 989, 596 B2, 2011.
41. Wei, Z.; Feng, J.; Lin, H.; Mullapudi, S.; Bishop, E.; Tous, G. I.; Casas-Finet, J.; Hakki, F.; Strouse, R.; Schenerman, M. A. Identification of a Single Tryptophan Residue as Critical for Binding Activity in a Humanized Monoclonal Antibody against Respiratory Syncytial Virus. *Anal. Chem.* **2007,** *79* (7), 2797–2805.
42. Zamani, L.; Andersson, F. O.; Edebrink, P.; Yang, Y.; Jacobsson, S. P. Conformational Studies of a Monoclonal Antibody, IgG1, by Chemical Oxidation: Structural Analysis

by Ultrahigh-Pressure LC-electrospray Ionization Time-of-flight MS and Multivariate Data Analysis. *Anal. Biochem.* **2008,** *380* (2), 155–163.

43. Caballero, D.; Samitier, J.; Bausells, J.; Errachid, A. Direct Patterning of Anti-Human Serum Albumin Antibodies on Aldehyde-Terminated Silicon Nitride Surfaces for HSA Protein Detection. *Small* **2009,** *5* (13), 1531–1534.
44. Shen, G.; Zhang, Y. Highly Sensitive Electrochemical Stripping Detection of Hepatitis B Surface Antigen Based on Copper-Enhanced Gold Nanoparticle Tags and Magnetic Nanoparticles. *Anal. Chim. Acta.* **2010,** *674* (1), 27.
45. Elia, G.; Silacci, M.; Scheurer, S.; Scheuermann, J.; Neri, D. Affinity-Capture Reagents for Protein Arrays. *Trends Biotechnol.* **2002,** *20* (12), s19–s22.
46. Ehlers, J.; Rondan, N. G.; Huynh, L. K.; Pham, H.; Marks, M.; Truong, T. N. Theoretical Study on Mechanisms of the Epoxy-Amine Curing Reaction. *Macromolecules* **2007,** *40* (12), 4370–4377.
47. Albayrak, N.; Yang, S. Immobilization of *Aspergillus oryzae* β-galactosidase on Tosylated Cotton Cloth. *Enzyme. Microb. Technol.* **2002,** *31* (4), 371–383.
48. Hermanson, G. T. *Immobilized Affinity Ligand Techniques,* 2nd ed. Academic Press: San Diego, 2008.
49. Kim, H. S.; Hage, D. S. Immobilization Methods for Affinity Chromatography. In *Handbook of Affinity Chromatography*; Hage, D. S., Ed.; CRC Press: Florida, 2006, 1st edn, pp 37–78.
50. Uttamchandani, M.; Wang, J.; Yao, S. Q. Protein and Small Molecule Microarrays: Powerful Tool for High-Throughput Proteomics. *Mol. BioSyst.* **2006**, 2, 58–68.
51. Sun, X.; Stabler, C. L.; Cazalis, C. S.; Chaikof, E. L. Carbohydrate and Protein Immobilization onto Solid Surfaces by Sequential Diel-Alder and Azide-Alkyne Cycloadditions. *Bioconjug. Chem.* **2006,** *17* (1), 52–57.
52. Colombo, M.; Bianchi, A. Click Chemistry for the Synthesis of RGD-Containing Integrin Ligands. *Molecules* 2010, 15 (1), 178–197
53. Sun, X. L.; Yang, L.; Chaikof, E. L.; Chemoselective Immobilization of Biomolecules with Aqueous Diel-Alders and PEG Chemistry. *Tetrahedron Lett* **2008,** *49* (16), 2510–2513.
54. Jullian, V.; Courtois, F.; Bolbach, G.; Chassaing, G. Carbon-Carbon Bond Ligation between β-cyclodextrin and Peptide by Photo-Irradiation. *Tetrahedron. Lett.* **2003,** *44* (34), 6437–6440.
55. Jensen, R. L.; Staỉde, L. W.; Wimmer, R.; Stensballe, A.; Duroux, M.; Larsen, K. L.; Wingren, C.; Duroux, L. Direct Site-Directed Photocoupling of Proteins onto Surfaces Coated with β-Cyclodextrins. *Langmuir* **2010,** *26* (13), 11597–11604.
56. Hino, N.; Hayashi, A.; Sakamoto, K.; Yokoyama, S. Site-Specific Incorporation of Non-Natural Amino Acids into Proteins in Mammalian Cells with an Expanded Genetic Code. *Nat. Protocol.* **2007,** *1* (6), 2957–2962.
57. Wang, F.; Robbins, S.; Guo, J.; Shen, W.; Schultz, P. G. Genetic Incorporation of Unnatural Amino Acids into Proteins in *Mycobacterium tuberculosis*. *PLoS ONE* **2010,** *5* (2), e9354.
58. Thermo Scientific Pierce Protein Research Products: Cross-linking reagents: A technical handbook. **2006**.
59. Wong, S. S. Conjugation of Proteins to Solid Matrices. In *Chemistry of Protein and Nucliec Acid Conjugation and Cross-Linking*; Wong, S. S, Ed.; CRC Press: Florida; 2011, 2nd edn, pp 295–318.

60. Oyerokun, F. T.; Vaia, R. A.; Maguire, J. F.; Farmer, B. L. Role of Solvent Selectivity in the Equilibrium Surface Composition of Monolayers Formed from a Solution Containing Mixtures of Organic Thiols. *Langmuir* **2010,** *26* (14), 11991–11997.
61. Koncki, R. Recent Developments in Potentiometric Biosensors for Biomedical Analysis. *Analytica. Chimica. Acta.* **2007,** *599* (1), 7–15.
62. Morfill, J.; Blank, K.; Zahnd, C.; Luginbuhl, B.; Kuhner, F.; Gottschalk, K. E.; Pluckthun, A.; Gaub, H. E. Affinity-Matured Recombinant Antibody Fragments Analyzed by Single-Molecule Force Spectroscopy. *Biophys. J.* **2007,** *93* (10), 3583–3590.
63. Imui, H.; Takehara, A.; Doi, F.; Kosuke, N.; Takai, M.; Miyake, S.; Ohkawa, H. A scFv Antibody-Based Immunoaffinity Chromatography Column for Clean-Up of Bisphenol a-contaminated Water Samples. *J. Agric. Food. Chem.* **2009,** *57* (2), 353–358.
64. Sun, M. M. C.; Beam, K. S.; Cerveny, C. G.; Hamblett, K. J.; Blackmore, R. S.; Torgov, M. Y.; Handley, F. G. M.; Ihle, N. C.; Senter, P. D.; Alley, S. C. Reduction and Alkylation Strategies for the Modification of Specific Monoclonal Antibody Disulfides. *Bioconjug. Chem.* **2005,** *16* (5), 1282–1290.
65. Chen, L. L.; Rosa, J. J.; Turner, S.; Pepinsky, R. B. Production of Multimeric Forms of CD4 through a Sugar-Based Cross-Linking Strategy. *J. Biol. Chem.* **1991,** *266* (27), 18237–18243.
66. Lin, P.; Weinrich, D.; Waldmann, H. Protein Biochips, Oriented Surface Immobilization of Proteins. *Macromol. Chem. Phy.* **2010,** *211* (2), 136–144.
67. Brogan, K. L.; Wolfe, K. N.; Jones, P. A.; Schoenfisch, M. H. Direct Oriented Immobilization of F(ab′) Antibody Fragments on Gold. *Anal. Chim. Acta.* **2003,** *496* (1–2), 73–80.
68. Hedhammer, M.; Graslund, T.; Hober, S. Protein Engineering Strategies for Selective Protein Purification. *Chem. Engineer. Technol.* **2005,** *28* (11), 1315–1325.
69. Rusmini, F.; Zhong, Z.; Feijen, J. Protein Immobilization Strategies for Protein Biochips. *Biomacromolecules* **2007,** *8* (6), 1775–1789.
70. Camarero, J. A. Recent Development in the Site-Specific Immobilization of Proteins onto Solid Supports. *Biopolymers* **2008,** *90* (3), 450–458.
71. Darain, F.; Ban, C.; Shim, Y. B. Development of a New and Simple Method for the Detection of Histidine-Tagged Proteins. *Biosens. Bioelectron.* **2004,** *20* (4), 857–863.
72. Cha, T.; Guo, A.; Zhu, X. Y. Enzymatic Activity on a Chip: The Critical Role of Protein Orientation. *Proteomics* **2005,** *5* (2), 416–419.
73. Hochuli, E. Large-Scale Chromatography of Recombinant Proteins. *J. Chromatog.* **1988,** *444,* 293–302.
74. Gaberc-Porekar, V.; Menart, V. Potential for Using Histidine Tags in Purification of Protein at Large Scale. *Chem. Eng. Technol.* **2005,** *28* (11), 1306–1314.
75. Paramban, R. I.; Bugos, R. C.; Su, W. W. Engineering Green Fluorescent Protein as a Dual Functional Tag. *Biotechnol. Bioengineer.* **2004,** *86* (6), 687–697.
76. Halliwell, C. M.; Morgan, G.; Ou, C.; Cass, A. E. G. Introduction of a (Poly)histidine Tag in -Lactate Dehydrogenase Produces a Mixture of Active and Inactive Molecules. *Anal. Biochem.* **2001,** *295* (2), 257–261.
77. Lin, P.; Weinrich, D.; Waldmann, H. Protein Biochips: Oriented Surface Immobilization of Proteins. *Macromol. Chem. Phys.* **2010,** *211* (2), 136–144.
78. Wong, L. S.; Khan, F.; Micklefield. J. Selective Covalent Protein Immobilization: Strategies and Applications. *Chem. Rev.* **2009,** *109* (9), 4025–4053.

79. Khan, F.; He, M.; Taussig, M. J. Double-Hexahistidine Tag with High-Affinity Binding for Protein Immobilization, Purification, and Detection on Ni^{+2}-Nitrilotriacetic Acid Surfaces. *Anal. Chem.* **2006,** *78* (9), 3072–3079.
80. Steinhauer, C.; Wingren, C.; Khan, F.; He, M.; Taussig, M. J.; Borrebaeck, C. A. K. Improved Affinity Coupling for Antibody Microarrays: Engineering of Double-$(His)_6$-Tagged Single Framework Recombinant Antibody Fragments. *Proteomics* **2006,** *6* (15), 4227–4234.
81. Laitinen, O. H.; Nordlund, H. R.; Hytönen, V. P.; Kulomaa, M. S. Brave New (strept) avidins in Biotechnology. *Trends Biotechnol.* **2007,** *25* (6), 269–277.
82. Grunwald, C. A Brief Introduction to the Streptavidin-Biotin System and its Usage in Modern Surface-Based Assays. *Zeitschrift fur Physikalische Chemie* **2008,** *222* (5–6), 789–821.
83. Chalkias, N. G.; Giannelis, E. P. An Avidin-Biotin Immobilization Approach for Horseradish Peroxidase and Glucose Oxidase on Layered Silicates with High Catalytic Activity Retention and Improved Thermal Behavior. *Indust. Biotechnol.* **2007,** *3* (1), 82–88.
84. Torres-Rodríguez, L. M.; Billon, M.; Roget, A.; Bidan, G. A Polypyrrole-Biotin-Based Biosensor: Elaboration and Characterization. *Synth. Met.* **1999,** *102* (1–3), 1328–1329.
85. Morimoto, K.; Kim, S. J.; Tanei, T.; Shimazu, K.; Tanji, Y.; Taguchi, T.; Tamaki, Y.; Terada, N.; Naguchi, S. Stem Cell Marker Aldehyde Dehydrogenase1-Positive Breast Cancers are Characterized by Negative Estrogen Receptor, Positive Human Epidermal Growth Factor Receptor Type 2 and High Ki67 Expression. *Cancer Sci.* **2009,** *100* (6), 1062–1068.
86. Pavlickova, P.; Hug, H. A Streptavidin-Biotin-Based Microarray Platform for Immunoassays. *Methods Mol. Biol.* **2004,** *264,* 73–83.
87. Pradier, C. M.; Salmain, M.; Liu, Z.; Methivier, C. Comparison of Different Procedures of Biotin Immobilization on Gold for the Molecular Recognition of Avidin: An FT-IRRAS Study. *Surf. Inter. Anal.* **2002,** *34* (1), 67–71.
88. Kolenko, P.; Dohnalek, J.; Duskova, J.; Skalova, T.; Collard, R.; Hasek, J. New Insights into Intra- and Intermolecular Interactions of Immunoglobulins: Crystal Structure of Mouse IgG2b-Fc at 2·1-Å Resolution. *Immunology* **2009,** *126* (3), 378–385.
89. Li, Y. J.; Zhang, X. E.; Zhou, Y. F.; Zhang, J. B.; Chen, Y. Y.; Li, W.; Zhang, Z. P. Reversible Immobilization of Proteins with Streptavidin Affinity Tags on a Surface Plasmon Resonance Biosensor Chip. *Anal. Bioanal. Chem.* **2006,** *386* (5), 356–361.
90. Holmberg, A.; Blomstergren, A.; Nord, O.; Lukacs, M.; Lundeberg, J.; Uhlen, M. The Biotin-Streptavidin Interaction can be Reversibly Broken Using Water at Elevated Temperatures. *Electrophoresis* **2005,** *26* (3), 501–510.
91. Mitchell, J. Small Molecule Immunosensing Using Surface Plasmon Resonance. *Sensors* **2010,** *10*, 7323–7346.
92. Evans, L.; Hughes, M.; Waters, J.; Cameron, J.; Dodsworth, N.; Tooth, D.; Greenfield, A.; Sleep, D. The Production, Characterisation and Enhanced Pharmacokinetics of scFv-Albumin Fusions Expressed in *Saccharomyces cerevisiae*. *Protein. Expr. Purif.* **2010,** *73* (2), 113.
93. Hutsell, S. Q.; Kimple, R. J.; Siderovski, D. P.; Willard, F. S.; Kimple, A. J. High Affinity Immobilization of Proteins Using Biotin- and GST-based Coupling Strategies. *Method. Mol. Biol.* **2010,** *627*, 75–90.

94. Nielsen, P.; Egholm, M.; Berg, R.; Buchardt, O. Sequence-Selective Recognition of DNA by Strand Displacement with a Thymine-Substituted Polyamide. *Science* **1991,** *254* (5037), 1497–1500
95. Egholm, M.; Buchardt, O.; Christensen, L.; Behrens, C.; Freier, S. M.; Driver, D. A.; Berg, R. H.; Kim, S. K.; Norden, B.; Nielsen, P. E. PNA Hybridizes to Complementary Oligonucleotides Obeying the Watson–Crick Hydrogen-Bonding Rules. *Nature* **1993,** *365* (6446), 566–568.
96. Sheng, H.; Ye, B. C. Different Strategies of Covalent Attachment of Oligonucleotide Probe onto Glass Beads and the Hybridization Properties. *Appl. Biochem. Biotechnol.* **2009,** *152* (1), 54–65.
97. Silva, T. A. R.; Ferreira, L. F.; Souza, L. M.; Goulart, L. R.; Madurro, J. M.; Brito-Madurro, A. G. New Approach to Immobilization and Specific-Sequence Detection of Nucleic Acids Based on poly(4-hydroxyphenylacetic acid). *Mat. Sci. Engineer.: C.* **2009,** *29* (2), 539–545.
98. Hook, F.; Ray, A.; Norden, B.; Kasemo, B. Characterization of PNA and DNA Immobilization and Subsequent Hybridization with DNA Using Acoustic-Shear-Wave Attenuation Measurements. *Langmuir* **2001,** *17* (26), 8305–8312.
99. Masuko, M. Hybridization of an Immobilized PNA Probe with its Complementary Oligodeoxyribonucleotide on the Surface of Silica Glass. *Nucleic. Acid. Res. Suppl.* **2003,** *3,* 145–146.
100. Lim, S. Y.; Chung, W.; Lee, H. K.; Park, M. S.; Park, H. G. Direct and Nondestructive Verification of PNA Immobilization Using Click Chemistry. *Biochem. Biophys. Res. Commun.* 2008, *376* (4), 633.
101. Ray, A.; Norden, B. Peptide Nucleic Acid (PNA): Its Medical and Biotechnical Applications and Promise for the Future. *FASEB J.* 2000, *14* (9), 1041–1060.
102. Zhong, M.; Fang, J.; Wei, Y. Site Specific and Reversible Protein Immobilization Facilitated by a DNA Binding Fusion Tag. *Bioconjug. Chem.* **2010,** *21* (7), 1177–1182.
103. Meyer, D. E.; Trabbic-Carlson, K.; Chilkoti, A. Protein Purification by Fusion with an Environmentally Responsive Elastin-Like Polypeptide: Effect of Polypeptide Length on the Purification of Thioredoxin. *Biotechnol. Prog.* **2001,** *17* (4), 720–728.
104. Trabbic-Carlson, K.; Meyer, D. E.; Liu, L.; Piervincenzi, R.; Nath, N.; LaBean, T.; Chilkoti, A. Effect of Protein Fusion on the Transition Temperature of an Environmentally Responsive Elastin-Like Polypeptide: A Role for Surface Hydrophobicity? *Protein. Eng. Des. Sel.* **2004,** *17* (1), 57–66.
105. Ong, S. R.; Trabbic-Carlson, K. A.; Nettles, D. L.; Lim, D. W.; Chilkoti, A.; Setton, L. A. Epitope Tagging for Tracking Elastin-Like Polypeptides. *Biomaterials* **2006,** *27* (9), 1930–1935.
106. Floss, D. M.; Schallau, K.; Rose-John, S.; Conrad, U.; Scheller, J. Elastin-like Polypeptides Revolutionize Recombinant Protein Expression and Their Biomedical Application. *Trends Biotechnol.* **2010,** *28,* 37–45.
107. Gao, D.; McBean, N.; Schultz, J. S.; Yan, Y.; Mulchandani, A.; Chen, W. Fabrication of Antibody Arrays Using Thermally Responsive Elastin Fusion Proteins. *J. Am. Chem. Soc.* **2006,** *128* (3), 676–677.
108. Zhang, L.; Liu, Y.; Chen, T. Label-free Amperometric Immunosensor Based on Antibody Immobilized on a Positively Charged Gold Nanoparticle/L-cysteine-modified Gold Electrode. *Microchimica. Acta.* **2009,** *164* (1), 161–166.

109. Hopp, T. P.; Prickett, K. S.; Price, V. L.; Libby, R. T.; March, C. J.; Pat Cerretti, D.; Urdal, D. L.; Conlon, P. J. A Short Polypeptide Marker Sequence Useful for Recombinant Protein Identification and Purification. *Nat. Biotech.* **1988,** *6* (10), 1204–1210.
110. Nakajima, H.; Brindle, P. K.; Handa, M.; Ihle, J. N. Functional Interaction of STAT5 and Nuclear Receptor Co-Repressor SMRT: Implications in Negative Regulation of STAT5-Dependent Transcription. *EMBO J.* **2001,** *20* (23), 6836–6844.
111. Butterfield, D. A.; Bhattacharya, D. Biofunctional Membranes: Site Specifically Immobilized Enzyme Arrays. In *New Insights in Membrane Science and Technology: Polymeric and Biofunctional Membranes*; Butterfield, D. A., Bhattacharya, D., Eds.; Elsevier: Amsterdam/Boston, 2003, pp 233–240.
112. Dennler, P.; Schibli, R.; Fischer, E. Enzymatic Antibody Modification by Bacterial Transglutaminase. *Method. Mol. Biol.* **2013,** *1045,* 205–215.
113. Hodneland, C. D.; Lee, Y.; Min, D.; Mrksich, M. Selective Immobilization of Proteins to Self-Assembled Monolayers Presenting Active Site-Directed Capture Ligands. *Proc. Natl. Acad. Sci. USA.* **2002,** *99* (8), 5048–5052.
114. Juillerat, A.; Gronemeyer, T.; Keppler, A.; Gendreizig, S.; Pick, H.; Vogel, H.; Johnsson, K. Directed Evolution of O6-Alkylguanine-DNA Alkyltransferase for Efficient Labeling of Fusion Proteins with Small Molecules *In Vivo*. *Chem. Biol.* **2003,** *10,* 313–317.
115. Kindermann, M.; George, N.; Johnsson, N.; Johnsson, K. Covalent and Selective Immobilization of Fusion Proteins. *J. Am. Chem. Soc.* **2003,** *125* (26), 7810–7811.
116. Huber, W.; Perspicace, S.; Kohler, J.; Müller, F.; Schlatter, D. SPR-Based Interaction Studies with Small Molecular Weight Ligands Using hAGT Fusion Proteins. *Anal. Biochem.* **2004,** *333* (2), 280.
117. Engin, S.; Trouillet, V.; Franz, C. M.; Welle, A.; Bruns, M.; Wedlich, D. Benzylguanine Thiol Self-Assembled Monolayers for the Immobilization of SNAP-tag Proteins on Microcontact-Printed Surface Structures. *Langmuir* **2010,** *26* (9), 6097–6101.
118. Kufer, S.; Dietz, H.; Albrecht, C.; Blank, K.; Kardinal, A.; Rief, M.; Gaub, H. Covalent Immobilization of Recombinant Fusion Proteins with hAGT for Single Molecule Force Spectroscopy. *Eur. Biophys. J.* **2005,** *35* (1), 72–78.
119. Kampmeier, F.; Ribbert, M.; Nachreiner, T.; Dembski, S.; Beaufils, F.; Brecht, A.; Barth, S. Site-Specific, Covalent Labeling of Recombinant Antibody Fragments via Fusion to an Engineered Version of 6-O-Alkylguanine DNA Alkyltransferase. *Bioconjug. Chem.* **2009,** *20* (5), 1010–1015.
120. Iversen, L.; Cherouati, N.; Berthing, T.; Stamou, D.; Martinez, K. L. Templated Protein Assembly on Micro-Contact-Printed Surface Patterns. Use of the SNAP-tag Protein Functionality. *Langmuir* **2008,** *24* (12), 6375–6381.
121. Sugimura, Y.; Ueda, H.; Maki, M.; Hitomi, K. Novel Site-Specific Immobilization of a Functional Protein Using a Preferred Substrate Sequence for Transglutaminase 2. *J. Biotechnol.* **2007,** *131* (2), 121–127.
122. Sala, A.; Ehrbar, M.; Trentin, D.; Schoenmakers, R. G.; Voirois, J.; Weber, F. E. Enzyme Mediated Site-Specific Surface Modification. *Langmuir* **2010,** *26* (13), 11127–11134.
123. Kusnezow, W.; Hoheisel, J. D. Antibody Microarrays: Promises and Problems. *Biotechniques* **2002,** *33,* S14–S23.
124. Wang, J.; Chao, Y. Immobilization of Cells with Surface-Displayed Chitin-Binding Domain. *Appl. Environ. Microbiol.* **2006,** *72* (1), 927–931.

125. Blank, K.; Lindner, P.; Diefenbach, B.; Plückthun, A. Self-Immobilizing Recombinant Antibody Fragments for Immunoaffinity Chromatography: Generic, Parallel, and Scalable Protein Purification. *Protein. Expr. Purif.* **2002,** *24* (2), 313.
126. Reulen, S.; van Baal, I.; Raats, J.; Merkx, M. Efficient, Chemoselective Synthesis of Immunomicelles Using Single-Domain Antibodies with a C-terminal Thioester. *BMC Biotechno.* **2009,** *9* (1), 66.
127. Saerens, D.; Huang, L.; Bonroy, K.; Muyldermans, S. Antibody Fragments as Probe in Biosensor Development. *Sensors* **2008,** *8* (8), 4669–4686.
128. Lindner, P.; Blank, K.; Diefenbach, B.; Plückthun, A. In *Chitin Binding Domains for Immobilizing Antibody Fragments in Immunoaffinity Chromatography*. Proceedings of the 5th International Conference of the European Chitin Society. Volume VI. Norwegian University of Science and Technology, Trondheim, The European Chitin Society (EUCHIS), 2002; Edited by Varum, K. M., Domard, A., and Smirdsrod, O., Eds., pp 261–262.
129. Tjoelker, L. W.; Gosting, L.; Frey, S.; Hunter, C. L.; Trong, H. L.; Steiner, B.; Brammer, H.; Gray, P. W. Structural and Functional Definition of the Human Chitinase Chitin-Binding Domain. *J. Biol. Chem.* **2000,** *275* (1), 514–520.
130. Pavlickova, P.; Schneider, E. M.; Hug, H. Advances in Recombinant Antibody Microarrays. *Clinica. Chimica. Acta.* **2004,** *343* (1–2), 17–35.
131. Park, T.; Park, J.; Lee, S.; Hong, H.; Lee, S. Polyhydroxyalkanoate Chip for the Specific Immobilization of Recombinant Proteins and its Applications in Immunodiagnostics. *Biotechnol. Bioproc. Eng.* **2006,** *11* (2), 173–177.
132. Kausaite-Minkstimiene, A.; Ramanaviciene, A.; Kirlyte, J.; Ramanavicius, A. Comparative Study of Random and Oriented Antibody Immobilization Techniques on the Binding Capacity of Immunosensor. *Anal. Chem.* **2010,** *82* (15), 6401–6408.
133. Vashist, S. K.; Holthofer, H.; Leister, K. A Method of Immobilising Biological Molecules to a Support and Products Thereof. **2009,** PCT/IE2008/000112(WO/2009/066275).
134. Schmid, A. H.; Stanca, S. E.; Thakur, M. S.; Thampi, K. R.; Suri, C. R. Site-Directed Antibody Immobilization on Gold Substrate for Surface Plasmon Resonance Sensors. *Sensor. Actuators B: Chem.* **2006,** *113* (1), 297.
135. Carrigan, S. D.; Scott, G.; Tabrizian, M. Real-Time QCM-D Immunoassay through Oriented Antibody Immobilization Using Cross-Linked Hydrogel Biointerfaces. *Langmuir* **2005,** *21* (13), 5966–5973.
136. Yuan, Y.; He, H.; Lee, L. J. Protein A-based Antibody Immobilization onto Polymeric Microdevices for Enhanced Sensitivity of Enzyme-Linked Immunosorbent Assay. *Biotechnol. Bioeng.* **2009,** *102* (3), 891–901.
137. Beyer, N. H.; Hansen, M. Z.; Schou, C.; Honjrup, P.; Heegaard, N. H. H Optimization of Antibody Immobilization for On-Line or Off-Line Immunoaffinity Chromatography. *J. Sep. Sci.* **2009,** *32* (10), 1604.
138. Zhu, K.; Li, J.; Wang, Z.; Jiang, H.; Beier, R. C.; Xu, F.; Shen, J.; Ding, S. Simultaneous Detection of Multiple Chemical Residues in Milk Using Broad-Specificity Antibodies in a Hybrid Immunosorbent Assay. *Biosens. Bioelectron.* **2011,** *26* (5), 2716.
139. Fitzpatick, J.; Manning, B. M.; O'Kennedy, R. Development of ELISA and Sensor-based Assays for the Detection of Ethynyl Estradiol in Bile. *Food Agric. Immunol.* **2003,** *15* (1), 55–64.
140. Lee, K. G.; Pillai, S. R.; Singh, S. R.; Willing, G. A. The Investigation of Protein A and *Salmonella* Antibody Adsorption onto Biosensor Surfaces by Atomic Force Microscopy. *Biotechnol. Bioeng.* **2008,** *99* (4), 949–959.

141. Usui-Aoki, K.; Shimada, K.; Koga, H. A Novel Antibody Microarray Format Using Non-Covalent Antibody Immobilization with Chemiluminescent Detection. *Mol. BioSyst.* **2007,** *3* (1), 36–42.
142. Yan, H.; Ahmad-Tajudin, A.; Bengtsson, M.; Xiao, S.; Laurell, T.; Ekstroim, S. Noncovalent Antibody Immobilization on Porous Silicon Combined with Miniaturized Solid-Phase Extraction (SPE) for Array Based ImmunoMALDI Assays. *Ana.l Chem.* **2011,** *83* (12), 4942–4948.
143. Puertas, S.; Moros, M.; Fernández-Pacheco, R.; Ibarra, M. R.; Grazú, V.; de la Fuente, J. M. Designing Novel Nano-Immunoassays: Antibody Orientation Versus Sensitivity, *J. Phys. D: Appl. Phys.* **2010,** *43* (47).
144. Darby, R. A. J.; Hine, A. V. LacI-Mediated Sequence-Specific Affinity Purification of Plasmid DNA for Therapeutic Applications. *FASEB J.* **2005,** *19* (7), 801–803.
145. Ueda, E. K. M, Gout, P. W.; Morganti, L. Current and Prospective Applications of Metal Ion-Protein Binding. *J. Chromat. A.* **2003,** *988* (1), 1–23.
146. Vashist, S. K.; Dixit, C. K.; MacCraith, B. D.; O'Kennedy, R. Effect of Antibody Immobilization Strategies on the Analytical Performance of a Surface Plasmon Resonance-Based Immunoassay. *Analyst* **2011,** *136,* 4431–4436.
147. Cutler, P. Immunoaffinity Chromatography. In *Protein Purification Protocols: Methods in Molecular Biology (Book Series). Volume 244*; Cutler, P., Ed.; Humana Press Inc.: Totowa, 2004; 2nd edn, pp 167–177.
148. Mallik, R.; Hage, D. S. Affinity Monolith Chromatography. *J. Separation. Sci.* **2006,** *29* (12), 1686–1704.
149. Moser, A. C.; Hage, D. S. Immunoaffinity Chromatography: An Introduction to Applications and Recent Developments. *Bioanalysis* **2010,** *2* (4), 769–790.
150. Zhang, S.; Wang, J.; Li, D.; Huang, J.; Yang, H.; Deng, A. A Novel Antibody Immobilization and its Application in Immunoaffinity Chromatography. *Talanta* **2010,** *82* (2), 704.
151. Stoevesandt, O.; Taussig, M. J.; He, M. Protein Microarrays: High-Throughput Tools for Proteomics. *Expert. Rev. Proteomic.* **2009,** *6* (2), 145–157.
152. Ellington, A. A.; Kullo, I. J.; Bailey, K. R.; Klee, G. G. Antibody-Based Protein Multiplex Platforms: Technical and Operational Challenges. *Clin. Chem.* **2010,** *56* (2), 186–193.
153. Wojciechowski, J.; Danley, D.; Cooper, J.; Yazvenko, N.; Taitt, C. R. Multiplexed Electrochemical Detection of *Yersinia pestis* and *Staphylococcal Enterotoxin B* using an Antibody Microarray. *Sensors* **2010,** *10* (4), 3351–3362.
154. Kaur, J.; Boro, R. C.; Wangoo, N.; Singh, K. R.; Suri, C. R. Direct Hapten Coated Immunoassay Format for the Detection of Atrazine and 2,4-dichlorophenoxyacetic Acid Herbicides. *Anal. Chim. Acta.* **2008,** *607* (1), 92.

CHAPTER 4

ADVANCES IN THIN FILM AND 2D BIOSENSORS

RAJESH KUMAR*, DIVYA GOYAL, and GAGANPREET K. SIDHU

Department of Physics, Panjab University, Chandigarh, India

**E-mail: rajeshbaboria@gmail.com*

CONTENTS

A new class of sensors has been developed for the biological recognition of various analytes such as enzymes, proteins, DNA, antigens, and many more. In recent years, the growth of biosensors has been accelerated due to the use of nanostructures-modified electrodes, porous thin films, and two-dimensional (2D) materials in order to increase the detection of specific molecules due to the higher surface area of these materials. Moreover, the use of nanoporous thin films and 2D materials along with nanomaterials opens up new possibilities for the construction of various types of biosensors. The higher conductivity and stability offered by this class of materials gives them edge over conventional nanomaterials based biosensors. These biosensors can be used for the real-time monitoring of various analytes. Based on various applications and recent strategies focused on thin films and 2D materials for the development of biosensors, these materials have presented new tools which have great potential in future. In this chapter we have comprehensively reviewed the recent breakthrough advancements in the biosensing applications of porous thin films and 2D materials.

4.1 INTRODUCTION

Biosensors have several advantages such as high specificity, high sensitivity, and low cost. Biosensor is a device in which biological recognition of an element takes place with primary selectivity of the element.[1] They are classified as biochips, immune sensors, glucometers, biocomputers, optrodes, chemical canaries, resonant mirrors, and so on.[2] Enzymes like glucose oxidase (GO_x) and horseradish peroxide (HRP) have been widely used as biorecognition elements. In these sensors, recognition elements play very active role in the detection of electroactive species. Widely used biorecognition elements are enzymes, microorganisms, nucleic acids, cells, and antibodies. Among these, enzymes are widely used in electrochemical biosensors.[3] A biosensor not only includes one which is based on biological purpose, but also one which is based on various other factors such as temperature, pressure, electrocardiograms, pH, Ca^{2+}, catecholamines, and so on.[4]

In recent years, biosensors have been prepared using thin films and various nanostructures as shown in Figure 4-1. Although nanostructures possess high surface-to-volume ratio compared to thin films, their drawbacks include the problem of low conduction and problem regarding integration with electronic circuits used to extract the generated signals.[5] Further, contact formation is also a challenge in case of nanostructure-based sensors.[6] The problem of contact formation can be overcome using thin film for the preparation of biosensors. The nanoporous thin films attract great interest due to their unique properties including structural support and so on.[7] These thin films possess high surface area due to high porosity, good conductance, and widely tunable Fermi level. Due to these properties, films shows high sensitivity and short response time for biosensors.[8]

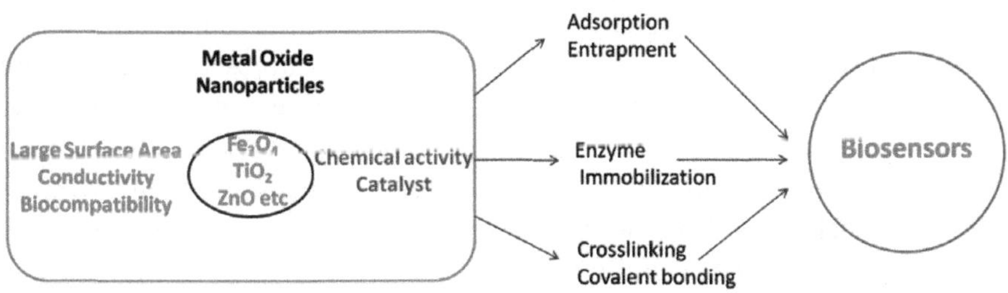

FIGURE 4-1 The systematic representation of bond between metal–oxides based thin films with analytes.

Similar to thin films, 2D materials also make better electric contact and allow more electrons to transfer. A huge number of elements or devices can be used as candidate for various sensing applications. Among them the devices with 2D channels are natural candidates for sensor applications due to the ultimately high surface-to-volume ratio and widely tunable Fermi level.[9] It is known that 2D materials play an important role in the designing of novel biosensor devices as they have unique electronic and optical properties along with large area available for biofunctionalization. These materials possess enhanced life time and improved durability. Further, increased carrier mobility and tunability of band structures as well as enhanced thermal and mechanical properties and biocompatibility are the key points for their use in biosensors to achieve selectivity and high sensitivity with very low detection limit. Such desirable aspects are achieved through the quantum effects, which are imposed at atomically thin dimensions.[10] While graphene may be the most studied among these materials, its lack of a band gap suggest that materials such as inorganic 2D transition metal dichalcogenide (TMDC) monolayered materials including molybdenum disulfide (MoS_2),[11] tungsten (IV) sulfide (WS_2),[12] molybdenum diselenide ($MoSe_2$),[13] tungsten diselenide (WSe_2),[14] and III–VI group semiconducting layered materials such as GaS and GaSe,[15] and so on are reported and identified for possible nanoelectronic device applications. The large surface area of 2D materials in various layered materials further depends on a number of layers. It means the sensitivity of these 2D materials can be tuned by changing the number of layers present in the materials. Among the various transduction techniques, electrochemical methods are so far widely studied and employed for the development of biosensors. Various 2D layered inorganic materials including MoS_2, WS_2, tin (IV) sulfide (SnS_2), and copper monosulfide (CuS) have been used in electrochemical biosensing for the detection of analytes such as glucose, dopamine, hydrogen peroxide, and DNA.[15,16] Recently biomolecules based on 2D MoS_2 have been used. They possess high fluorescence quenching efficiency of monolayered MoS_2 and different affinities toward single- and double-stranded DNA have been utilized in sensing processes.[17] Fine control over the TMDC film thickness, down to a few layers, is achieved by modifying the thickness of the predeposited metal layer. A direct band gap of a single-layer MoS_2 is 1.9 eV.[11d,12a]

4.2 FABRICATION OF THIN FILMS

A number of methods has been reported till now regarding the synthesis of thin films like spin-coating, electrostatic layer-by-layer self-assembly, and inkjet printing. Langmuir–Blodgett (LB) technique is also used to obtain high-quality nanofilms. In this method nanoporous films of different thickness are obtained by varying various parameters like molecule's nature, deposition pressure, and composition of subphase and counter ions.[18] This method is of great importance as via this method a single layer of molecules is formed on the liquid (water) surface and then this layer is transferred to the substrates. Different layers can be deposited on the substrate by repeating LB method. There are numerous biosensors based on nanothin films for the sensing purpose so as to improve their sensitivity and selectivity accordingly to the requirement, because sensitivity depends on the thickness of layers. The more the film is thin, the more it shows sensitivity. For this purpose, the use of LB method is the best.[19]

4.3 FABRICATION OF 2D MATERIALS

2D materials are fully compatible with semiconductor fabrication technology, which led to cheaper- and high-performance devices. For the synthesis of high-quality films of 2D materials at a large scale, chemical vapor deposition (CVD) and physical vapor deposition techniques are considered as one of the good methods of choice in a reproducible manner. In addition to these above mentioned techniques there are many other techniques used for the fabrication such as radio frequency sputtering, spin-coating, and so on.[20]

4.4 FACTORS INFLUENCING SENSITIVITY

There exist numerous factors that affect the sensitivity of sensors. One of them is resistance of thin film, which changes with thickness and number of deposition layers. In case of a polymer, a thin film of single layer shows more sensitivity compared to a film of more layers.[21] It is reported that the thin film of polyaniline shows faster response compared to the thin film of polyaniline and acetic acid because of thickness difference. In addition to thickness of a layer, there are many other factors that also change sensor sensitivity like porosity, doping, temperature, and so on.[22]

A sensor's performance is calculated in terms of sensor response (S) and response time. Further, S can be defined by relative resistance change as shown in eq 4.1:

$$S = \frac{\Delta R}{R_o} = \frac{R_{sensor} - R_o}{R_o} \times 100\% \quad (4.1)$$

where R_0 is the initial resistance of the sensor and R_{sensor} is the measured resistance.[23]

4.5 THIN FILM BASED BIOSENSORS

Thin films prepared with different chemical composition have unique properties in comparison to bulk materials. The surface of thin films plays an important and leading role in solving various problems regarding stability of different interfaces. Due to the dominating role of its substrate it creates the bridge between various molecular structures, which helps a lot in the development of biosensors. These unique properties and leading role of a substrate makes these thin films better materials for the formation and for the various applications of biosensors (Table 4.1).

Table 4-1 Summarization of thin films based electrodes

Biosensor	Detection of analytes	Specification
Carbon nanotube (CNT) thin film biosensors	CNT thin film was coated on poly(methyl methacrylate). CNT thin films device makes selective detection of viruses.	Films exhibit 5 orders of higher sensitivity compared to electrical impedance sensors
Polymer thin film biosensors	Creatinine biosensors are based on multienzyme sequences. They catalyze the conversion of creatinine via creatine and sarcosine to glycine, formaldehyde, etc.	
ZnO thin film biosensor		ZnO nanostructures are suitable candidate for small integrated biosensor devices
ZnO surface acoustic wave (SAW) biosensor	ZnO/Si SAW device detects prostate specific antigen (PSA) antibody–antigen immunoreaction as a function of PSA concentrations.	Sensitivity of $ZnO/SiO_2/Si$ SAW Love mode sensors is 8.64 $m^2 mg^{-1}$ which is 2 to 5 times higher that of ZnO/ lithium tantalate and SiO_2/ quartz Love sensors
ZnO Lamb biosensor	A $Si/SiO_2/Si_3N_4/Cr/Au/ZnO$ Lamb wave device detects human immunoglobulin E based on conventional cystamine self-assembled monolayers technology	Sensitivity of 8.52×10^7 $cm^2 g^{-1}$
ZnO thin film bulk acoustic resonator biosensor	ZnO film was used to detect DNA and protein molecules.	Possesses a sensitivity of 2400 Hz $cm^2 ng^{-1}$ which is about 2500 times higher than that of the conventional quartz crystal microbalance device
Other thin film based biosensors		3(aminopropyl) triethoxysilane having wide dynamic range (up to 60 mM) and rapid response time

4.5.1 CARBON NANOTUBE THIN FILM BIOSENSORS

These biosensors were fabricated by depositing a homogenous and transferable carbon nanotube (CNT) thin film fabricated by the CVD technique. There are many reports related to CNT thin film biosensors. Here we mention one of the examples of such biosensors, in which CNT thin film was coated on poly(methyl methacrylate) (PMMA) for support and growth of substrate, which makes contact with source and drain. This CNT thin film device makes selective detection of viruses. Such films exhibits five-fold higher sensitivity compared to electrical impedance sensors with identical microelectrode dimensions (no CNT).[24]

4.5.2 POLYMER THIN FILM BIOSENSORS

Polymer thin films embedded with metal nanoparticles provided the suitable microenvironment for immobilization of biomolecules retaining their biological activity with desired orientation, to facilitate electron transfer between the immobilized enzymes and electrode surfaces, better conformation, and high biological activity, thus resulting in enhanced sensing performance.[25] The sensitivity of these materials depends on the method of signal transduction, which in turn depends on the strength of binding of polymer and metal nanoparticles. A number of biosensors based on metal nanoparticles–polymers frameworks has been reported. We now discuss few of them. Eggins et al[25a] describe about polymer–metal nanoparticles for the electrochemical biosensor for monitoring and control of blood glucose levels. It is very easy to use and cheaper. There is another biosensor proposed by Tsuchida and Yoda known as creatinine biosensors which are based on multienzyme sequences.[25b] It consists of creatinine amidohydrolase, creatine amidinohydrolase, and sarcosine oxidase. They catalyze the conversion of creatinine via creatine and sarcosine to glycine, formaldehyde, and so on. There are many more biosensors based on polymer–metal nanoparticles that have been reported.[25c]

4.5.3 ZNO THIN FILM BIOSENSOR

Zinc oxide (ZnO) nanostructures and thin films have attracted much interest as materials for biosensors due to their biocompatibility, chemical stability, high isoelectric point, electrochemical activity, high electron mobility, and ease of synthesis via diverse methods. Importantly, ZnO nanostructures have shown the binding of biomolecules in desired orientations with improved conformation and high biological activity, resulting in enhanced sensing characteristics. Furthermore, compatibility with complementary metal–oxide semiconductor technology for constructing integrated circuits (ICs) makes ZnO nanostructures a suitable candidate for small integrated biosensor devices.[26]

4.5.4 ZNO SURFACE ACOUSTIC WAVE BIOSENSOR

It is reported that ZnO acoustic wave devices can be successfully used as biosensors, based on a biomolecule recognition system. Among these biosensors, Love wave, surface acoustic wave (SAW), and film bulk acoustic resonator (FBAR) devices using inclined ZnO films are the promising waves for applications in highly sensitive biodetection systems.[27] A ZnO/Si SAW device has been successfully used in the detection of prostate specific antigen (PSA) antibody–antigen immuno-reaction as a function of PSA concentrations.[27] The immobilization of PSA on the surface of ZnO SAW shifted the resonance frequencies to a lower value. A linear dependence was found between the resonance frequency change and the PSA/alpha 1-antichymotrypsin complex concentrations over the broad dynamic range of 2–10,000 ng mL^{-1}.[28] It is reasonable to use ZnO as a guiding layer on substrates of ST-cut quartz to form Love-mode biosensors. In addition to ZnO other substrate materials used for Love-mode ZnO sensors include lithium tantalate ($LiTaO_3$), lithium niobate ($LiNbO_3$), and sapphire. It can also be prepared by using a polymer film (such as PMMA, polyimide, SU-8, or parylene C) on top of the ZnO layer as the guiding layer.[29] A Love-mode ZnO device shows maximum sensitivity up to 18.77×10^{-8} m^2 s kg^{-1}, which is comparatively much higher than that of SiO_2/quartz Love-mode SAW device. Most of the above mentioned ZnO Love-mode sensors are based on a bulk piezoelectric substrate (e.g., quartz, $LiNbO_3$, and $LiTaO_3$), which are expensive and incompatible with IC fabrication.[30]

4.5.5 ZNO LAMB WAVE BIOSENSOR

In Lamb wave sensors, the wave propagation velocity in the membrane is slower than the acoustic wave velocity in the fluids on the surface; thus, the energy is not easily dissipated and can be used in liquid samples.[31] A ZnO based Lamb wave device has been used for monitoring the growth of the bacterium *Pseudomonas putida* in a bolus of toluene, as well as the reaction of antibodies in an immunoassay for an antigen present in breast cancer patients.[32] A Si/SiO_2/Si_3N_4/Cr/Au/ZnO Lamb wave device has been used for detecting human immunoglobulin E based on conventional cystamine self-assembled monolayers technology, with a sensitivity of 8.52×10^7 cm^2 g^{-1} at a wave frequency of 9 MHz.[33] However, the Lamb wave biosensor has not been widely studied because of (1) low sensitivity due to the low operation frequency and (2) difficulties in fabrication of thin and fragile membrane structures.

4.5.6 ZNO THIN FILM BULK ACOUSTIC RESONATOR BIOSENSOR

Currently, thin film bulk acoustic resonator (FBAR) biosensors attract more attention compared to the SAW and quartz crystal microbalance (QCM) biosensors due to their better properties like high sensitivity, low insertion loss and high power handling capability.[34] High-frequency ZnO FBAR sensors have good sensitivity and high energy densities owing to the trapping of the standing wave between the two electrodes. A label free FBAR gravimetric biosensor based on a ZnO film was used to detect DNA and protein molecules.[35] It possesses a sensitivity of 2400 Hz cm^2 ng^{-1} which is about 2500 times higher than that of the conventional QCM device. A recent study shows the use of Al/ZnO/Pt/Ti FBAR design which gives the sensitivity of 3654 kHz cm^2 ng^{-1} with a good thermal stability.[36] ZnO shear mode FBAR device has been used in a water–glycerol solution, with sensitivity of 1000 Hz cm^2 ng^{-1}.[37] Weber et al.[38] have fabricated a ZnO FBAR device, which operates in a transversal shear mode, using a ZnO film. The fabricated device has a high sensitivity of 585 Hz cm^2 ng^{-1} and detection limit of 2.3 ng/cm^2 (for an avidin/anti-avidin biorecognition system).

4.5.7 OTHER THIN FILM BASED BIOSENSORS

Thin film glucose biosensor based on plasma-polymerized films (PPFs) reported that the films deposited on glucose biosensor are extremely thin, deposited well onto the substrate, and have a highly cross-linked network structure and functional groups, such as amino groups, which enable a large amount of enzyme to be immobilized. Sensors fabricated by using this type of thin films are more reproducible, exhibit lower noise, and reduce the effect of interference to a greater degree than that of conventional fabricated sensors, for example, via 3(aminopropyl) triethoxysilane (APTES) having wide dynamic range (up to 60 mM) and rapid response time. The amperometric response observed by the interference of ascorbic acid and acetaminophen, due to its highly cross-linking network structure, was reduced by size discrimination of PPFs.[39] Gooding et al. studied the variation in the sensitivity of biosensor with the surface roughness.[40] The dependence of film's thickness has also been studied and a short response time (11.5–0.8 s; glucose concentration, 5–20 mM) was observed for lower thickness.[41]

The flawless and highly cross-linking network of PPF reduces the noise level of baseline of PPF-based sensors by 2 to 5 times compared to the APTES-based sensors; however, the APTES-based sensors have high detection limit of the order of 0.25 mM compared to 0.12 mM of PPF-based sensors which is due to their pinhole defects. The oxidation current of PPF-based sensors observed to be stable for 30 days, whereas in the case of APTES-based sensors oxidation current disappears just after 10 days thus enabling PPF form a stable thin film. This stable nature of the film immobilizes the enzyme, and hence prevents its denaturation.

4.6 EFFECT OF POROSITY

Porous oxide patterns can be controlled (e.g., pore size, pore pitch, and wall size) by tuning the various parameters such as anodization voltage and/or current, electrolyte concentration, anodization temperature, post-fabrication etching, and so on.

The pattern of porosity improves the surface area and allows more receptor molecules per unit area, and hence ensures effective functioning of sensors. Thust et al and Drott et al reported for porous silicon/silicon dioxide or porous gold. However, Lin et al and van Noort et al reported about the efficacy of porous silicon dioxide used in combination with interferometric and ellipsometric biomolecular interactions. It was observed that the sensitivity of optical biosensors can be improved at the detection place on a planar gold surface by making the surface porous. The more the porosity of the surface, the more is the attachment of ligands per unit surface area. Consequently, this results in larger optical shifts when interacting with specifically binding analyte molecules. It was reported by Danny van Noort et al[42] that when porous gold layer of 500 nm thickness was deposited on a planar gold surface by electrochemical deposition, a six-fold increase in the ellipsometric response was observed compared to the normal planar gold surface.[43]

Drott et al. reported an improvement in enzyme activity with porous size of porous silicon reactor compared to that of a nonporous reference reactor.[44] A GO_x was immobilized onto three porous microreactors and a non-porous reactor, to observe the glucose monitoring. It was observed that the system showed linear response of glucose up to 15 mM with 0.5 μl injection volume. The enzyme activity also improved by a factor of 100 compared to the nonporous reference reactor when the reactor anodized at

50 mA cm^{-2}. These results demonstrate the potential of porous silicon as a surface enlarging matrix for microenzyme reactors.[44]

The electrochemical deposition of thin film polymers involves one step covalent graft of various functional groups on metallic or semiconducting surfaces. The pictorial representation of the attachment of a polymer to the metallic or semiconducting surface is shown in Figure 4-2. Author studied the feasibility of a sensor based on surfaces of anodic porous alumina (APA) with enhanced Raman spectroscopy as the sensing principle. They used APA (Anodic porous alumina) pores reservoirs for release of active principles triggering for possible cell reactions and SERS (surface enhanced Raman spectroscopy) to use its high sensitivity and localization properties so as to set a high quality spectroscopy of living cells on flat 2-D substrates.

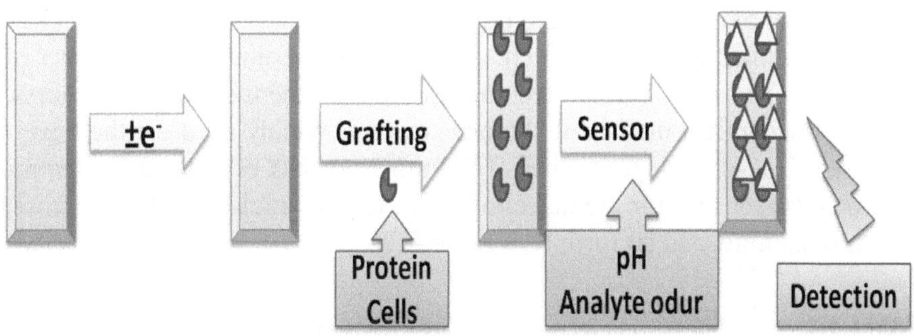

FIGURE 4-2 General scheme of a thin film based biosensor.

Another experiment with Au-coated APA substrate was performed using SERS to check the effect of pore size in the range of 20–50 nm , and it was observed that the best results were obtained for largest pores with smallest wall thickness (~50 nm).[45] Au coated on the substrate, to make measurements of SERS measurements performed on APA surfaces with varying pore size with adsorption of R6G (Rhodamine 6G) molecules. It was observed that there was maximum SERS intensity for the highest value of pore size. A small *d* (high pore wall), that is, more number of hot spots occurring at the gap between the adjacent Au–APA pore walls and a low *w*, that is, small gap between such pores are required for the optimum SERS effect. A cell culture experiment performed on fibroblast-like cells proves the biocompatibility of a material. No significant difference in cell adhesion and proliferation was observed for the different pore size in the range of 50–130 nm.

The effect of porosity was also studied in the case of nanoporous ZnO films, which were fabricated by using sol–gel method and deposited on glass substrate using dip coating technique. The prepared film was crack-free and highly porous in nature with 13.6 nm surface roughness. The high porosity of ZnO thin films is appropriate for loading high quantities of enzymes for the fabrication of electrochemical biosensor devices. Due to high porosity of the film, large number of oxygen vacancies is available at grain boundaries and interstitial sites and this changes the microstructure of the film.[46]

4.7 2D MATERIALS BASED BIOSENSORS

Although thin films play a leading role in the fabrications and applications of biosensors, the more layered materials also play an important role in the development of biosensors. More layered materials possess remarkable electrical and optical properties which can be tuned by adjusting the number of layers.[11a,11b] There are various techniques used for the transduction process, but electrochemical method is the one which is widely used for the development of biosensors. A number of 2D materials has been used in electrochemical biosensing for the detection of various biomolecules like glucose, DNA, dopamine, and so forth.[15,16] (Table 4.2).

TABLE 4-2 Summarization of 2D materials based electrodes

Enzymatic biosensor based on 2D materials		
Biosensor	Detection of analytes	Specification
Glucose biosensor	Sensor observed using PEDOT: PSS (poly(3,4-ethylenedioxythiophene) polystyrene sulfonate) semiconductor	Sensitivity of 1.65 μA per 1 mM of glucose concentration Linear response range from 1.1 mM to 16.5 mM of glucose. Response time of 10–20 s.
Cholesterol biosensor	ChO_x/Glu/PANI–Stearic acid LB film/ITO electrodes	Sensitivity of 88.9 nA mg^{-1} dL Linear range from 25 mg dL^{-1} to 400 mg dL^{-1} of cholesterol concentration
H_2O_2 sensor	Graphene–AgNPLs hybrid thin films exhibit remarkable electro-catalytical activity toward H_2O_2 electrochemical reduction.	Linear range from 2×10^{-5} M to 1×10^{-2} M Response time of less than 2 s The biosensor possess high selectivity and high sensitivity of 183.5 A cm^{-2} mM^{-1}. Detection limit of 3×10^{-6} M

TABLE 4-2 *(Continued)*

Enzymatic biosensor based on 2D materials		
Other enzymatic sensors		
Encapsulating urea into a single-layer PPy membrane	Detection of urea present in blood	Linear detection of the analyte from 0.1 μM to 1 mM
Solution-gated organic field effect transistors with a penicillinase-functionalized α-sexithiophene semiconductor	Detection of penicillin	The sensitivity is good as inorganic sensors (as high as 80 $\mu V \cdot \mu M^{-1}$)
Nonenzymatic biosensors		
C-reactive protein (based on single-walled carbon nanotubes)	Detecting field effect transistors	Shows sensitivity in the concentration range from 10^{-4} to 10^{2} $\mu g \cdot mL^{-1}$
Organic field effect devices incorporating single-stranded DNA as the recognition element	Detection of DNA	
MoS_2 thin films based DNA sensors	Detection of DNA molecules	Low detection limit of 10 fMto large range upto 10^6. High sensitivity of 17 mV/dec
Ferrocene (Fc)-modified thin films based biosensors	Fc derivatives serve as redox-active tags for the electrochemical determination of analytes	
pH sensor		An MoS_2 based pH sensor achieved sensitivity as high as 713 for a pH change by 1 unit [23]. Ultrasensitive and specific protein sensing is also achieved with a sensitivity of 196 even at 100 femtomolar concentration.

4.7.1 ENZYMATIC BIOSENSOR BASED ON 2D MATERIALS

Recently, 2D materials have received considerable popularity in the simple detection of various analytes like proteins and enzymes because of their high sensitivity, low cost, portability, simpler in nature, and low power requirements. All these unique features of 2D materials make them an excellent candidate for detection of analytes and open new doors for various biomedical applications such as clinical diagnosis, environment monitoring, and so on.

4.7.1.1 GLUCOSE BIOSENSOR

Glucose biosensors are of extensive use due to their simple and known sensing mechanism. Their sensing mechanism involves the catalyzes of glucose using GO_x enzyme which is an ideal enzyme due to its high stability.[47] The resultant byproduct H_2O_2 can electrochemically react with (oxidized) conjugated polymers, causing a change in conductivity.[48] Further, during this process an electrochemical break down upon exposure to electrical potential results in the generation of charge carriers which could be used to generate an electrical signal (see the equations following the text in this regard). In addition to these factors, the sensor can also used for various medical applications such as monitoring of blood glucose levels in diabetic individuals.

$$\text{Glucose} \xrightarrow[\text{GO}_x]{} \text{Gluconic acid} + H_2O_2$$

$$H_2O_2 \xrightarrow[0.7\text{V vs SCE}]{} O_2 + 2H^+ + 2e^- \qquad \text{SCE (saturated calomel electrode)}$$

Another glucose-based sensor was observed using PEDOT: PSS (poly(3,4-ethylenedioxythiophene) polystyrene sulfonate) semiconductor.[49] In this sensor GO_x was immobilized on the PEDOT–PSS conducting polymer film using a simple cost-effective spin-coating technique, and was entrapped in the polymer matrix during electrochemical polymerization. The device so formed shows the sensitivity of 1.65 μA per 1 mM of glucose concentration, with the linear response range from 1.1 mM to 16.5 mM of glucose and response time of 10–20 s.[50]

Recently, Liao et al.[51] have used GO_x embedded into graphene and reduced graphene oxide (rGO)-modified gate electrodes with which a glucose sensor of linear range from 10 nM to 1 μM was achieved. The high surface area of the graphene/rGO improved the magnitude of detection limit by a factor of two. This sensor shows great selectivity toward both uric acid and L-ascorbic acid, as well as to glucose devices with and without the enzyme. Liao et al also reported an organic electrochemical transistors (OECTs) sensor known as dopamine sensors with high sensitivity and selectivity coated with polymer nafion or chitosan based and are e suitable for low-cost and disposable sensing application.[53]

4.7.1.2 ENZYMETIC CHOLESTEROL BIOSENSOR

Cholesterol oxidase /Glu/polyaniline–stearic acid LB film/indium tin oxide (ChO_x/Glu/PANI–Stearic acid LB film/ITO) electrodes have been synthesized by depositing mixed monolayers of PANI- on ITO using the LB technique.[54] In this biosensor ChO_x is linked covalently to LB monolayers of PANI–Stearic acid. This sensor possesses a sensitivity of 88.9 nA mg^{-1} dL with the linear range from 25 mg dL^{-1} to 400 mg dL^{-1} of cholesterol concentration. The low value of Michaelis–Menten constant (K_M = 1.21 mM) indicates the increased interaction between ChO_x and cholesterol in the PANI–Stearic acid LB film for the immobilized enzyme. This low value of K_M compared to the free enzyme (3.71 mM) indicates a great advantage of PANI–Stearic acid LB film as a potential electrode for the development of an efficient cholesterol sensor.[54]

4.7.1.3 H_2O_2 SENSOR

A high performance H_2O_2 biosensor was fabricated by depositing a layer of silver nanocrystals on graphene thin film electrode by double pulse electrochemical method. A mechanism for biosensing based on the inducing effect between silver nanoparticles and 2D graphene template has been proposed. Graphene–silver nanoplates hybrid thin films exhibit remarkable electrocatalytical activity toward H_2O_2 electrochemical reduction. This hybrid possesses good linear range from 2×10^{-5} M to 1×10^{-2} M with fast amperometric response time of less than 2 s. The biosensor possesses high selectivity and high sensitivity of 183.5 A cm^{-2} mM^{-1} with estimated detection limit of 3×10^{-6} M. These results imply that double pulse potential electrochemical method provides new opportunities in controllable preparation of multifarious nanomaterials on the graphene substrate.[55]

4.7.1.4 OTHER ENZYMATIC SENSORS BASED ON 2D MATERIALS

Although the reaction between glucose and GO_x has attracted majority of attention, when comes under organic thin film transistor (OTFT)-based enzymatic biosensors. However, nowadays, many reports based on other enzymes based sensors are also available under this category.

Wang et al reported a method of encapsulating urea into a single-layer pancreatic polypeptide (PPy) membrane for the detection of urea present in blood that causes some serious illnesses in humans and other animals, depending on its concentration. In this sensor, a PPy film was formed by electropolymerization in the presence of the enzyme, which was used as the recognition element in a bio-field effect transistor (Bio-FET). This sensor shows the linear detection of the analyte from 0.1 μM to 1 mM in a phosphate buffer solution.[56]

Buth et al. reported the detection of penicillin using solution-gated organic field effect transistors with a penicillinase-functionalized α-sexithiophene semiconductor, which rely on the pH-sensitive nature of the α-sexithiophene combined with the proton liberating enzymatic reaction for the operation. The sensitivity is claimed to be almost as good as inorganic sensors (as high as 80 $\mu V \mu M^{-1}$).[57] Khodagholy et al.[58] demonstrated biosensor with lactose oxidase incorporated into an ionogel on top of the devices. It can detect lactate approximately to 10 mM.

There are some recent publications reporting enzymatic sensors, which although not employing transistors in their architecture, involve fabrication techniques that could be applicable for the use in sensors that do incorporate OTFTs. For example, Phongphut et al. reported the sensing of triglyceride, a biomarker for some serious diseases, by immobilizing three different enzymes (lipase, glycerol kinase, and glycerol-3-phosphate oxidase) onto a working electrode fabricated using a PEDOT: PSS/gold nanocomposites material. All these three enzymes act on triglycerides in order to liberate H_2O_2, with low detection limit of 7.88 $mg{\cdot}dL^{-1}$.[59] Another recent example was reported in 2013 by Jia et al.[60] in which a novel electrochemical sensor was fabricated in the form of a temporary-transfer tatto capable of continuous sensing lactose levels in human sweat. Lactate is a biomarker for tissue oxygenation, critical in assessing physical performance in sports and other health-care areas. Sensor shows high selectivity with very small or no response to uric acid, glucose, creatinine. or ascorbic acid. It does not incorporate an OTFT, so it requires an external potentiostat in the same way as conventional electrochemical cell requires. It was fabricated using printing techniques and solution-based materials, making it compatible and improving it by using organic transistors. A submillimolar sensor based on urea using urease-functionalized polyaniline and poly(vinylsulfonic acid) pH-sensitive membranes was deposited on the working electrodes in an electrochemical cell was recently reported by Vieira and co-workers.[61] One of the major drawbacks of most of the OTFT based enzymatic glucose sensors is that they have slow response time, that is, of the order of tens or hundreds of seconds.

4.7.2 NONENZYMATIC BIOSENSORS BASED ON 2D MATERIALS

Recently, many highly effective nonenzymatic sensors have been reported. Some of the most common recognition elements used in this category of sensors include proteins and antibodies (nonenzyme). In addition of the above mentioned sensors, there are some other sensors which do not have specific recognition elements; rather, they mainly rely on electrochemical reduction and/or oxidation reactions induced by an electrical potential. These types of sensors have high sensitivity, but suffer from poor selectivity due to their nature. These sensors work well if their sensitivity to the target analyte is relatively high compared with other moieties. There are many types of nonenzymatic biosensors that have been reported in the literature.

For example, lactic acid was detected at concentrations of 10 μM using organic transistors with α-sexithiophene (α6T) and copper phthalocyanine (CuPc).[62] Similarly, Magliulo et al. reported the detection of biotin at the concentration level of 15 pM using OFET-based sensors with functional biointerlayer (FBI). This FBI connects the source with the drain.[63] The FBI consists of two materials: streptavidin (SA) a protein that has an extremely high affinity to biotin deposited by spin-coating technique and self-assembly process, and the second material is the organic semiconductor, P3HT. Successful sensing trials were also conducted in which the SA layer was replaced firstly with a biotin antibody and secondly with the enzyme, HRP, showing that the FBI–OFET may be a promising platform for a variety of biological recognition elements. In another paper, authors reported the immobilization of biotin on an organic semiconductor layer in a FET, and SA was used as the target analyte.[64]

Justino et al. recently reported a C-reactive protein (CRP) for detecting N-Tunneling field effect transistors (NTFET) based on single-walled carbon nanotubes. It shows sensitivity in the concentration range from 10^{-4} to 10^{2} $\mu g \cdot mL^{-1}$,[65] which covers the critical range of around 1–10 $\mu g \cdot mL^{-1}$, in which CRP behaves as an important biomarker, such as cardiovascular disease. The use of a simple OTFT-based sensor is good for the CRP detection compared to many other conventional detection methods (immunoassays) which are expensive. These NTFET -based devices utilize CRP antibodies by immobilizing them on the surface of the CNT joining the source and drain electrodes, and change the current between the source and drain by depositing the different concentrations of CRP on top of the device.

Chartuprayoon et al. reported another organic-based biosensor which uses antibodies as recognition element applied on thin strips of polypyrrole as the semiconductor layer.[56] Here, the cucumber mosaic virus (CMV) was detected with CMV antibodies.

The detection of DNA, which is a very important part of various medical applications, is also a popular area of research for various organic transistor based biosensors. Lin et al. showed the detection of DNA using organic field effect devices incorporating single-stranded DNA (ssDNA) as the recognition element. It detects DNA successfully to the concentration level of 1 μM.[66] Furthermore, Kahn et al. exhibited DNA detection at the level as low as 1 nM concentration using peptide nucleic acid as a recognition element.[67] Several other authors also reported the detection of DNA using similar detection mechanism incorporated onto different types of transistors.[68]

In an another recent study, Tarabella et al. demonstrated the effectiveness and reliable sensing of OECTs devices for detecting liposome-based nanoparticles for a wider dynamic range down to 10^{-5} mg·mL^{-1} with a lowest detection limit, assessed in real-time monitoring, of 10^{-7} mg·mL^{-1}. Even more recently, Huang et al. have shown that glial fibrillary acidic protein, a biomarker which is associated with brain injury, can be detected with an OTFT-based biosensor in which its antibodies were immobilized on a polymer film on top of the device.[69] Tang et al. presented the detection of dopamine down to 5 nM concentrations. The d0etection mechanism was standard electrochemistry using a platinum gate electrode suspended in the analyzed solution.[70]

The sensing of bovine serum albumin (BSA) and BSA antibodies (anti-BSA) has been studied by Khan et al.[71] The author observed that in the sensing of BSA, anti-BSA was used as the detection element and vice versa. Pentacene was used as the semiconductor in such devices, which after passivation with perfluoropolymer, was found to be air- and water stable, and these properties broaden their potential applications.

4.7.3 MOS_2 THIN FILMS BASED DNA SENSORS

The 2D nature provides a large sensing area for high responsivity, provides good ease of fabrication compared to 1D nanobiosensors,[72] low noise level in solution,[73] and high sensitivity to biomolecules.[74] In this regard, an MoS_2 was fabricated by selective chemical synthesis process that facilitates fabrication process and also increases surface area. It was successfully designed for the sensitive detection of DNA hybridization by immobilizing DNA on the MoS_2 surface, which provides the coupling of surface charges with the channel conductance. The sensor achieved a detection limit of 10 fM, a high sensitivity of 17 mV/dec and a large dynamic range of 10^6 for the detection of DNA molecules. The reference figure of MoS_2 transistor with sensitivity that is 74-fold higher than that of graphene FET biosensors.

The biosensor related to the detection of biomolecules is shown in Figure 4-3. This label-free, highly sensitive, and scalable MoS_2 Bio-FET works well at low voltage value with low power consumption, which makes it a suitable material for various applications in disease diagnostics, environmental monitoring, food safety, and public security based on detection of DNA molecules.[75]

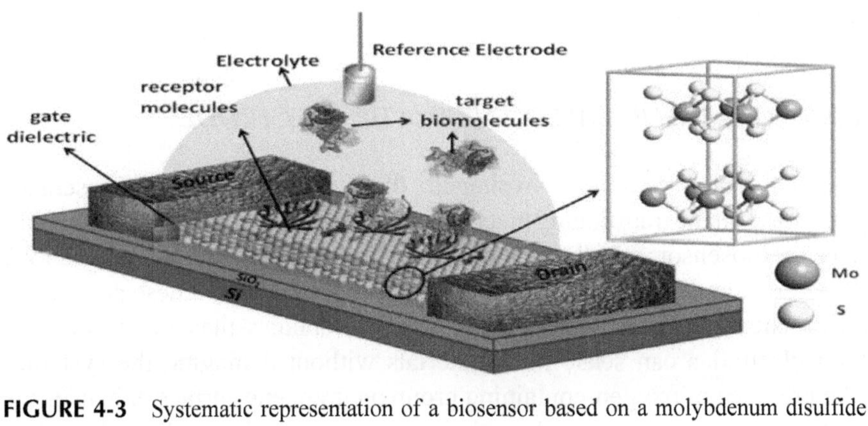

FIGURE 4-3 Systematic representation of a biosensor based on a molybdenum disulfide field-effect.

4.7.4 FERROCENE-MODIFIED THIN FILMS BASED BIOSENSORS

Ferrocene (Fc) derivatives serve as redox-active tags for the electrochemical determination of analytes.[76] Typically, DNA-based sensors with ssDNA chains are constructed by modifying the electrode surface with the F tag.[77] Redox reaction of the Fc tag depends on DNA chain hybridization, thus yielding a hybridization-dependent redox signal. An Fc-tagged aptamer has also been attached to the electrode surface.[78]

4.7.5 pH SENSOR

An MoS_2-based pH sensor achieved sensitivity as high as 713 with 1 unit change in pH value. It operates efficiently over a wide range of pH (3–9). The pH sensing is based on the protonation/deprotonation of the OH groups on the gate dielectric, thereby changing the dielectric surface charge. Low value of pH tends to protonate the surface of the OH groups, and thereby generating positive surface charges on the dielectric. On the other side high, pH value would tend to deprotonate the surface OH groups and hence generating negative charges. This pH-dependent surface charge together with the electrolyte gate voltage applied through the reference electrode determines the effective surface potential of the dielectric. A significant increase in current is obtained at a particular applied bias with decrease in pH value (or higher positive charge on the dielectric surface that causes lowering of the threshold voltage of the FET), leading to the successful demonstration of the MoS_2 pH sensor. The shift in threshold voltage is found to be 59 mV/pH.[79]

4.7.6 GRAPHENE BASED ENZYMATIC ELECTRODES

Chaubey and Malhotra[80] explained the categorization of biosensors depending on the transducing mechanism: (1) resonant biosensor, (2) optical-detector biosensors, (3) thermal detection biosensors, (4) ion-sensitive FET biosensors, and (5) electrochemical biosensors. Among all these categories, electrochemical biosensors possess more advantages than others because their electrodes can sense host materials without damaging the system.[80] The presence of oxygen-containing group on graphene surface enhances the electron transfer rate. Based on their previous experimental results, Chen et al.[81] observed the biocompatibility of graphene paper and hence reported that it is very suitable for biomedical and various other applications as shown in Figure 4-4.[82] Till now it is understood that biomolecules undergo better redox reaction with 3D structures, but recent research has shown that graphene enhances direct electron transfer between enzymes and electrodes. There are a number of graphene-based biosensors (Table 4.3) which we now discuss.

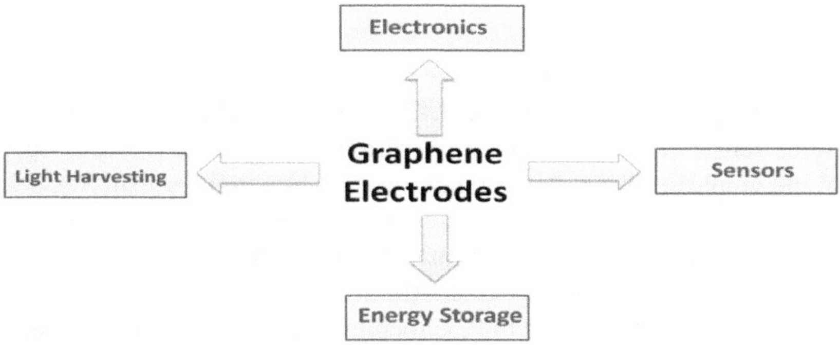

FIGURE 4-4 The use of graphene-based electrode for various applications.[82]

TABLE 4-3 Summarization of Graphene-based electrodes

Graphene based enzymatic electrodes		
Biosensor	Detection of analytes	Specifications
Glucose oxidase biosensor	Organically modified graphene for detection of glucose and maltose	Detection limit = 0.168 mM (S/N = 3) Sensitivity = 0.261 A mM^{-1} cm^{-2}
Cytochrome *c* biosensor	Reduction of nitric oxide	
Nicotinamide adenine dinucleotide (NADH) biosensor	Electrochemical determination of NADH	Linearity of electrode from 0.25 mM to 2 mM Sensitivity of 37.43 Am M^{-1} cm^{-2}
Hemoglobin biosensor	Chitosan–graphene-modified electrode for the detection of hemoglobin	Increased linearly from 30 mV s^{-1} to 150 mV s^{-1}
Horseradish peroxidase (HRP) biosensor	Detection of H_2O_2	Linear response range to H_2O_2 was from 5×10^{-6} M to 5.13×10^{-3} M Detection limit of 1.7×10^{-6} M (S/N = 3)
Cholesterol biosensor		Sensitivity toward cholesterol ester was 2.07 ± 0.1 A M^{-1} cm^{-2} Detection limit of 0.2 M
Graphene based nonenzymatic electrodes		
Hydrogen peroxide	Electrochemical detection of H_2O_2	Wider linear range, low detection limit, good reproducibility, and long-term stability
Ascorbic acid (AA), uric acid (UA), and dopamine (DA)	Simultaneous detection of AA, DA, and UA	

4.7.6.1 GO_X BIOSENSOR

The use of graphene is very effective in curing various biomedical diseases due to its highly sensitive and cost-effective nature. The use of graphene in various sensors has been studied by various researchers. It was observed that the graphene based glucose biosensors exhibit good sensitivity, selectivity, and reproducibility. This excellent performance was attributed to the large surface-area-to-volume ratio and high conductivity of graphene. Chitosan shows very good biocompatibility with GO_x. Wu et al.[83] reported that the chitosan–graphene/PtNP nanocomposite film can also be used for the detection of blood glucose. Zeng et al.[84] reported the detection limit and sensitivity of organically modified graphene for enzyme-based glucose and maltose biosensing. The detection limit of the biosensor is 0.168 mM (S/N=3) with sensitivity at 0.261 Am M^{-1} cm^{-2}. Such a similar type of sensor is shown in Figure 4-5.[85] Figure 4-5 shows an electrochemical biosensor fabricated by Dai et al. by entrapping both GO_x and HRP in the mesopores of SBA-15.[85,86]

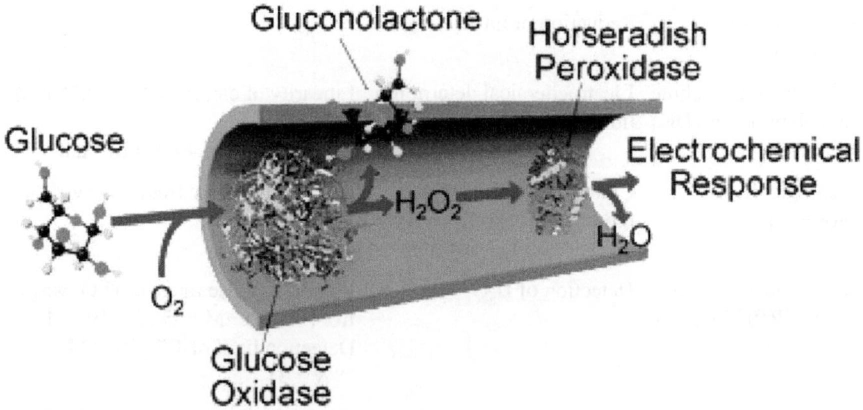

FIGURE 4-5 An electrochemical biosensor formed by entrapping both glucose oxidase and horseradish peroxidase in the mesopores of SBA-15.[85,86]

4.7.6.2 CYTOCHROME C BIOSENSOR

Cytochrome *c* (Cyt-*c*) is also one of the sensors based on graphene which shows high transfer rate. The electron transfer rate constant of Cyt-*c* at the graphene electrode was observed to be 1.95 s^{-1}, which is much higher than conventional electrodes. Due to high transfer rate, Cyt-*c* maintains its bioactivity on the surface of the electrode and shows an enzyme-like activity for the reduction of nitric oxide.[87]

4.7.6.3 NADH BIOSENSOR

The development of nicotinamide adenine dinucleotide (NADH)-based biosensors is due to the effective oxidation of NADH at low potentials. It is reported by one of the scientists, Shan et al.[88], who designed a biosensor based on an interleukin(IL)-functionalized graphene for the electrochemical determination of NADH. The authors have reported good linearity of IL–chitosan–graphene modified electrode from 0.25 mM to 2 mM and a high sensitivity of 37.43 Am M^{-1} cm^{-2}.

4.7.6.4 HEMOGLOBIN BIOSENSOR

Chitosan–graphene-modified electrode for the detection of hemoglobin has been reported by Xu et al.[89]. The current response of hemoglobin at the chitosan–graphene/glassy carbon electrode showed a well resolved redox peak with increased linearly from 30 mV s^{-1} to 150 mV s^{-1}.

4.7.6.5 HRP BIOSENSOR

Similarly like graphene-based sensors discussed, same graphene based sensor was studied by Zhou et al.[90] for the detection of H_2O_2. Chitosan–graphene/gold nanoparticle/HRP electrode fabricated with linear response range to H_2O_2 was from 5×10^{-6} M to 5.13×10^{-3} M, and a detection limit of 1.7×10^{-6} M ($S/N = 3$).

Figure 4-6[91] shows the mechanism of H_2O_2 biosensor, in which H_2O_2 is reduced by the immobilization of HRP. Moreover, HRP gets reduced with the aid of a mediator, which itself gets oxidized. The oxidized mediator gets electrochemically reduced on the electrode, leading to an increase in the current.[91]

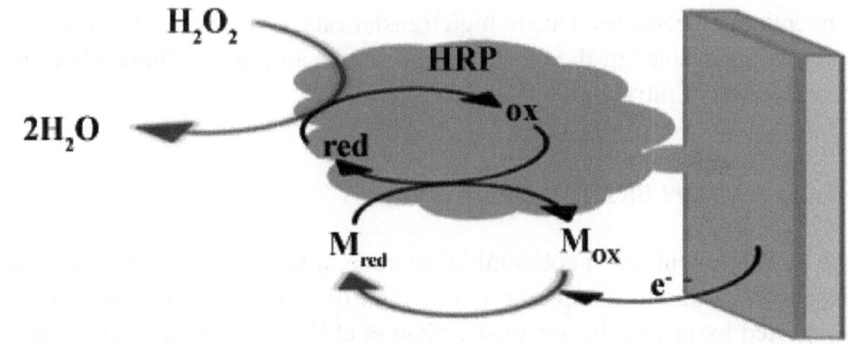

FIGURE 4-6 Schematic diagram of the reaction sequence within the enzyme electrode.[91]

4.7.6.6 NON-ENZYMATIC CHOLESTEROL BIOSENSOR

The surface of graphene/Pt–nanoparticles hybrid material was immobilized by cholesterol oxidase and cholesterol esterase to develop the cholesterol biosensor, with sensitivity and detection limit of the electrode toward cholesterol ester being 2.07 ± 0.1 AM^{-1} cm^{-2} and 0.2 M, respectively. These biosensors can be used for various purposes. Figure 4-6 shows the generation of H_2O_2 as a byproduct by using cholesterol biosensor. Cholesterol gets oxidized to cholest-4-en-3-one catalyzed by ChO_x to generate H_2O_2 as a byproduct. Figure 4-7[92] illustrates the cholesterol biosensor.[93]

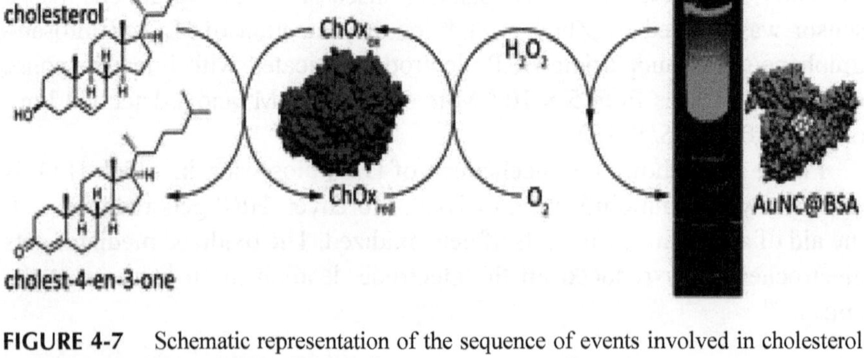

FIGURE 4-7 Schematic representation of the sequence of events involved in cholesterol biosensing employing AuNCs–BSA as fluorescent reporters.[92]

4.7.7 GRAPHENE BASED NONENZYMATIC ELECTRODES

The introduction of graphene in sensing areas offers an exciting approach for various nonenzymatic biosensors. The use of nonenzymatic electrodes as glucose sensors as an ideal system was first investigated by Loeb.[94] It comes under the fourth generation to analytical glucose oxidation. These sensors play a vital role in the detection of glucose, as they avoid the expensive and fragile enzymes. There is a wide literature in the light of this field, as one of the researcher group, that is, Kong et al.[95] developed Au nanoparticles of high density via chemical reduction of $HAuCl_4$ with sodium citrate by using thionine functionalized graphene oxide. Luo et al.[96] also reported a nonenzymatic glucose sensor, with metallic Cu nanoparticles on graphene sheets. Xiao et al. also reported a nonenzymatic glucose sensor. They used one-step electrochemical synthesis method for the fabrication of Pt–Ni nanoparticle–graphene nanocomposites.[97] Chen and co-workers have developed electronic graphene biosensors. Such biosensors have been developed to detect the building blocks of living beings. Chen et al. fabricated a CVD-grown graphene sensor to detect glucose and glutamate electrically, with a lower limit of detection of ~0.1 mM and ~5 Mm, respectively. Figures 4-8[98,99] and 4-9[100] show graphene sensors and the fabricated platform for the detection of GO_x.

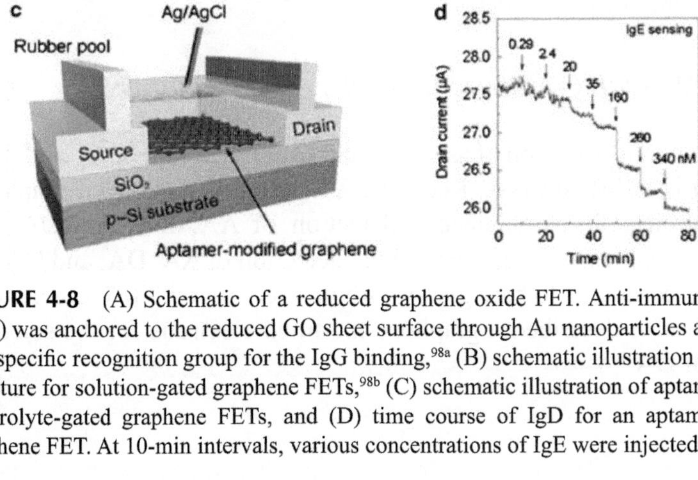

FIGURE 4-8 (A) Schematic of a reduced graphene oxide FET. Anti-immunoglobulin G (IgG) was anchored to the reduced GO sheet surface through Au nanoparticles and functions as a specific recognition group for the IgG binding,[98a] (B) schematic illustration of the device structure for solution-gated graphene FETs,[98b] (C) schematic illustration of aptamer-modified electrolyte-gated graphene FETs, and (D) time course of IgD for an aptamer-modified-graphene FET. At 10-min intervals, various concentrations of IgE were injected.[99a,99b]

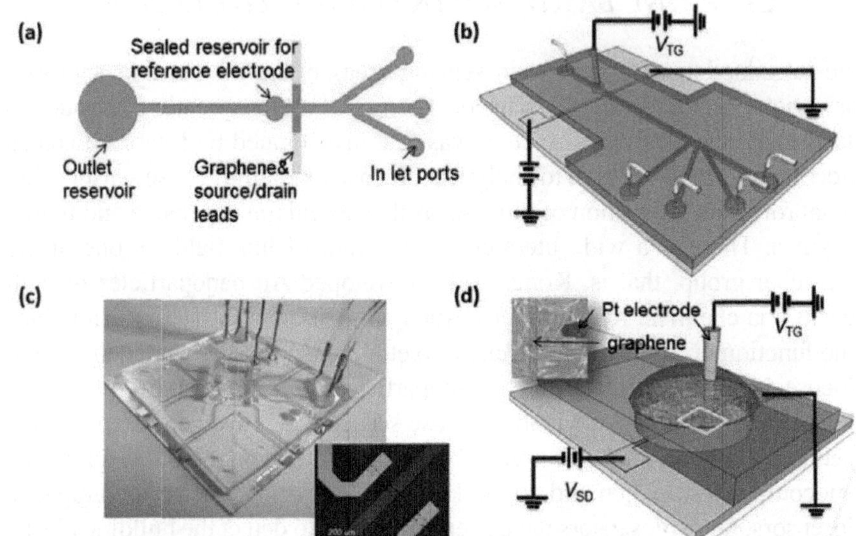

FIGURE 4-9 Solution-gated graphene field-effect transistor (FET) fabrications, (A) path of the microfluidic channel, (B) structure of the microfluidic graphene FET and measurement scheme, (C) optical image of an assembled microfluidic graphene FET, and (D) structure and measurement scheme of the solution-gated FET in an inverted cell structure.[101]

4.7.7.1 HYDROGEN PEROXIDE

Zhang et al.[101] designed a PSSA–g–PPY functionalized graphene for the electrochemical detection of H_2O_2. The biosensor showed a wide linear range, low detection limit, good reproducibility, and long-term stability.

4.7.7.2 ASCORBIC ACID, URIC ACID, AND DOPAMINE

The simultaneous detection of ascorbic acid (AA), uric acid (UA), and dopamine (DA) is a difficult task. Recently, Zhou et al.[102] fabricated graphene-based electrodes for simultaneous detection of AA, DA, and UA. The electroanalytical performance toward the detection of AA, DA, and UA was much better compared to bare glassy carbon (GC) or graphite/GC electrode. Wang et al. also described a graphene-modified electrode for the selective detection of DA.[103]

4.7.8 2D BASED OTHER BIOSENSING APPLICATIONS

It is known that 2D molybdenum trioxide (MoO_3) FET based biosensing platform use BSA as a model protein. The conduction channel is a nano-structured film made of 2D α- MoO_3 nanoflakes, with majority of the nano-flake thickness being either equal or less than 2.8 nm. The response time is impressively low (less than 10 s) which is due to the high permittivity of the 2D α-MoO_3. The MoO_3 film resistance increased on an average by ~1.0% for low BSA concentrations of 1 mg mL^{-1} (~15 μM). The average change in the resistance percentage of ~24.5% was observed for maximum BSA concentrations of 25 mg mL^{-1}. The response times were observed to be less than 10 s for all BSA concentrations. The detection limit of 12 was obtained using the standard deviations in the signal/noise fluctuation which was ~0.23%, that is, equal to 250 μg mL^{-1}. Biocompatibility of MoO_3, its high permittivity, and large electron mobility along with relative abundance and ease of synthesis, makes 2D α-MoO_3 favorable for future nanoelectronic and sensor applications.[104]

The application of MoS_2 FET based biosensors can become even more promising due to recent progress in large-area synthesis of 2D MoS_2 using CVD methods. Further, MoS_2 biosensor electrically detects PSA with high sensitivity and in a label-free manner. The fabrication of this type of biosensors does not require a dielectric layer of oxides such as HfO_2 due to the hydrophobic nature of MoS_2. Such an oxide-free operation improves sensitivity and simplifies sensor design. For a quantitative and selective detection of PSA antigen, anti-PSA antibody was immobilized on the sensor's surface. Measured off-state current of the device showed a significant decrease with an increase in applied PSA concentration. The minimum detectable concentration of PSA is 1 pg mL^{-1}, which is of several orders of magnitude below the clinical cut-off level of 4 ng mL^{-1}.[105] Figure 4-10[106] shows how a label-free MoS_2 nanosheet based field-effect biosensor detects cancer marker protein.

Nanocrystalline diamond is also a very promising candidate for future biosensor applications. Diamond is particularly suitable for various biofunctional and biosensor applications due to its several unique properties, like good biocompatibility and a large electrochemical potential window. Protein molecules attached covalently to the hydrogen terminated nanocrystalline diamond thin films which were modified by the layer of the amino group. In these films, immobilized biomolecules are fully functional and are equally active. A direct electron transfer takes place between the enzyme redox center and diamond electrode after the modification of electrodes with the enzyme catalyase. These electrodes are also very sensitive toward detection of H_2O_2 due to their dual role as a substrate for biofunctionalization and as an electrode.[107]

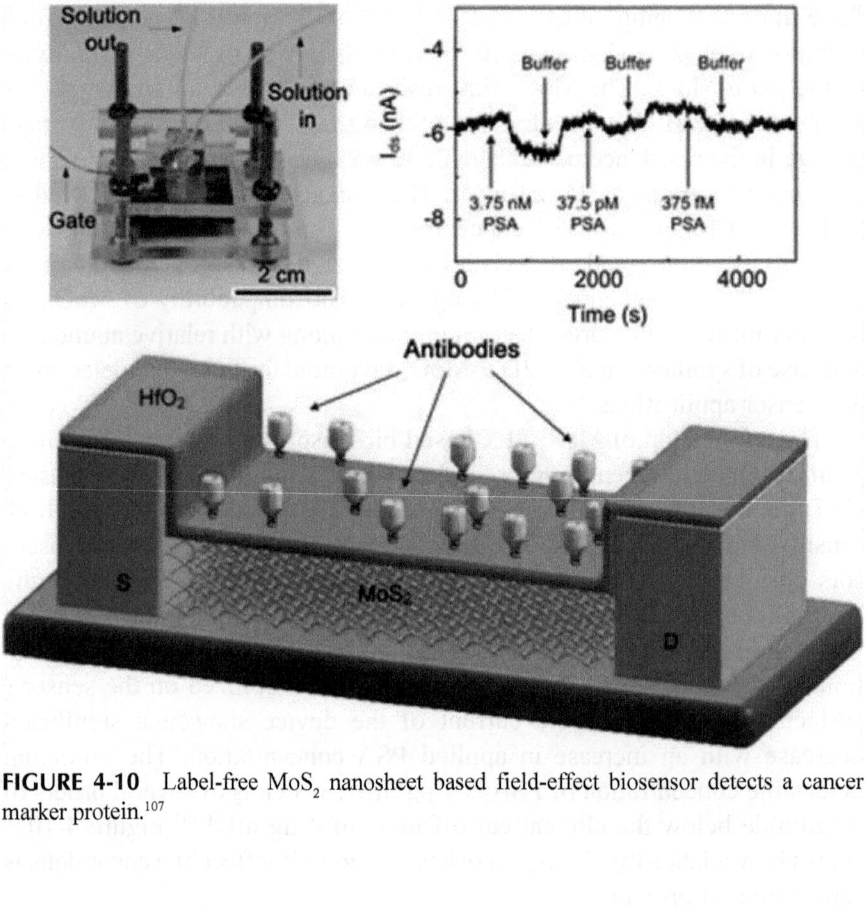

FIGURE 4-10 Label-free MoS_2 nanosheet based field-effect biosensor detects a cancer marker protein.[107]

4.8 CONCLUSION

Currently, research in the area of biosensing is more attracted toward the increase in the integration of various electronic and biological systems compared to previous points of interest like faster, cheaper, and efficient working of sensors. Therefore, future development of highly sensitive and specific biosensors and bioelectronics devices will require the combination of multidisciplinary areas like quantum, solid state physics, biology, electrical engineering, and so on. Advancement in any of these above mentioned fields will have significant effect on the future of medical field.

This chapter summarized the recent progress in porous thin films and 2D materials for biosensing applications. These porous thin films have great advantages over conventional materials and play an important role in designing biosensor devices for future due to their unique properties like large surface area (due to high porosity), tunable band gap, and tunable physical and chemical properties depending on the requirements. It is expected that in future combination of thin films and 2D materials with DNA/RNA and various other biomolecules like enzymes, proteins, antigens, and so on will be very useful for developing new molecular designs, programming, and so on. These layered materials will be in great demand in future due to their great potential for biosensing applications. More work will have to be done in the biosensing application of thin films and 2D materials as it is still in the premature stage and also due to the fact that there are several unsolved problems related to these materials for their enhanced application in medical and industrial purposes.

ACKNOWLEDGMENT

The authors are thankful to the University Grant Commission and Department of Science and Technology, India for the financial assistance and fellowship.

KEYWORDS

- **Biosensors**
- **Porosity**
- **Nanomaterials**
- **Sensitivity**
- **CNT**
- **Biomolecules**
- **Nanotubes**

REFERENCES

1. Mateescu, A.; Wang, Y.; Dostalek, J.; Jonas, U. Thin Hydrogel Films for Optical Biosensor Applications. *Membranes* **2012,** *2* (1), 40–69.
2. Varma, M. M. Polymer Thin Film Structures for Ultra-low Cost Biosensing. *Optik-Int. J. Light Elect. Optic.* **2012,** *123* (15), 1400–1403.
3. (a) Koydemir, H. C.; Külah, H.; Özgen, C. Thin Film Biosensors. In *Thin Films and Coatings in Biology*, Springer: 2013; pp 265–300; (b) Singh, A.; Kaushik, A.; Kumar, R.; Nair, M.; Bhansali, S., Electrochemical Sensing of Cortisol: A Recent Update. *Appl. Biochem. Biotechnol.* **2014,** *174* (3), 1115–1126.
4. Kissinger, P.T. Biosensors—A Perspective. *Biosens Bioelectron* **2005,** *20* (12), 2512–2516.
5. Kumar, R.; Sidhu, G. K.; Goyal, N.; Nair, M.; Kaushik, A. Cerium Oxide Nanostructures for Bio-Sensing Application. *ScienceJet* **2015,** *4* (161), 1–21.
6. (a) Lin, V. S.-Y.; Motesharei, K.; Dancil, K.-P. S.; Sailor, M. J.; Ghadiri, M. R. A Porous Silicon-Based Optical Interferometric Biosensor. *Science* **1997,** *278* (5339), 840–843; (b) Kaushik, A.; Kumar, R.; Huey, E.; Bhansali, S.; Nair, N.; Nair, M. Silica Nanowires: Growth, Integration, and Sensing Applications. *Microchim. Acta.* **2014,** *181* (15–16), 1759–1780.
7. Bi, Z.; Anderoglu, O.; Zhang, X.; MacManus-Driscoll, J. L.; Yang, H.; Jia, Q.; Wang, H., Nanoporous Thin Films with Controllable Nanopores Processed from Vertically Aligned Nanocomposites. *Nanotechnology* **2010,** *21* (28), 285606.
8. Singh, S.; Arya, S. K.; Pandey, P.; Malhotra, B.; Saha, S.; Sreenivas, K.; Gupta, V. Cholesterol Biosensor Based on rf Sputtered Zinc Oxide Nanoporous Thin Film. *Appl. Phys. Lett.* **2007,** *91* (6), 063901–063901-3.
9. Zhao, H.; Guo, Q.; Xia, F.; Wang, H. Two-Dimensional Materials for Nanophotonics Application. *Nanophotonics* **2015,** *4,* 128–142.
10. Balendhran, S.; Walia, S.; Nili, H.; Ou, J. Z.; Zhuiykov, S.; Kaner, R. B.; Sriram, S.; Bhaskaran, M.; Kalantar-zadeh, K., Two-Dimensional Molybdenum Trioxide and Dichalcogenides. *Adv. Funct. Mater.* **2013,** *23* (32), 3952–3970.
11. (a) Ramakrishna Matte, H.; Gomathi, A.; Manna, A. K.; Late, D. J.; Datta, R.; Pati, S. K.; Rao, C. MoS2 and WS2 Analogues of Graphene. *Angew. Chem.* **2010,** *122* (24), 4153–4156; (b) Late, D. J.; Liu, B.; Matte, H.; Rao, C.; Dravid, V. P. Rapid Characterization of Ultrathin Layers of Chalcogenides on SiO2/Si Substrates. *Adv. Funct. Mater.* **2012,** *22* (9), 1894–1905; (c) Late, D. J.; Liu, B.; Matte, H. R.; Dravid, V. P.; Rao, C. Hysteresis in Single-Layer MoS2 Field Effect Transistors. *ACS Nano.* **2012,** *6* (6), 5635–5641; (d) Late, D. J.; Huang, Y.-K.; Liu, B.; Acharya, J.; Shirodkar, S. N.; Luo, J.; Yan, A.; Charles, D.; Waghmare, U. V.; Dravid, V. P. Sensing Behavior of Atomically Thin-Layered MoS2 Transistors. *ACS Nano.* **2013,** *7* (6), 4879–4891.
12. (a) Allen, M. J.; Fowler, J. D.; Tung, V. C.; Yang, Y.; Weiller, B. H.; Kaner, R. B. Temperature Dependent Raman Spectroscopy of Chemically Derived Graphene. *Appl. Phys. Lett.* **2008,** *93* (19), 193119; (b) Late, D. J. Temperature Dependent Phonon Shifts in Single-Layer WS2. *ACS Appl. Mater. Interfaces* **2014,** *6* (2), 1158–1163.
13. (a) Late, D. J.; Shirodkar, S. N.; Waghmare, U. V.; Dravid, V. P.; Rao, C. Thermal Expansion, Anharmonicity and Temperature-Dependent Raman Spectra of Single-and Few-Layer MoSe2 and WSe2. *ChemPhysChem* **2014,** *15* (8), 1592–1598; (b) Late, D. J.; Doneux, T.; Bougouma, M. Single-Layer MoSe2 Based NH3 Gas Sensor. *Appl. Phys. Lett.* **2014,** *105* (23), 233103.

14. Late, D. J.; Liu, B.; Luo, J.; Yan, A.; Matte, H.; Grayson, M.; Rao, C.; Dravid, V. P. GaS and GaSe Ultrathin Layer Transistors. *Adv. Mater.* **2012,** *24* (26), 3549–3554.
15. Xi, Q.; Zhou, D.-M.; Kan, Y.-Y.; Ge, J.; Wu, Z.-K.; Yu, R.-Q.; Jiang, J.-H. Highly Sensitive And Selective Strategy for microRNA Detection Based on WS2 Nanosheet Mediated Fluorescence Quenching and Duplex-Specific Nuclease Signal Amplification. *Anal. Chem.* **2014,** *86* (3), 1361–1365.
16. Yuan, Y.; Li, R.; Liu, Z. Establishing Water-Soluble Layered ws2 Nanosheet as a Platform for Biosensing. *Anal. Chem.* **2014,** *86* (7), 3610–3615.
17. Zhu, C.; Zeng, Z.; Li, H.; Li, F.; Fan, C.; Zhang, H., Single-Layer MoS2-Based Nanoprobes for Homogeneous Detection of Biomolecules. *J. Am. Chem. Soc.* **2013,** *135* (16), 5998–6001.
18. Zhavnerko, G.; Marletta, G. Developing Langmuir–Blodgett Strategies Towards Practical Devices. *Mater. Sci. Eng., B.* **2010,** *169* (1), 43–48.
19. Luz, R. A.; Iost, R. M.; Crespilho, F. N. Nanomaterials for Biosensors and Implantable Biodevices. In *Nanobioelectrochemistry*, Springer: 2013; pp 27–48.
20. (a) Wang, H.; Liu, F.; Fu, W.; Fang, Z.; Zhou, W.; Liu, Z. Two-Dimensional Heterostructures: Fabrication, Characterization, and Application. *Nanoscale* **2014,** *6* (21), 12250–12272; (b) Salvatore, G. A.; Münzenrieder, N.; Barraud, C.; Petti, L.; Zysset, C.; Büthe, L.; Ensslin, K.; Tröster, G. Fabrication and Transfer of Flexible Few-Layers MoS_2 Thin Film Transistors to Any Arbitrary Substrate. ACS Nano **2013,** *7* (10), 8809–8815; (c) Gupta A.; Sakthivel T.; Seal S., Recent development in 2D materials beyond graphene. *Prog. Mater. Sci.* **2015,** 73, 44–126.
21. Xie, D.; Jiang, Y.; Pan, W.; Li, D.; Wu, Z.; Li, Y. Fabrication and Characterization of Polyaniline-Based Gas Sensor by Ultra-Thin Film Technology. *Sens. Actuators, B: Chem.* **2002,** *81* (2), 158–164.
22. Comini, E. Metal Oxide Nano-Crystals for Gas Sensing. *Anal. Chim. Acta* **2006,** *568* (1), 28–40.
23. Im J.; Sengupta S. K.; Barucha M. F.; Granz C. D.; Ammu s.; Manohar S. K.; whitten J. E. A Hybrid Chemiresistive Sensor System for the Detection of Organic Vapors. *Sens. Actuators B* **2011**, *156*, 715–722.
24. Mandal, H. S.; Su, Z.; Ward, A.; Tang, X. S. Carbon Nanotube Thin Film Biosensors for Sensitive and Reproducible Whole Virus Detection. *Theranostics* **2012,** *2* (3), 251.
25. (a) Eggins B. R., Chemical Sensors and Electrochemical Biosensors: Analytical Techniques in the Sciences. Wiley, England **2002**; (b) Tsuchida, T.; Yoda K., Multi-enzyme Membrane Electrodes for Determination of Creatinine and Creatine in Serum. *Clin. Chem.* **1983**, *26*, 51–55; (c) Prakash, S.; Chakrabarty, T.; Singh, A. K.; Shahi, V. K. Polymer Thin Films Embedded with Metal Nanoparticles for Electrochemical Biosensors Applications. *Biosens. Bioelectron.* **2013,** *41*, 43–53.
26. Arya, S. K.; Saha, S.; Ramirez-Vick, J. E.; Gupta, V.; Bhansali, S.; Singh, S. P. Recent Advances in ZnO Nanostructures and Thin Films for Biosensor Applications: Review. *Anal. Chim. Acta.* **2012,** *737*, 1–21.
27. Fu, Y.; Luo, J.; Du, X.; Flewitt, A.; Li, Y.; Markx, G.; Walton, A.; Milne, W. Recent Developments on ZnO Films for Acoustic Wave Based Bio-Sensing and Microfluidic Applications: A Review. *Sens. Actuators, B: Chem.* **2010,** *143* (2), 606–619.
28. Wood, C.; Evans, S.; Cunningham, J.; O'Rorke, R.; Wälti, C.; Davies, A. Alignment of Particles In Microfluidic Systems Using Standing Surface Acoustic Waves. *Appl. Phys. Lett.* **2008,** *92* (4), 044104.

29. (a) Chu, S.-Y.; Water, W.; Liaw, J.-T. An Investigation of the Dependence of ZnO Film on the Sensitivity of Love Mode Sensor in ZnO/quartz Structure. *Ultrasonics* **2003,** *41* (2), 133–139; (b) Jian, S.-J.; Chu, S.-Y.; Huang, T.-Y.; Water, W. Study of Preferred Orientation Of Zinc Oxide Films on the 64° LiNbO 3 Substrates and Their Applications as Liquid Sensors. J. *Vacu. Sci. Technol. A: Vacuum, Surfaces, and Films* **2004,** *22* (6), 2424–2430; (c) Krishnamoorthy, S.; Iliadis, A. Development of High Frequency ZnO/SiO 2/Si Love Mode Surface Acoustic Wave Devices. *Solid-State Electron.* **2006,** *50* (6), 1113–1118.
30. (a) Powell, D. A.; Kalantar-zadeh, K.; Wlodarski, W. Numerical Calculation of SAW Sensitivity: Application to ZnO/LiTaO3 Transducers. *Sens. Actuators, A: Phy.* **2004,** *115* (2–3), 456–461; (b) Du, J.; Harding, G. L. A Multilayer Structure for Love-Mode Acoustic Sensors. *Sens. Actuators, A: Phy.* **1998,** *65* (2), 152–159.
31. Wenzel, S. W.; White, R. M. Flexural Plate-Wave Gravimetric Chemical Sensor. *Sens. Actuators, A: Phy.* **1990,** *22* (1), 700–703.
32. White, R. Introductory Lecture Acoustic interactions from Faraday's crispations to MEMS. *Faraday Discuss.* **1997,** *107*, 1–13.
33. Huang, I.-Y.; Lee, M.-C. Development of a FPW Allergy Biosensor for Human IgE Detection by MEMS and Cystamine-Based SAM Technologies. *Sens. Actuators, B: Chem.* **2008,** *132* (1), 340–348.
34. (a) Bjurstrom, J.; Rosén, D.; Katardjiev, I.; Yanchev, V. M.; Petrov, I. Dependence of the Electromechanical Coupling on the Degree of Orientation of c-textured Thin AlN Films. *IEEE Trans. Ultrason. Ferroelectr. Freq. Control IEEE Transactions on* **2004,** *51* (10), 1347–1353; (b) Kang, Y.-R.; Kang, S.-C.; Paek, K.-K.; Kim, Y.-K.; Kim, S.-W.; Ju, B.-K. Air-Gap Type Film Bulk Acoustic Resonator Using Flexible Thin Substrate. *Sens. Actuators, A: Phy.* **2005,** *117* (1), 62–70.
35. Gabl, R.; Feucht, H.-D.; Zeininger, H.; Eckstein, G.; Schreiter, M.; Primig, R.; Pitzer, D.; Wersing, W. First Results on Label-Free Detection of DNA and Protein Molecules Using a Novel Integrated Sensor Technology Based on Gravimetric Detection Principles. Biosens. Bioelectron. **2004,** *19* (6), 615–620.
36. (a) Lin, R.-C.; Chen, Y.-C.; Chang, W.-T.; Cheng, C.-C.; Kao, K.-S. Highly Sensitive Mass Sensor Using Film Bulk Acoustic Resonator. *Sens. Actuators, A: Phy.* **2008,** *147* (2), 425–429; (b) Yan, Z.; Zhou, X.; Pang, G.; Zhang, T.; Liu, W.; Cheng, J.; Song, Z.; Feng, S.; Lai, L.; Chen, J. ZnO-Based Film Bulk Acoustic Resonator for High Sensitivity Biosensor Applications. *Appl. Phys. Lett.* **2007,** *90* (14), 143503–143503-3.
37. Link, M.; Weber, J.; Schreiter, M.; Wersing, W.; Elmazria, O.; Alnot, P. Sensing Characteristics of High-Frequency Shear Mode Resonators in Glycerol Solutions. *Sens. Actuators, B: Chem.* **2007,** *121* (2), 372–378.
38. Weber, J.; Albers, W. M.; Tuppurainen, J.; Link, M.; Gabl, R.; Wersing, W.; Schreiter, M. Shear Mode FBARs as Highly Sensitive Liquid Biosensors. *Sens. Actuators, A: Phy.* **2006,** *128* (1), 84–88.
39. Muguruma, H.; Hiratsuka, A.; Karube, I. Thin-Film Glucose Biosensor Based on Plasma-Polymerized Film: Simple Design for Mass Production. *Anal. Chem.* **2000,** *72* (11), 2671–2675.
40. Gooding, J.; Praig, V.; Hall, E. Platinum-Catalyzed Enzyme Electrodes Immobilized on Gold Using Self-Assembled Layers. *Anal. Chem.* **1998,** *70* (11), 2396–2402.
41. Hiratsuka, A.; Muguruma, H.; Nagata, R.; Nakamura, R.; Sato, K.; Uchiyama, S.; Karube, I. Mass Transport Behavior of Electrochemical Species Through Plasma-Polymerized Thin Film on Platinum Electrode. *J. Membr. Sci.* **2000,** *175* (1), 25–34.

42. van Noort, D.; Mandenius, C.-F. Porous Gold Surfaces for Biosensor Applications. *Biosens. Bioelectron.* **2000,** *15* (3), 203–209.
43. Grieshaber, D.; MacKenzie, R.; Voeroes, J.; Reimhult, E. Electrochemical Biosensors-Sensor Principles and Architectures. *Sensors* **2008,** *8* (3), 1400–1458.
44. Drott, J.; Lindström, K.; Rosengren, L.; Laurell, T. Porous Silicon as the Carrier Matrix in Microstructured Enzyme Reactors Yielding High Enzyme Activities. *J. Micromech. Microeng.* **1997,** *7* (1), 14.
45. Toccafondi, C.; Thorat, S.; La Rocca, R.; Scarpellini, A.; Salerno, M.; Dante, S.; Das, G. Multifunctional Substrates of Thin Porous Alumina for Cell Biosensors. *J. Mater. Sci.: Mater. Med.* **2014,** *25* (10), 2411–2420.
46. Ansari, A. A.; Khan, M.; Alhoshan, M.; Alrokayan, S.; Alsalhi, M. Nanoporous Characteristics Of Sol—Gel-Derived ZnO Thin Film. *J. Semicond.* **2012,** *33* (4), 042002.
47. Wilson, R.; Turner, A. Glucose Oxidase: An Ideal Enzyme. *Biosens. Bioelectron.* **1992,** *7* (3), 165–185.
48. Macaya, D. J.; Nikolou, M.; Takamatsu, S.; Mabeck, J. T.; Owens, R. M.; Malliaras, G. G. Simple Glucose Sensors with Micromolar Sensitivity Based on Organic Electrochemical Transistors. *Sens. Actuators, B: Chem.* **2007,** *123* (1), 374–378.
49. Syritski, V.; Idla, K.; Öpik, A. Synthesis and Redox Behavior of PEDOT/PSS and PPy/DBS Structures. *Synth. Met.* **2004,** *144* (3), 235–239.
50. Liu, J.; Agarwal, M.; Varahramyan, K. Glucose Sensor Based on Organic Thin Film Transistor Using Glucose Oxidase and Conducting Polymer. *Sens. Actuators, B: Chem.* **2008,** *135* (1), 195–199.
51. Liao, C.; Zhang, M.; Niu, L.; Zheng, Z.; Yan, F. Highly Selective and Sensitive Glucose Sensors Based on Organic Electrochemical Transistors with Graphene-Modified Gate Electrodes. *J. Mater. Chem. B.* **2013,** *1* (31), 3820–3829.
52. Liao, C.; Zhang, M.; Niu, L.; Zheng, Z.; Yan, F. Organic Electrochemical Transistors with Graphene-Modified Gate Electrodes for Highly Sensitive and Selective Dopamine Sensors. *J. Mater. Chem. B.* **2014,** *2* (2), 191–200.
53. Tang, H.; Yan, F.; Lin, P.; Xu, J.; Chan, H. L. Highly Sensitive Glucose Biosensors Based on Organic Electrochemical Transistors Using Platinum Gate Electrodes Modified with Enzyme and Nanomaterials. *Adv. Funct. Mater.* **2011,** *21* (12), 2264–2272.
54. Matharu, Z.; Sumana, G.; Arya, S. K.; Singh, S.; Gupta, V.; Malhotra, B. Polyaniline Langmuir-Blodgett Film Based Cholesterol Biosensor. *Langmuir* **2007,** *23* (26), 13188–13192.
55. Zhong, L.; Gan, S.; Fu, X.; Li, F.; Han, D.; Guo, L.; Niu, L. Electrochemically Controlled Growth of Silver Nanocrystals on Graphene Thin Film and Applications for Efficient Nonenzymatic H_2O_2 Biosensor. *Electrochim. Acta* **2013,** *89*, 222–228.
56. Elkington, D.; Cooling, N.; Belcher, W.; Dastoor, P. C.; Zhou, X. Organic Thin-Film Transistor (OTFT)-Based Sensors. *Electronics* **2014,** *3* (2), 234–254.
57. Buth, F.; Donner, A.; Sachsenhauser, M.; Stutzmann, M.; Garrido, J. A. Biofunctional Electrolyte-Gated Organic Field-Effect Transistors. *Adv. Mater.* **2012,** *24* (33), 4511–4517.
58. Khodagholy, D.; Curto, V. F.; Fraser, K. J.; Gurfinkel, M.; Byrne, R.; Diamond, D.; Malliaras, G. G.; Benito-Lopez, F.; Owens, R. M. Organic Electrochemical Transistor Incorporating an Ionogel as a Solid State Electrolyte for Lactate Sensing. *J. Mater. Chem.* **2012,** *22* (10), 4440–4443.

59. Phongphut, A.; Sriprachuabwong, C.; Wisitsoraat, A.; Tuantranont, A.; Prichanont, S.; Sritongkham, P. A Disposable Amperometric Biosensor Based on Inkjet-Printed Au/PEDOT-PSS Nanocomposite for Triglyceride Determination. *Sens. Actuators, B: Chem.* **2013,** *178*, 501–507.
60. Jia, W.; Bandodkar, A. J.; Valdés-Ramírez, G.; Windmiller, J. R.; Yang, Z.; Ramírez, J.; Chan, G.; Wang, J. Electrochemical Tattoo Biosensors for Real-Time Noninvasive Lactate Monitoring in Human Perspiration. *Anal. Chem.* **2013,** *85* (14), 6553–6560.
61. Vieira, N. C.; Figueiredo, A.; Fernandes, E. G.; Guimaraes, F. E.; Zucolotto, V. Nanostructured Polyaniline Thin Films as Urea-Sensing Membranes in Field-Effect Devices. *Synth. Met.* **2013,** *175*, 108–111.
62. Someya, T.; Dodabalapur, A.; Gelperin, A.; Katz, H. E.; Bao, Z. Integration and Response of Organic Electronics with Aqueous Microfluidics. *Langmuir* **2002,** *18* (13), 5299–5302.
63. Magliulo, M.; Mallardi, A.; Gristina, R.; Ridi, F.; Sabbatini, L.; Cioffi, N.; Palazzo, G.; Torsi, L. Part per Trillion Label-Free Electronic Bioanalytical Detection. *Anal. Chem.* **2013,** *85* (8), 3849–3857.
64. Magliulo, M.; Mallardi, A.; Mulla, M. Y.; Cotrone, S.; Pistillo, B. R.; Favia, P.; Vikholm-Lundin, I.; Palazzo, G.; Torsi, L. Electrolyte-Gated Organic Field-Effect Transistor Sensors Based on Supported Biotinylated Phospholipid Bilayer. *Adv. Mater.* **2013,** *25* (14), 2090–2094.
65. James, C. Polypyrrole Nanoribbon Based Chemiresistive Immunosensors for Viral Plant Pathogen Detection. *Anal. Methods* **2013,** *5* (14), 3497–3502.
66. Lin, T.-W.; Kekuda, D.; Chu, C.-W. Label-Free Detection of DNA Using Novel Organic-Based Electrolyte-Insulator-Semiconductor. *Biosens. Bioelectron.* **2010,** *25* (12), 2706–2710.
67. Khan, H. U.; Roberts, M. E.; Johnson, O.; Förch, R.; Knoll, W.; Bao, Z. In Situ, Label-Free DNA Detection Using Organic Transistor Sensors. *Adv. Mater.* **2010,** *22* (40), 4452–4456.
68. (a) Kergoat, L.; Piro, B.; Berggren, M.; Pham, M.-C.; Yassar, A.; Horowitz, G. DNA Detection with a Water-Gated Organic Field-Effect Transistor. *Org. Electron.* **2012,** *13* (1), 1–6; (b) Khan, H. U.; Roberts, M. E.; Johnson, O.; Knoll, W.; Bao, Z. The Effect of ph and DNA Concentration on Organic Thin-Film Transistor Biosensors. *Org. Electron.* **2012,** *13* (3), 519–524; (c) Lin, P.; Luo, X.; Hsing, I.; Yan, F. Organic Electrochemical Transistors Integrated in Flexible Microfluidic Systems and Used for Label-Free DNA Sensing. *Adv. Mater.* **2011,** *23* (35), 4035–4040.
69. Tarabella, G.; Balducci, A. G.; Coppedè, N.; Marasso, S.; D'Angelo, P.; Barbieri, S.; Cocuzza, M.; Colombo, P.; Sonvico, F.; Mosca, R. Liposome Sensing and Monitoring by Organic Electrochemical Transistors Integrated in Microfluidics. *Biochim. Biophys. Acta (BBA)-General Subjects* **2013,** *1830* (9), 4374–4380.
70. Huang, W.; Besar, K.; LeCover, R.; Dulloor, P.; Sinha, J.; Hardigree, J. F. M.; Pick, C.; Swavola, J.; Everett, A. D.; Frechette, J. Label-Free Brain Injury Biomarker Detection Based on Highly Sensitive Large Area Organic Thin Film Transistor with Hybrid Coupling Layer. *Chem. Sci.* **2014,** *5* (1), 416–426.
71. (a) Tang, H.; Lin, P.; Chan, H. L.; Yan, F. Highly Sensitive Dopamine Biosensors Based on Organic Electrochemical Transistors. *Biosens. Bioelectron.* **2011,** *26* (11), 4559–4563; (b) Khan, H. U.; Jang, J.; Kim, J.-J.; Knoll, W. Effect of Passivation on the Sensitivity and Stability of Pentacene Transistor Sensors in Aqueous Media. *Biosens.*

Bioelectron. **2011,** *26* (10), 4217–4221; (c) Khan, H. U.; Jang, J.; Kim, J.-J.; Knoll, W. In Situ Antibody Detection and Charge Discrimination Using Aqueous Stable Pentacene Transistor Biosensors. *J. Am. Chem. Soc.* **2011,** *133* (7), 2170–2176.

72. (a) Cheng, Z.; Hou, J.; Zhou, Q.; Li, T.; Li, H.; Yang, L.; Jiang, K.; Wang, C.; Li, Y.; Fang, Y. Sensitivity Limits and Scaling of Bioelectronic Graphene Transducers. *Nano Lett.* **2013,** *13* (6), 2902–2907; (b) Kim, D.-J.; Sohn, I. Y.; Jung, J.-H.; Yoon, O. J.; Lee, N.-E.; Park, J.-S. Reduced Graphene Oxide Field-Effect Transistor for Label-Free Femtomolar Protein Detection. *Biosens. Bioelectron.* **2013,** *41*, 621–626; (c) Kim, D. J.; Park, H. C.; Sohn, I. Y.; Jung, J. H.; Yoon, O. J.; Park, J. S.; Yoon, M. Y.; Lee, N. E. Electrical Graphene Aptasensor for Ultra-Sensitive Detection of Anthrax Toxin with Amplified Signal Transduction. *Small* **2013,** *9* (19), 3352–3360.
73. (a) Cheng, Z.; Li, Q.; Li, Z.; Zhou, Q.; Fang, Y. Suspended Graphene Sensors with Improved Signal and Reduced Noise. *Nano Lett.* **2010,** *10* (5), 1864–1868; (b) Heller, I.; Chatoor, S.; Männik, J.; Zevenbergen, M. A.; Dekker, C.; Lemay, S. G. Influence of Electrolyte Composition on Liquid-Gated Carbon Nanotube and Graphene Transistors. *J. Am. Chem. Soc.* **2010,** *132* (48), 17149–17156; (c) Heller, I.; Chatoor, S.; Männik, J.; Zevenbergen, M. A.; Oostinga, J. B.; Morpurgo, A. F.; Dekker, C.; Lemay, S. G. Charge Noise in Graphene Transistors. *Nano Lett.* **2010,** *10* (5), 1563–1567.
74. (a) Dong, X.; Shi, Y.; Huang, W.; Chen, P.; Li, L. J. Electrical Detection of DNA Hybridization with Single-Base Specificity Using Transistors Based on CVD-Grown Graphene Sheets. *Adv. Mater.* **2010,** *22* (14), 1649–1653; (b) He, Q.; Sudibya, H. G.; Yin, Z.; Wu, S.; Li, H.; Boey, F.; Huang, W.; Chen, P.; Zhang, H. Centimeter-Long and Large-Scale Micropatterns of Reduced Graphene Oxide Films: Fabrication and Sensing Applications. *ACS Nano.* **2010,** *4* (6), 3201–3208; (c) Ohno, Y.; Maehashi, K.; Yamashiro, Y.; Matsumoto, K. Electrolyte-Gated Graphene Field-Effect Transistors for Detecting ph and Protein Adsorption. *Nano Lett.* **2009,** *9* (9), 3318–3322.
75. Lee, D.-W.; Lee, J.; Sohn, I. Y.; Kim, B.-Y.; Son, Y. M.; Bark, H.; Jung, J.; Choi, M.; Kim, T. H.; Lee, C. Field-Effect Transistor with a Chemically Synthesized MoS2 Sensing Channel for Label-Free and Highly Sensitive Electrical Detection of DNA Hybridization. *Nano Res.* **2015,** *8*(7), 2340–2350.
76. Takahashi, S.; Anzai, J.-I. Recent Progress in Ferrocene-Modified Thin Films and Nanoparticles For Biosensors. *Materials* **2013,** *6* (12), 5742–5762.
77. Kang, D.; Zuo, X.; Yang, R.; Xia, F.; Plaxco, K. W.; White, R. J. Comparing the Properties of Electrochemical-Based DNA Sensors Employing Different Redox Tags. *Anal. Chem.* **2009,** *81* (21), 9109–9113.
78. Song, M.-J.; Lee, S.-K.; Lee, J.-Y.; Kim, J.-H.; Lim, D.-S. Electrochemical Sensor Based on Au Nanoparticles Decorated Boron-Doped Diamond Electrode Using Ferrocene-Tagged Aptamer for Proton Detection. *J. Electroanal. Chem.* **2012,** *677*, 139–144.
79. Sarkar, D.; Liu, W.; Xie, X.; Anselmo, A. C.; Mitragotri, S.; Banerjee, K. MoS2 Field-Effect Transistor for Next-Generation Label-Free Biosensors. *ACS Nano.* **2014,** *8* (4), 3992–4003.
80. Chaubey, A.; Malhotra, B. D. Mediated Biosensors. *Biosens. Bioelectron.* **2002,** *17* (6–7), 441–456.
81. Chen, H.; Müller, M. B.; Gilmore, K. J.; Wallace, G. G.; Li, D. Mechanically Strong, Electrically Conductive, and Biocompatible Graphene Paper. *Adv. Mater.* **2008,** *20* (18), 3557–3561.
82. Huang, X.; Zeng, Z.; Fan, Z.; Liu, J.; Zhang, H. Graphene-Based Electrodes. *Adv. Mater.* **2012,** *24* (45), 5979–6004.

83. Wu, H.; Wang, J.; Kang, X.; Wang, C.; Wang, D.; Liu, J.; Aksay, I. A.; Lin, Y. Glucose Biosensor Based on Immobilization of Glucose Oxidase in Platinum Nanoparticles/Graphene/Chitosan Nanocomposite Film. *Talanta* **2009,** *80* (1), 403–406.
84. Zeng, G.; Xing, Y.; Gao, J.; Wang, Z.; Zhang, X. Unconventional Layer-by-Layer Assembly Of Graphene Multilayer Films for Enzyme-Based Glucose and Maltose Biosensing. *Langmuir* **2010,** *26* (18), 15022–15026.
85. Dai, Z.; Bao, J.; Yang, X.; Ju, H. A Bienzyme Channeling Glucose Sensor with a Wide Concentration Range Based on Co-Entrapment of Enzymes in SBA-15 Mesopores. *Biosens. Bioelectron.* **2008,** *23* (7), 1070–1076.
86. Lee, C.-H.; Lin, T.-S.; Mou, C.-Y. Mesoporous Materials for Encapsulating Enzymes. *Nano Today* **2009,** *4* (2), 165–179.
87. Shan, C.; Yang, H.; Han, D.; Zhang, Q.; Ivaska, A.; Niu, L. Graphene/AuNPs/chitosan Nanocomposites Film for Glucose Biosensing. *Biosens. Bioelectron.* **2010,** *25* (5), 1070–1074.
88. Shan, C.; Yang, H.; Han, D.; Zhang, Q.; Ivaska, A.; Niu, L. Electrochemical Determination of NADH and Ethanol Based on Ionic Liquid-Functionalized Graphene. *Biosens. Bioelectron.* **2010,** *25* (6), 1504–1508.
89. Xu, H.; Dai, H.; Chen, G. Direct Electrochemistry and Electrocatalysis of Hemoglobin Protein Entrapped in Graphene and Chitosan Composite Film. *Talanta* **2010,** *81* (1–2), 334–338.
90. Zhou, K.; Zhu, Y.; Yang, X.; Luo, J.; Li, C.; Luan, S. A Novel Hydrogen Peroxide Biosensor Based on Au–Graphene–HRP–Chitosan Biocomposites. *Electrochim Acta* **2010,** *55* (9), 3055–3060.
91. Ahammad, A. S. Hydrogen Peroxide Biosensors Based on Horseradish Peroxidase and Hemoglobin. *J. Biosens. Bioelectron* **2013,** *S9*, 1–12.
92. Chen, X.; Baker, G. A. Cholesterol Determination Using Protein-Templated Fluorescent Gold Nanocluster Probes. *Analyst* **2013,** *138* (24), 7299–7302.
93. Dey, R. S.; Raj, C. R. Development of an Amperometric Cholesterol Biosensor Based on Graphene–Pt Nanoparticle Hybrid Material. *J. Phys. Chem. C.* **2010,** *114* (49), 21427–21433.
94. Loeb, W. Sugar Decomposition III. Electrolysis of Dextrose. *Biochemische Zeitschrift Biochemistry* **1909**, 132–144.
95. Kong, F.-Y.; Li, X.-R.; Zhao, W.-W.; Xu, J.-J.; Chen, H.-Y. Graphene Oxide–Thionine–Au Nanostructure Composites: Preparation and Applications in Non-Enzymatic Glucose Sensing. *Electrochem. Commun.* **2012,** *14* (1), 59–62.
96. Luo, J.; Jiang, S.; Zhang, H.; Jiang, J.; Liu, X. A Novel Non-Enzymatic Glucose Sensor Based on Cu Nanoparticle Modified Graphene Sheets Electrode. *Anal. Chim. Acta.* **2012,** *709*, 47–53.
97. Gao, H.; Xiao, F.; Ching, C. B.; Duan, H. One-Step Electrochemical Synthesis of PtNi Nanoparticle-Graphene Nanocomposites for Nonenzymatic Amperometric Glucose Detection. *ACS Appl. Mater. Interfaces* **2011,** *3* (8), 3049–3057.
98. (a) Mao, S.; Lu, G.; Yu, K.; Bo, Z.; Chen, J. Specific Protein Detection Using Thermally Reduced Graphene Oxide Sheet Decorated with Gold Nanoparticle-Antibody Conjugates. *Adv. Mater.* **2010,** *22* (32), 3521–3526; (b) Ang, P. K.; Chen, W.; Wee, A. T. S.; Loh, K. P. Solution-Gated Epitaxial Graphene as pH Sensor. *J. Am. Chem. Soc.* **2008,** *130* (44), 14392–14393.
99. (a) Ohno, Y.; Maehashi, K.; Matsumoto, K. Label-Free Biosensors Based on Aptamer-Modified Graphene Field-Effect Transistors. *J. Am. Chem. Soc.* **2010,** *132* (51),

18012–18013; (b) Liu, S.; Guo, X. Carbon Nanomaterials Field-Effect-Transistor-Based Biosensors. *NPG Asia Mater.* **2012,** *4* (8), e23.

100. Ritzert, N. L.; Li, W.; Tan, C.; Rodríguez-Calero, G. G.; Rodríguez-López, J.; Hernández-Burgos, K.; Conte, S.; Parks, J. J.; Ralph, D. C.; Abruña, H. D. Single Layer Graphene as an Electrochemical Platform. Faraday Discussions **2014.**
101. Zhang, J.; Lei, J.; Pan, R.; Xue, Y.; Ju, H. Highly Sensitive Electrocatalytic Biosensing of Hypoxanthine Based on Functionalization of Graphene Sheets with Water-Soluble Conducting Graft Copolymer. *Biosens. Bioelectron.* **2010,** *26* (2), 371–376.
102. Rand, K. D.; Zehl, M.; Jensen, O. N.; Jørgensen, T. J. Protein Hydrogen Exchange Measured at Single-Residue Resolution by Electron Transfer Dissociation Mass Spectrometry. *Anal. Chem.* **2009,** *81* (14), 5577–5584.
103. Ping, J.; Wu, J.; Wang, Y.; Ying, Y. Simultaneous Determination of Ascorbic Acid, Dopamine and Uric Acid Using High-Performance Screen-Printed Graphene Electrode. *Biosens. Bioelectron.* **2012,** *34* (1), 70–76.
104. Balendhran, S.; Walia, S.; Alsaif, M.; Nguyen, E. P.; Ou, J. Z.; Zhuiykov, S.; Sriram, S.; Bhaskaran, M.; Kalantar-zadeh, K. Field Effect Biosensing Platform Based on 2D α-MoO3. *ACS Nano.* **2013,** *7* (11), 9753–9760.
105. (a) Liu, K.-K.; Zhang, W.; Lee, Y.-H.; Lin, Y.-C.; Chang, M.-T.; Su, C.-Y.; Chang, C.-S.; Li, H.; Shi, Y.; Zhang, H. Growth of Large-Area and Highly Crystalline MoS2 Thin Layers on Insulating Substrates. *Nano Lett.* **2012,** *12* (3), 1538–1544; (b) Zhan, Y.; Liu, Z.; Najmaei, S.; Ajayan, P. M.; Lou, J. Large-Area Vapor-Phase Growth and Characterization of MoS2 Atomic Layers on a SiO2 Substrate. *Small* **2012,** *8* (7), 966–971.
106. Wang, L.; Wang, Y.; Wong, J. I.; Palacios, T.; Kong, J.; Yang, H. Y. Functionalized MoS2 Nanosheet-Based Field-Effect Biosensor for Label-Free Sensitive Detection of Cancer Marker Proteins in Solution. *Small* **2014,** *10* (6), 1101–1105.
107. Härtl, A.; Schmich, E.; Garrido, J. A.; Hernando, J.; Catharino, S. C.; Walter, S.; Feulner, P.; Kromka, A.; Steinmüller, D.; Stutzmann, M. Protein-Modified Nanocrystalline Diamond Thin Films for Biosensor Applications. *Nat. Mater.* **2004,** *3* (10), 736–742.

CHAPTER 5

INTEGRATED ELECTRONICS, ANALYTICAL TRANSDUCERS, AND SIGNAL PROCESSING

PANDIARAJ MANICKAM[1,*], VAIRAMANI KANAGAVEL[2], ROBSON BENJAMIN ALBY[3], KARUNAKARAN CHANDRAN[4], and SHEKHAR BHANSALI[1]

[1]*Bio-MEMS and Microsystems Laboratory, Department of Electrical and Computer Engineering, Florida International University, Miami, USA*

[2]*University Science Instrumentation Centre, Madurai Kamaraj University, Madurai, Tamil Nadu, India*

[3]*Department of Physics, The American College, Madurai, Tamil Nadu, India*

[4]*Biomedical Research Laboratory, Department of Chemistry, VHNSN College (Autonomous), Virudhunagar, Tamil Nadu, India*

**E-mail: pmanicka@fiu.edu*

CONTENTS

Biosensor is a self-contained integrated receptor transducer device, and consists of a biological recognition element in intimate contact or integrated with the transducer. Biosensor technology has been envisioned to play a significant analytical role in medicine, agriculture, food safety, bioprocessing, environmental, and industrial monitoring. However, bulky and complex bench-top instruments used for detection and signal processing restrict the commercialization of biosensor technology. This also affects long-term implementation of biosensor system in point-of-care clinical applications. Over the past decade, the integration of biosensors with miniaturized electronic circuits to develop hand-held biosensor devices has garnered considerable attention in the scientific community and industry. This chapter outlines some recent advances made in interfacing the biosensors with miniaturized electronic circuits to develop compact biosensor devices for point-of-care diagnosis.

5.1 INTRODUCTION

Over the past decade many important technological advances have led to the development of tools and materials needed for the construction of new biosensor devices. Starting with Clark's oxygen electrode sensor, there has been much advancement in improving the sensitivity, selectivity, and multiplexing capacity of the modern biosensor. According to the International Union of Pure and Applied Chemistry recommendations 1999, a biosensor is an independently integrated receptor transducer device, which is capable of providing selective quantitative or semiquantitative analytical information using a biological recognition element.[1,2] More clearly a biosensor is an analytical device used for the detection of analytes with three major constituents:

i. the bioreceptor,
ii. the transducer or detection element, and
iii. the reading device.

The bioreceptor is a biological recognition element that serves to respond to a target analyte of interest. The transducer serves to convert the biorecognition event into a measurable electrical signal. The amplifier within the biosensor is a key component in the whole endeavor as it serves to amplify small input signals from the transducer and deliver an amplified signal that consists of the essential features of the measured waveform. The amplified signal is then processed by the signal processor, at which point it can be routed to storage locations in memory, displayed on a variety of visualization platforms, or sent straight to analysis either onboard or as a postprocessing task. The working principle of the biosensor is schematically represented in Figure 5-1.

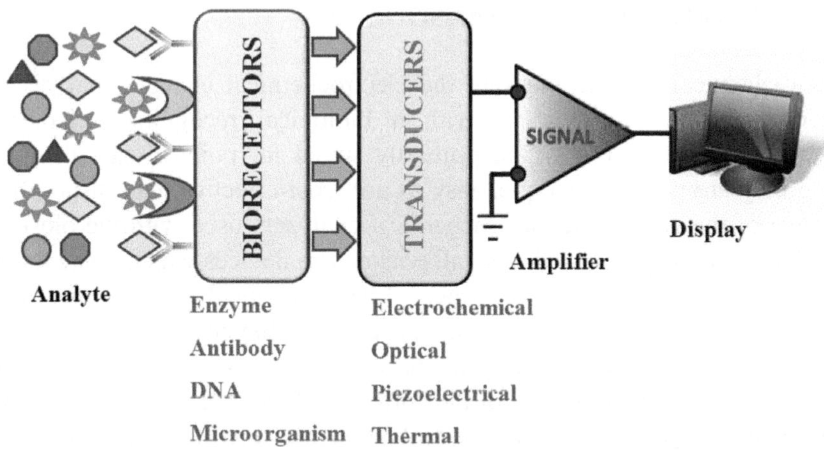

FIGURE 5-1 Schematic representation of working principle of biosensors.

Today, a large number of biosensors can be found in laboratories around the world and there is growing number of biosensors being developed as diagnostic tools in point-of-care testing. However, the analytical instruments used for biosensing are generally bulky and complex. This limits the long-term implementation of biosensor systems in point-of-care/clinical diagnosis. Recent technological advancements in microelectronics allow us to replace the traditional cumbersome instruments with easy to use miniaturized electrochemical systems. Such portable (hand-held), battery-powered, electrochemical analyzers offer tremendous potential for obtaining the desired analytical information in a faster, simpler ("user friendly"), and cheaper manner than do traditional laboratory based transducers. Generally, biosensor techniques can be classified either by the type of biological signaling mechanism they utilize or by the type of signal transduction they employ. In many cases, the main limitation in realizing point-of-care testing/sensing devices is the ability to miniaturize the analytical transducer and the lack of a cost-effective production method. Biosensor signal transduction can be accomplished via a great variety of methods. Although a variety of transducer methods have been feasible toward the development of biosensor technology, the most common methods are electrochemical and optical followed by the piezoelectric type.[3–6] This chapter highlights recent advances, trends, and applications of miniaturized electrochemical transducer systems for biosensor applications.

5.2 ELECTROCHEMICAL TRANSDUCERS

Electrochemical sensors measure the electrochemical changes that occur when target analyte interacts with a biological recognition element. Electrochemical biosensors are primarily suited for point-of-care devices since they are portable, simple, easy to use, cost-effective, and disposable in most of the cases. The electrochemical analyzers used with the biosensors have been miniaturized to small pocket size devices which make them applicable for home use or in surgery. Electrochemical sensors constitute the largest and oldest group of chemical sensors. They are further classified based on the electro-analytical principles employed. Voltammetric sensors are based on the measurement of the current–voltage relationship. A potential is applied to the sensor and a current proportional to the concentration of the electro-active species of interest is measured (amperometry is a special case of voltammetry, where the potential is kept constant). Potentiometric sensors are based on the measurement of the potential at an electrode at equilibrium state, that is, no current is allowed to flow during the measurement. The measured potential is proportional to the logarithm of the concentration of the electro-active species (Nernst equation). Conductometric sensors are based on the measurement of a conductance by applying an alternating current potential with small amplitude to a pair of electrodes in order to prevent polarization. The presence of charge carriers determines the sample conductance. Based on their high sensitivity, simplicity, and cost competitiveness, more than half of the biosensors reported in the literature are based on electrochemical transducers.[7]

5.3 ELECTROCHEMICAL ANALYZER THEORY

Electrochemical analyzer is an instrument used in the analytical chemistry to detect an analyte by measuring the potential and/or current in an electrochemical cell. The electrochemical cell, where the experiments are carried out, consists of a working electrode (WE), a reference electrode (RE), and a counter electrode (CE). The WE is an electrode where the reaction of interest is taking place.[8,9] The RE is used to produce a constant potential at the electrochemical cell. The CE (also known as auxiliary electrode), is an electrode which is used to close the current circuit in the electrochemical cell. It is usually made of an inert material (e.g., platinum, gold, graphite, and glassy carbon) and it does not participate in the electrochemical reaction. The key component of any biosensor based electrochemical analyzer is its potentiostat, an electronic circuit that precisely controls the potential of the WE, and therefore the electrochemical reaction at the electrode's surface. Figure 5-2 represents the basic components of a modern electrochemical analyzer. It comprises a biosensor, interfaced with potentiostat, analog-to-digital converter (ADC), digital-to-analog converter (DAC), microcontroller, and a computer.

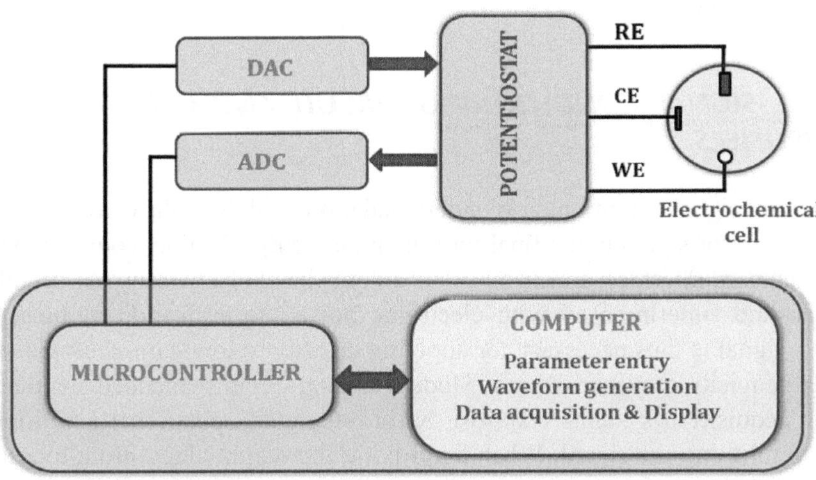

RE = Reference electrode CE = Counter electrode WE= Working electrode
DAC = Digital to analog converter ADC = Analog to digital converter

FIGURE 5-2 Basic components of a modern electrochemical analyzer system.

In a traditional electrochemical biosensor, the front end is the biorecognition or transducer unit which produces an electrical signal that depends on the measured parameter. This signal is sent over a transmission link to the signal processing section of the system which could include amplification, data conversion, or any number of other processing schemes. This can then be transferred to a computer for readout, data storage, or some other further processing. The advantages of this setup are that the sensor can be quite cheap and is separated from the data acquisition and processing. Indeed, the conventional electrochemical analyzers are capable of performing many different kinds of electrochemical analysis. However, there are disadvantages; the unamplified signal from the sensor will typically be quite weak and prone to the introduction of noise during transmission. Moreover, the total cost and size of the commercial instruments limit their applicability in point-of-care diagnosis. With a significant increase in the advancement in electronic technology, the interfacing of biosensor platform with microelectronics provides a new paradigm in designing inexpensive smart biosensing devices. A "smart" sensor system adds some form of signal processing, feedback, or control to the front end of the device, integrated with the sensor itself.[10] A key advantage of the electronics integrated biosensor platform is that the biosensing will be done using low-cost- and low power portable data acquisition systems while information processing will be done by proper digital circuitry.

5.3.1 SIGNAL CONDITIONING CIRCUIT AND OPERATIONAL AMPLIFIERS

The term signal conditioning is understood more widely as the conversion of the biosensor signal to the final form including amplification, conversion to the standard interface, and so on.[11] The output signal of a biosensor is usually weak and superimposed with electronic noise. Amplification of biosensor's signal is thus necessary for applying directly to low-gain, multiplexed data acquisition system inputs. Modern analog circuits intended for these data acquisition systems comprise basic integrated operational amplifiers for amplifying the signal. When amplifying the output signal from biosensors and electrochemical sensor, the amplifier must have a high frequency response, to avoid distortion and drift in the output.[12] The operational amplifiers normally are required to have high input impedance so that its loading effect on the transducer output signal is minimized. The signal processing is one of the major parts of the biosensor device. Signal will be processed in two phases: the first phase is analog signal processing coming from the transducer of the biosensors and the second phase is digital signal processing and data acquisition system. Data acquisition system will send digital numeric value to the microcontroller. The success of the design of any biosensor analytical system depends heavily on the design and performance of the signal conditioning circuits. Even a costly and accurate transducer may fail to deliver good performance if the signal conditioning circuit is not designed properly.

5.3.2 POTENTIOSTAT

Basically, a potentiostat is an electronic device that controls the voltage difference between a WE and RE. By injecting currents into the sensor through a CE, the potentiostat circuit also controls the electrochemical reaction at the electrode surface.[13] In other words, the potentiostat in an electrochemical analyzer has two tasks. Firstly, it measures the potential difference between the WE and RE without polarizing the RE, and compares the potential difference with a preset voltage. Secondly, it injects a current flowing from the CE to the WE in order to counteract the difference between the preset voltage and the existing WE potential. After completing the two tasks, the output of the potentiostat is the current flowing from the CE to the WE. A current to voltage converter (i/V) unit of the potentiostat is used to convert the current signal from the WE into a voltage signal which is then transferred to the computer via the ADC interface of the microcontroller unit.

Figure 5-3 shows a simple potentiostat circuit made up of two operational amplifiers (op-amp A and op-amp B) with distinct functions. The potential chosen for the WE (V_{bias}) is generated through the DAC and is fed to the op-amp A and to the RE. The electrochemical current generated at the WE as a result of applied potential is then converted into an equivalent voltage by another op-amp which is a current-to-voltage converter as per eq 5-1. The i/V gain can be adjusted by selecting the feedback resistor, R_f.

$$V_{out} = -I_{sensor} \times R_f \tag{5-1}$$

The potentiostat can be designed using one of many amplifiers in different configurations. Depending on the type of the electro-analytical techniques used, a fixed or a variable voltage is needed to apply to the electrochemical cell. For cyclic voltammetry (CV), this circuit sweeps repeatedly within the voltage range of interest. For amperometry, a fixed voltage value has to be applied and the voltage chosen on the basis of the electrochemical reaction. A band gap reference circuit can be used to generate a fixed voltage.[14] To design a very slow triangular waveform voltage, a mixed signal design approach can be used. Sometimes, a digitally programmable waveform generator will be used for biasing by employing an up-down counter, comparator, latch, and a DAC.[15]

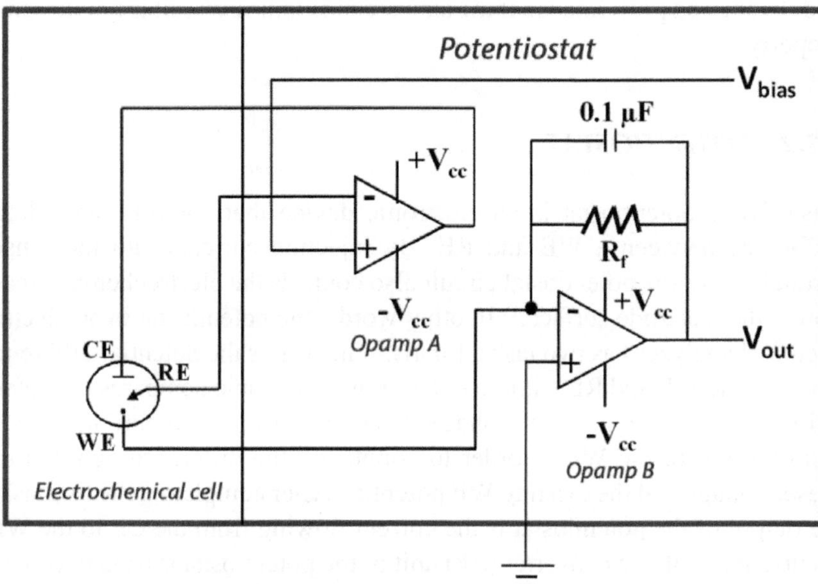

FIGURE 5-3 Block diagram of a simple potentiostat circuit.

Conventional potentiostats are designed for research purpose and are capable of performing many different kinds of complex electrochemical analysis. As such they typically are bulky and therefore inappropriate for miniaturized point-of-care biosensing assays. For field use and on-site diagnostics, the design and testing of miniaturized electronics interfaced electrochemical analyzers have been an active research topic in the past decade. The advantage of this custom-made potentiostat is that they can be made into very small, of the order of millimeters, and customized at fabrication specifically for specific experimental requirements such as current magnitude and frequency.

5.4 ELECTRONIC INTERFACING OF VOLTAMMETRIC BIOSENSORS

Cyclic voltammetry is a very important electrochemical technique used to study the mechanism of charge transfer reaction of redox species, in particular to determine the concentration of biomolecules using its oxidation/reduction reaction pathways. In the CV experiment, the potential of an electrode is cycled from an initial potential (E_i) to a final potential (E_f) and then back to E_i. During the potential sweep, the potentiostat measures the current (between the WE and CE) resulting from the applied potential.[16] The resulting current–potential plot is termed a cyclic voltammogram. The cyclic voltammogram is a complicated, time-dependent function of a large number of physical and chemical parameters. A voltammetric sensor can operate in other modes such as linear, square wave, and differential pulse voltammetric modes. Consequently, the respective current–potential response for each mode will be different. Several groups have designed and fabricated miniaturized single chip potentiostat that can be integrated to the sensor platform; this includes the development of a battery operated palm sized amperometric analyzer and a portable system that uses stripping voltammetry to perform trace metal analysis.[16,17]

A microcontroller-based potentiostat was developed to perform voltammetric technique for the analysis.[18] During this decade, by using the advantages of electronic technologies, more such advanced potentiostat circuits with more attractive performance have been reported widely. For example, Loncaric et al.[19] have designed a portable, universal serial bus (USB)-based electrochemical biosensor device for the detection of lysozyme. The device implements CV measurements by means of a portable potentiostat in conjunction with a miniaturized electrochemical cell. The interfaced microelectronic components, viz. potentiostat, ADC, and microcontroller unit of the electrochemical device are powered and controlled entirely by the USB. This system is compatible with any computer that has a USB port and is equipped with the Microsoft Excel program.

As shown in Figure 5-4, the USB-powered device was designed using a simple scanning potentiostat circuit governed by a microcontroller. This microcontroller uses USB bus power and does not require any additional power source. The op-amp used for op-amps 1–6 was the AD823 (analog devices) in a single-supply operation. Op-amp 1 functions as a buffering unit, whereas op-amps 2 and 3 serve as the potentiostat (Figure 5-4). In order to produce a scanning output voltage for generating a cyclic voltammogram, the microcontroller was programmed to output a ramp signal whose voltage levels and rate of change can be easily adjusted through user controlled potentiometers.

In 2011, Rowe and co-workers[20] reported an open source potentiostat named "*CheapStat-do it yourself*", will easily be constructed by anyone proficient at assembling circuits. The cost of making a CheapStat device is under US\$100—significantly less than the cost of commercial potentiostat instruments which can cost thousands of dollars. This low-cost, hand-held

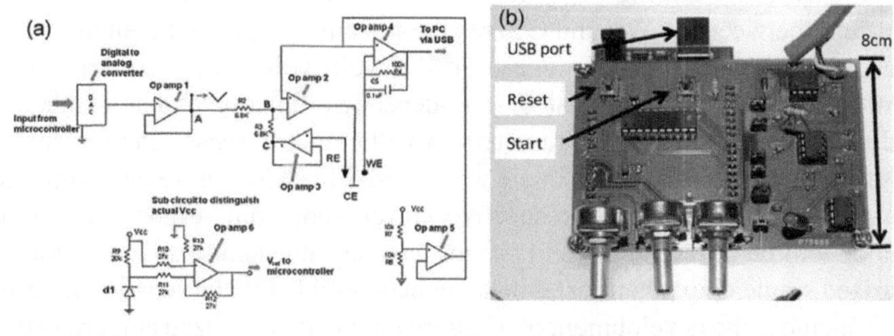

FIGURE 5-4 (A) Analog circuit of the universal serial bus-based potentiostat and (B) photograph of the potentiostat prototype showing the function keys/ports (Reproduced from Ref. [19])

instrument (Figure 5-5) is an open source and can be used in many different aspects of analytical chemistry research and education. This device is capable of carrying out various electrochemical protocols (e.g., CV, square-wave voltammetry, and anodic stripping voltammetry) and providing sufficient performance in chemical and biomarker detections.

FIGURE 5-5 The CheapStat an inexpensive "do-it-yourself" potentiostat (Reproduced from Ref. [20]).

In one study, Cruz et al.[21] developed a low-cost miniaturized ultra-low-power potentiostat LMP91000 interfacing with the BeagleBone board. The apparatus carries out CV measurements on three-electrode systems, and stores the data. Our group has also recently reported a cost-effective, USB based portable electrochemical analyzer performing CV for the determination of cytochrome *c* (cyt *c*) a heme-containing metalloprotein.[22] The developed electrochemical analyzer consists of a potentiostat, current-to-voltage converter constructed by the LM358 op-amp, a MAX5354 DAC, and a PIC18F4550 microcontroller based data acquisition system integrated with a screen printed electrode based cyt *c* biosensor. The electrochemical analyzer is connected through a standard USB port of the computer for data analysis. The application software is developed using Visual Basic 6.0. The dimensions of the whole device are 10 cm × 13 cm on printed circuit boards with total cost of less than US$50 and making it cost-effective and portable (Figure 5-6). The developed portable electrochemical analyzer belongs to the human interface device (HID) class. The HID class is well supported by all the operating systems. Hence, the need for an external device driver is avoided. Graphical user interface (GUI) based system implemented in the current version of the device provides the use of graphical widgets like

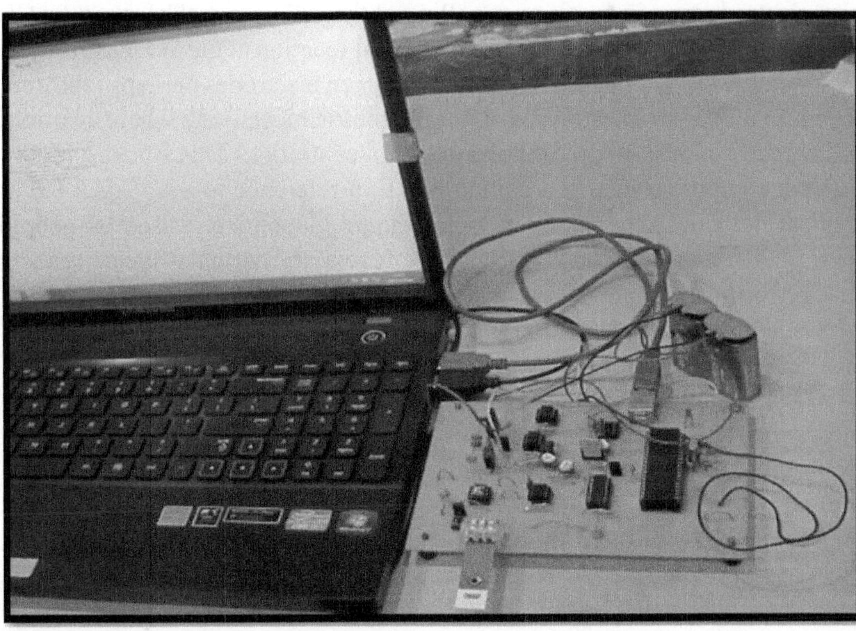

FIGURE 5-6 Photograph of the universal serial bus based electrochemical analyzer showing the interface with cyt c biosensor and laptop (Reproduced from Ref. [22]).

required voltage range setup, push buttons, charts, displays, and so on. It creates a user friendly GUI so that an inexperienced user can become quickly conversant with a complex application.

5.5 ELECTRONIC INTERFACING OF AMPEROMETRIC BIOSENSORS

Amperometric sensors are considered to be a subclass of voltammetric sensors and mostly used for continuous measurement of target analytes. In amperometric sensors, a fixed potential is applied to the electrochemical cell, and a corresponding current, due to a reduction or oxidation reaction, is then obtained. This current can be used to quantify the species involved in the reaction. The key consideration of an amperometric sensor is that it operates at a fixed potential.

Amperometric electrochemical sensor structures can be divided into two types: two-electrode and three-electrode structures (Figure 5-7). The former comprises a WE and an RE. A lot of research has been devoted to the development of potentiostats for two-electrode amperometric sensor applications. The latter comprises a WE, an RE, and a CE. This type is preferred over the two-electrode type in precise controlling of the cell potential because the CE supplies current required for electrochemical reaction at the WE electrode to maintain the stability of the RE.[23] Clark oxygen electrodes perhaps represent the basis for the simplest forms of amperometric biosensors, where a current is produced in proportion to the oxygen concentration. This is measured by the reduction of oxygen at a platinum WE in reference to a Ag/AgCl RE at a given potential. The commonly used home glucometers, used by people with diabetes to monitor blood glucose levels are portable device readers that perform amperometry for glucose detection.[24]

Using integrated circuit (IC) techniques a number of successful potentiostat designs performing amperometric techniques have been proposed.[25–27] The design and development of a palm-sized (9 cm × 11 cm × 3 cm), cost-effective, microcontroller operated amperometric analyzer was reported by Avdikos et al.[28] The low-power-consumption electronics used allow 8 h of autonomous operation with a 9 V battery (110 mAh), thus making this unit suitable for on-field measurements. Its operation is based mainly on the simple two-electrode potentiostatic mode. This approach allows the production of small size and low-power-consumption potentiostats, which are commercially available and allow even micro-electrode applications. An embedded system designed to provide miniaturized electrochemical

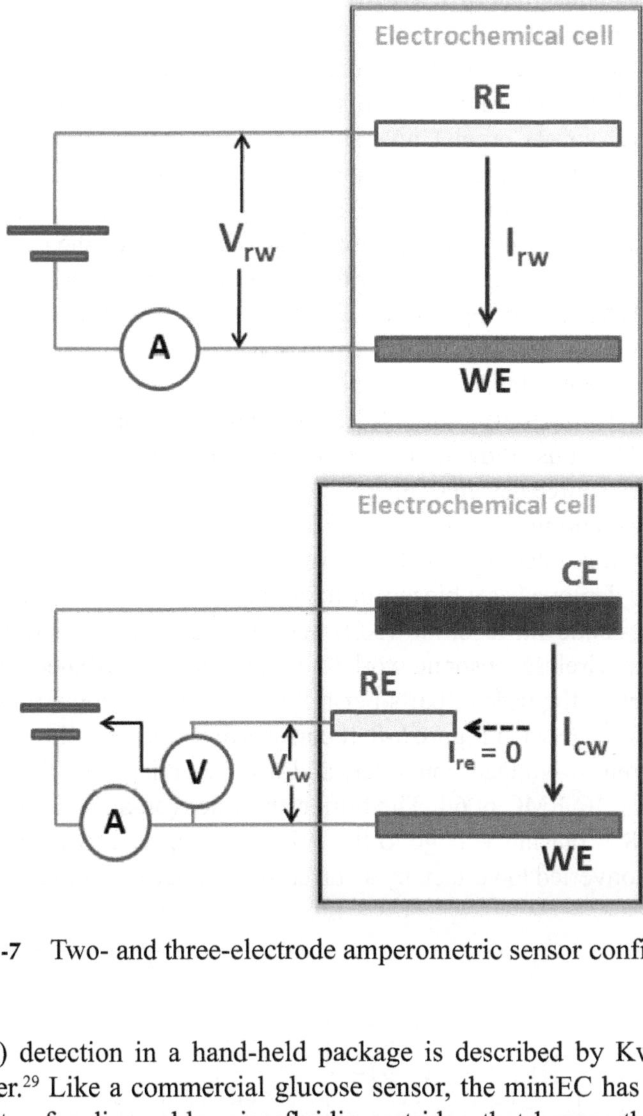

FIGURE 5-7 Two- and three-electrode amperometric sensor configuration.

(miniEC) detection in a hand-held package is described by Kwakye and Baeumner.[29] Like a commercial glucose sensor, the miniEC has two parts. It consists of a disposable microfluidic cartridge that houses the amperometric biosensor and an embedded system to power the sensor and measure, display, and store the sensor results. The potentiostat circuit was built with low power, low noise MAX407 op-amps and performs two-electrode amperometry. Transimpedance amplifier circuit is implemented in the circuit to amplify the signal generated at the WE. This device uses MSP430 microcontroller together with low-power-electronic components and can run for months on a single AA or AAA battery. The total cost of the device is around US$50 thus making it a really cost-effective platform for commercial biosensing application.

Design and operation of an accurate, field-ready, cost-effective, and microcontroller-based amperometric potentiostat was demonstrated by Angenent's group at Cornell University and used in novel subsurface bioelectrochemical system based biosensor.[30] They presented a design for this amperometric potentiostat using a ATMega644 microcontroller, an external DAC, and a series of operational amplifiers. They utilized the serial peripheral interface of the microcontroller to communicate between the microcontroller and DAC and the operational amplifiers to process the signal from the DAC and apply potential to the electrochemical cell, and the internal ADC to record the current at the WE. This robust and field-ready potentiostat, which can withstand temperatures of −30 °C, can be manufactured at relatively low cost (US$600), thus, allowing for *en masse* deployment at field sites. This is the first open source design of an inexpensive potentiostat that is field ready for long-term chronoamperometry. The micro controller unit (MCU)-based potentiostat was integrated with electrodes and a solar panel based power system, and deployed as a biosensor to monitor microbial respiration.

A home-made nitric oxide (NO) sensor was designed and prototype connected to wireless sensor network for determining the exhaled nitric oxide (ENO) level of the user in parts per billion.[31] The amperometric measurement circuit is shown in Figure 5-8. It consists of a voltage follower, a potentiostat, current-to-voltage converter, and a differential amplifier. It uses a quad op-amp IC LMC 6064. The buffer amplifier (op-amp A) supplies the required 0.8 V biasing voltage to the WE. The amperometric current from the WE is converted to voltage by a current-to-voltage converter (op-amp C)

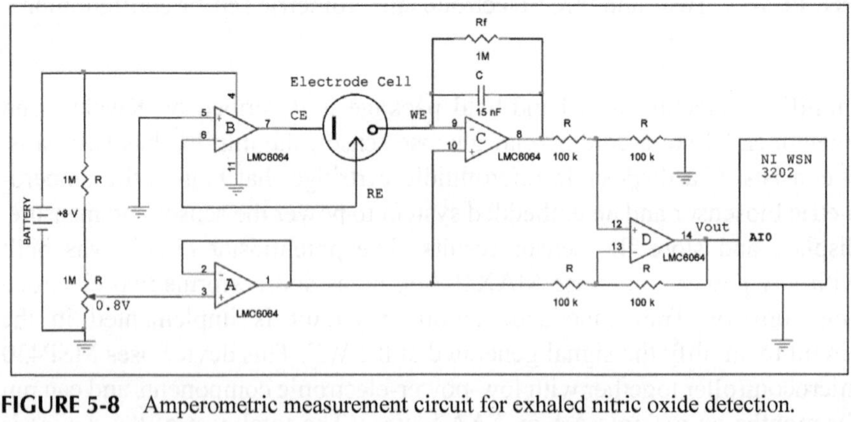

FIGURE 5-8 Amperometric measurement circuit for exhaled nitric oxide detection.

and applied as an inverting input of the differential amplifier (op-amp D). The output of the current-to-voltage converter is given by eq 5-2:

$$V_o = - (i_{ox} R_f) + V_{bias} \tag{5-2}$$

where V_o is the output voltage of I/V converter, i_{ox} is the oxidation current flowing through the WE, Rf is the feedback resistor, and V_{bias} is the bias voltage applied to the WE.

5.6 LABVIEW BASED VIRTUAL INSTRUMENT FOR CV

Programmers develop embedded software applications every day in order to increase efficiency and productivity of the analytical devices. Laboratory Virtual Instrument Engineering Workbench (LabVIEW), as a programming language, is a powerful tool that can be used to help achieve these goals. LabVIEW is a graphically based programming language developed by National Instruments (NI). Its graphical nature makes it ideal for test and measurement (T and M), automation, instrument control, data acquisition, and data analysis applications. This results in significant productivity improvements over conventional programming languages.[32] Recently, virtual instrumentation is rapidly replacing the costly bench top instrumentation because it offers flexible, fast, and cost-effective solutions.[33] Virtual Instrument (VI) is a LabVIEW programming element.

A VI consists of a front panel, block diagram, and an icon that represents the program. The front panel is used to display controls and indicators for the user, and the block diagram contains the code for the VI. The icon, which is a visual representation of the VI, has connectors for program inputs and outputs. Programming languages such as C and BASIC uses functions and subroutines as programming elements. LabVIEW uses the VI. The front panel of a VI handles the functions inputs and outputs, and the code diagram performs the work of the VI. Multiple VIs can be used to create large-scale applications; in fact, large-scale applications may have several hundred VIs. A VI may be used as the user interface or as a subroutine in an application. User interface elements such as graphs are easily accessed, as drag-and-drop units in LabVIEW. The VIs were developed to demonstrate the instrumentation principles, study the oxidase reactions, detection of trace metal, morphine, and so on.[34–36] Economou et al.[37] developed a VI to perform square wave voltammetry for the determination of riboflavin. Madasamy et al.[38] have developed LabVIEW based electrochemical analyzer for detection of NO from human samples. The potentiostat circuit for the virtual electrochemical analyzer was constructed using an op-amp LM 358. The potentiostat circuit was connected to the NI MyDAQ for the sensor data acquisition (Figure 5-9). The GUI software was developed using LabVIEW 10.0 to control the potentiostat, acquire the current response, and process the acquired current signal. Commercially available NO analyzers are functioning based on the chemical reaction of NO with ozone (O_3) and the concentration of NO is measured with respect to the luminescent intensity. These analyzers involve high cost, corrosive chemicals, and also not a selective method for the determination of NO. In an effort to diminish the cost of the measuring unit, make it environment and user friendly and also to achieve the specific determination of NO, we have introduced here

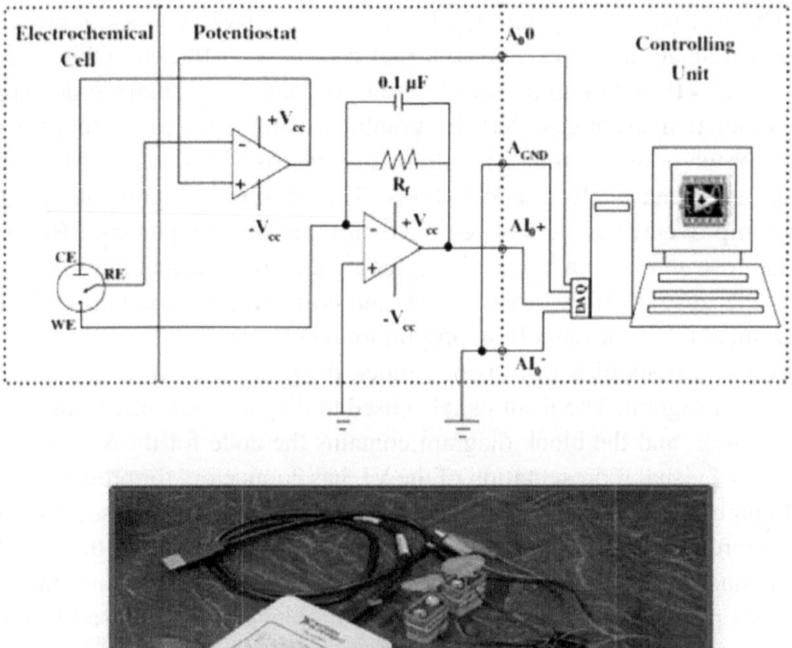

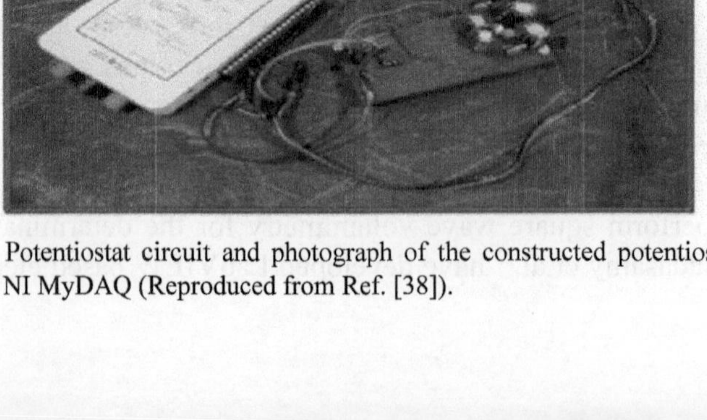

FIGURE 5-9 Potentiostat circuit and photograph of the constructed potentiostat circuit connected with NI MyDAQ (Reproduced from Ref. [38]).

a compact, flexible, and low cost electrochemical NO analyzer. The main advantages of developed virtual electrochemical is its user-friendly software that draws the linear plot and measures the concentration of NO present in the unknown sample. The mathematical parameters like slope, intercept, correlation coefficient, and best linear fit values are displayed as soon as the linear graph plotted.

5.7 ELECTRONIC INTERFACING OF POTENTIOMETRIC BIOSENSORS

Potentiometric measurement involves determination of the electrical potential between two electrodes at zero current flow. A standard three-electrode cell is built with the three-electrode setup. These biosensors are based on ion-selective electrodes and ion-sensitive field effect transistors (FETs) (Figure 5-10).[39] The primary outputting signal is possibly due to ions accumulated at the ion-selective membrane interface. Current flowing through the electrode is equal to or near zero. The electrode follows the presence of the monitored ion resulting from the enzyme reaction. A FET based immunosensor device is composed of a semiconductor channel and source, drain, and gate electrodes, all of which are located on a substrate (commonly a Si wafer) with an insulating (dielectric) surface layer. The source and drain electrodes communicate with each other through the semiconductive channel, while the gate electrode is used to modulate the channel conductance via an applied electrical potential.

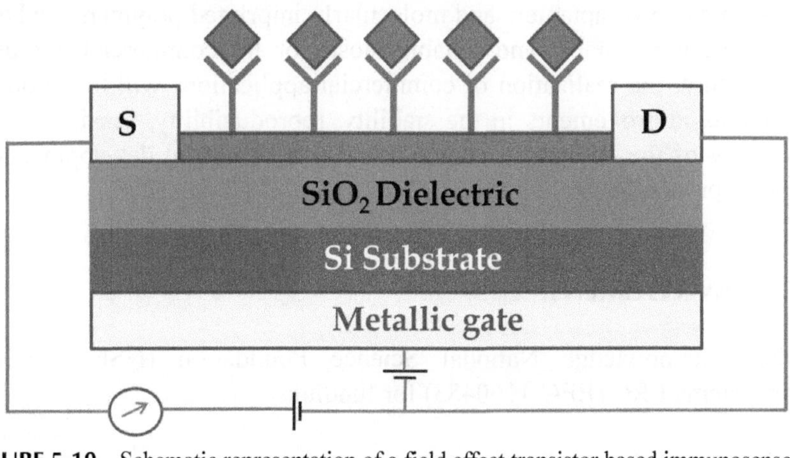

FIGURE 5-10 Schematic representation of a field effect transistor based immunosensor.

The rapid development of IC technology during the past decades has initiated many initiatives to fabricate potentiometric biosensors on silicon or complementary metal–oxide–semiconductor (CMOS) substrates. However, CMOS provides excellent means to meet some of the key criteria of potentiometric biosensors such as miniaturization of the devices, low power consumption, rapid sensor response characteristics, or batch fabrication at industrial standards and low costs. Additional advantages come from the possibility of monolithic co-integration of circuitry and transducers. These include improved sensor signal-to-noise characteristics due to on-chip signal processing and analog/digital conversion, or the realization of smart features on the sensor chip.

5.8 KEY TRENDS, FUTURE PERSPECTIVES, AND CONCLUSION

Because of its simplicity, the electrochemical biosensors are the perfect candidate for point-of-care diagnostics and health-care monitoring. The interfacing of low-cost electronics has brought the electrochemical biosensor to new levels in clinical diagnostics. The advancements in universal electronic modules and open source platforms enable the design of new generation of biosensor devices integrated with smartphones. Although the progress in the field of biosensor development is steadily increasing, the success rate of such biosensors in the market is very low due to biosensor's stability and repeatability along with power requirement. To overcome this issue, the current approaches are focusing on the design of a self-powered electronic interfaced biosensor for implantable application. New types of biorecognition elements, viz. aptamers and molecularly imprinted polymers are being used to design a stable and reliable biosensor for commercial purposes. In the future, the realization of commercial applications will hinge on the continuous improvements in the stability, reproducibility, sensitivity, and specificity of the devices in combination with industrial developments of low-cost processing techniques.

ACKNOWLEDGMENT

Authors acknowledge National Science Foundation (NSF), ASSIST Nanosystems ERC (EEC-1160483) for funding.

KEYWORDS

- **Biosensor**
- **Electrochemical transducers**
- **Electrode**
- **Potentiostat**
- **Integrated circuit**
- **Converter**

REFERENCES

1. Marks, R. S.; Cullen, D. C.; Karube, I.; Lowe, C. R.; Weetall, H. H.; Eds. *Handbook of Biosensors and Biochips*; John Wiley & Sons, Ltd: Chichester, 2008.
2. Thévenot, D. R.; Toth, K.; Durst, R. A.; Wilson, G. S. Electrochemical Biosensors: Recommended Definitions and Classification. *Biosens. Bioelectron.* **2001,** *16* (1–2), 121–131.
3. Buerk, D. G. *Biosensors: Theory and Applications*; Technomic Publishing Company: Lancaster, UK., 1993.
4. Collings, A. F.; Caruso, F. Biosensors: Recent Advances. *Reports Prog. Phys.* **1997,** *60* (11), 1397–1445.
5. Gonsalves, K. E.; Halberstadt, C. R.; Laurencin, C. T.; Nair, L. S., Eds. *Biomedical Nanostructures*; John Wiley & Sons, Inc.: Hoboken, NJ, 2007.
6. Wang, J. *Analytical Electrochemistry. 3rd Ed*; John Wiley & Sons: New Jersey, 2006.
7. Mutlu, M. *Biosensors in Food Processing, Safety, and Quality Control*; CRC Press: Boca Raton, 2012; Vol. 12.
8. Sawyer, Donald T.; Andrzej Sobkowiak, J. L. R. *Electrochemistry for Chemists*; Wiley-Interscience: New York; 2nd ed, 1995.
9. Bard, Allen J., Faulkner, L. R. *Electrochemical Methods: Fundamentals and Applications*; Wiley: New York; 2nd edn, 2000.
10. Ponmozhi, J.; Frias, C.; Marques, T.; Frazão, O. Smart Sensors/Actuators for Biomedical Applications: Review. *Measurement* **2012,** *45* (7), 1675–1688.
11. Mathivanan, N. *PC-Based Instrumentation: Concepts and Practice*; Prentice-Hall of India Pvt. Ltd: New Delhi, 2007.
12. Villagrasa, J. P.; Colomer-Farrarons, J.; Miribel, P. L. *Bioelectronics for Amperometric Biosensors, State of the Art in Biosensors – General Aspects*; Rinken, T., Ed.; InTech, 2013. ISBN: 978-953-51-1004-0; DOI: 10.5772/52248.
13. Colomer-Farrarons, J.; Miribel-Català, P. L.; Rodríguez-Villarreal, A. I.; Samitier, J. *Portable Bio-Devices: Design of Electrochemical Instruments from Miniaturized to Implantable Devices, New Perspectives in Biosensors Technology and Applications*; Serra, A., (Ed.).; InTech, 2011. ISBN: 978-953-307-448-1; DOI: 10.5772/17212.

14. Carrara, S.; Ghoreishizadeh, S.; Olivo, J.; Taurino, I.; Baj-Rossi, C.; Cavallini, A.; de Beeck, M. O.; Dehollain, C.; Burleson, W.; Moussy, F. G.; et al. Fully Integrated Biochip Platforms for Advanced Healthcare. *Sensors (Basel).* **2012,** *12* (8), 11013–11060.
15. Li, L.; Qureshi, W. A.; Liu, X.; Mason, A. J. Amperometric Instrumentation System with On-Chip Electrode Array for Biosensor Application. In *2010 Biomedical Circuits and Systems Conference (BioCAS)*; IEEE, **2010**; pp 294–297.
16. Settle, F. A. *Handbook of Instrumental Techniques for Analytical Chemistry*; Prentice Hall: Upper Saddle River, NJ, 1997.
17. Heinze, J. Cyclic Voltammetry—"Electrochemical Spectroscopy". New Analytical Methods(25). *Angew. Chemie Int. Ed. English* **1984,** *23* (11), 831–847.
18. Gopinath, A. V.; Russell, D. An Inexpensive Field-Portable Programmable Potentiostat. *Chem. Educ.* **2006,** *11* (1), 23–28.
19. Loncaric, C.; Tang, Y.; Ho, C.; Parameswaran, M. A.; Yu, H.-Z. A USB-Based Electrochemical Biosensor Prototype for Point-of-Care Diagnosis. *Sens. Actuators B Chem.* **2012,** *161* (1), 908–913.
20. Rowe, A. A.; Bonham, A. J.; White, R. J.; Zimmer, M. P.; Yadgar, R. J.; Hobza, T. M.; Honea, J. W.; Ben-Yaacov, I.; Plaxco, K. W. CheapStat: An Open-Source, "Do-It-Yourself" Potentiostat for Analytical and Educational Applications. *PLoS One* **2011,** *6* (9), e23783.
21. Cruz, A. F. D.; Norena, N.; Kaushik, A.; Bhansali, S. A Low-Cost Miniaturized Potentiostat for Point-of-Care Diagnosis. *Biosens. Bioelectron.* **2014,** *62*, 249–254.
22. Pandiaraj, M.; Benjamin, A. R.; Madasamy, T.; Vairamani, K.; Arya, A.; Sethy, N. K.; Bhargava, K.; Karunakaran, C. A Cost-Effective Volume Miniaturized and Microcontroller Based Cytochrome c Assay. *Sens. Actuators A Phys.* **2014,** *220*, 290–297.
23. Thomas, F. G.; Henze, G. *Introduction to Voltammetric Analysis: Theory and Practice*; CSIRO Publishing: Collingwood, 2001.
24. Nie, Z.; Deiss, F.; Liu, X.; Akbulut, O.; Whitesides, G. M. Integration of Paper-Based Microfluidic Devices with Commercial Electrochemical Readers. *Lab Chip.* **2010,** *10* (22), 3163–3169.
25. Serra, P. A.; Rocchitta, G.; Bazzu, G.; Manca, A.; Puggioni, G. M.; Lowry, J. P.; O'Neill, R. D. Design and Construction of a Low Cost Single-Supply Embedded Telemetry System for Amperometric Biosensor Applications. *Sens. Actuators B Chem.* **2007,** *122* (1), 118–126.
26. Huang, C.-Y.; Syu, M.-J.; Chang, Y.-S.; Chang, C.-H.; Chou, T.-C.; Liu, B.-D. A Portable Potentiostat for the Bilirubin-Specific Sensor Prepared from Molecular Imprinting. *Biosens. Bioelectron.* **2007,** *22* (8), 1694–1699.
27. Ramfos, I.; Vassiliadis, N.; Blionas, S.; Efstathiou, K.; Fragoso, A.; O'Sullivan, C. K.; Birbas, A. A Compact Hybrid-Multiplexed Potentiostat for Real-Time Electrochemical Biosensing Applications. *Biosens. Bioelectron.* **2013,** *47*, 482–489.
28. Avdikos, E. M.; Prodromidis, M. I.; Efstathiou, C. E. Construction and Analytical Applications of a Palm-Sized Microcontroller-Based Amperometric Analyzer. *Sens. Actuators B Chem.* **2005,** *107* (1), 372–378.
29. Kwakye, S.; Baeumner, A. An Embedded System for Portable Electrochemical Detection. *Sens. Actuators B Chem.* **2007,** *123* (1), 336–343.
30. Friedman, E. S.; Rosenbaum, M. A.; Lee, A. W.; Lipson, D. A.; Land, B. R.; Angenent, L. T. A Cost-Effective and Field-Ready Potentiostat That Poises Subsurface Electrodes to Monitor Bacterial Respiration. *Biosens. Bioelectron.* **2012,** *32* (1), 309–313.

31. Surendran, E.; Natarajan, M.; Krishna, A. V.; Kanagavel, V.; Chandran, K.; Thangamuthu, M.; Pandiaraj, M. Deployment of Wireless Sensor Network for the Measurement of Exhaled Nitric Oxide in In-Home Healthcare. *Sens. Transducers J.* **2012,** *142* (7), 87–94.
32. Bitter, R.; Mohiuddin, T.; Nawrocki, M. *LabVIEW: Advanced Programming Techniques*; CRC Press: Hoboken; Har/Cdr ed, 2000.
33. Johnson, G. W.; Jennings, R. *LabVIEW Graphical Programming*; McGraw-Hill Professional: New York, 2006.
34. Economou, A.; Voulgaropoulos, A. LabVIEW-Based Sequential-Injection Analysis System for the Determination of Trace Metals by Square-Wave Anodic and Adsorptive Stripping Voltammetry on Mercury-Film Electrodes. *J. Autom. Methods Manag. Chem.* **2003,** *25* (6), 133–140.
35. Lenehan, C. E.; Barnett, N. W.; Lewis, S. W. Design of LabVIEW-Based Software for the Control of Sequential Injection Analysis Instrumentation for the Determination of Morphine. *J. Autom. Methods Manag. Chem.* **2002,** *24* (4), 99–103.
36. Jensen, M. B. Using LabVIEW to Demonstrate Instrumentation Principles. *Anal. Bioanal. Chem.* **2011,** *400* (9), 2673–2676.
37. Economou, A.; Bolis, S.; Efstathiou, C.; Volikakis, G. A. "virtual" Electroanalytical Instrument for Square Wave Voltammetry. *Anal. Chim. Acta.* **2002,** *467* (1–2), 179–188.
38. Madasamy, T.; Pandiaraj, M.; Balamurugan, M.; Karnewar, S.; Benjamin, A. R.; Venkatesh, K. A.; Vairamani, K.; Kotamraju, S.; Karunakaran, C. Virtual Electrochemical Nitric Oxide Analyzer Using Copper, Zinc Superoxide Dismutase Immobilized on Carbon Nanotubes in Polypyrrole Matrix. *Talanta* **2012,** *100*, 168–174.
39. Hierlemann, A.; Baltes, H. CMOS-Based Chemical Microsensors. *Analyst* **2003,** *128* (1), 15–28.

CHAPTER 6

NANOBIOTECHNOLOGY FOR ENZYMATIC SENSORS

MD. AZAHAR ALI[1]* and CHANDAN SINGH[2]

[1]*Department of Electrical and Computer Engineering, Iowa State University, Ames, USA*

[2]*Department of Science & Technology Centre on Biomolecular Electronics, Biomedical Instrumentation Section, CSIR-National Physical Laboratory, New Delhi, India*

**E-mail: azahar@iastate.edu*

CONTENTS

An enzymatic biosensor is known to utilize an enzyme as a recognition element and considered as an established and reliable tool for clinical diagnostics. It is an integration of the biocatalyst (catalytic reaction with a target analyte) and transducer material (transduced signal via optically or electrically). The enzymatic biosensors provide easy, simple, cost-effective, and fast detection of various target analytes including glucose, cholesterol, urea, and so on in serum, urine, body fluids, and so on. A large number of enzyme-based biosensors has been developed using nanostructured materials. This chapter deals with the basic principle of enzymatic biosensor, techniques for enzyme immobilization, role of nanomaterials, and also application of microfluidics for development of miniaturized enzymatic biosensors.

6.1 INTRODUCTION TO ENZYMATIC BIOSENSORS

An enzymatic biosensor is an analytical device consisting an enzyme as a biological sensing element that recognize the specific target analyte and the generated biological responses were converted into a measurable signal using a transducer.[1] In our daily life, the enzymatic biosensors are can be potentially used for health-care diagnostics, food safety, defense, and environmental monitoring. The human health care is one of the potential areas where enzymatic biosensors find various applications including monitoring of glucose, cholesterol, urea, lactate, and so on in human blood serum, saliva, or sweat. For practical and commercial applications, several types of enzymatic sensors have been widely used for glucose (clinical diagnostics/treatment for diabetic patients), urea (clinical applications), lactate (sports medicine, food science, biotechnology, etc.), and glutamate (food science, biotechnology, etc.).[1] In this chapter, our aim is to describe the prospects, and state-of-the-art enzyme-based biosensors using nanomaterials, conducting polymers (CPs), microfluidics, and so on for clinical diagnostics.

The genesis of the concept of enzyme-based biosensor was presented by Leland C. Clark Jr. in 1962 at the New York Academy of Sciences Symposium and was commercialized as a clinical analyzer (Yellow Springs Instruments) in 1975.[2] This is the first glucose enzyme electrode that relied on a thin layer of glucose oxidase (GO_x) entrapped over an oxygen electrode via a semipermeable dialysis membrane. In this device, the electrochemical detection has been employed to monitor the oxygen consumed by the enzyme-catalyzed reaction between glucose and GO_x in the presence of oxygen. A negative potential was applied to the platinum cathode for a reductive detection of the oxygen consumption. The method of detection is based on the decrease in oxygen concentration that is proportional to glucose concentration. This proposed oxygen electrode is known as the first generation of biosensor which is oxygen dependent.[3] In the first-generation biosensors, the normal product of the reaction diffuses to the transducer and causes the electrical response. The second-generation biosensors involved the specific "mediators" between the reaction and the transducer in order to generate improved response. Thus, this generation of biosensor depends on a mediator but does not depend on oxygen for biochemical reaction. In 1980, efforts have been focused on the development of mediator-based "second-generation" biosensors by introducing the commercial strips for self-monitoring of blood glucose. In the third-generation biosensors, the reaction causes the response and does not involve any mediator diffusion. Thus, in these biosensors, the direct electron transfer (ET) between the biomolecules (including the enzyme) and the electrode surface excludes any intermediate ET reactions with redox mediators. Most enzymatic biosensors employ a class of enzymes such as oxidoreductases, and the two subclasses most known are oxidases and dehydrogenases.[4] Their biochemical reaction sequences can be given as[4]

$$S_{red} + E_{ox} \rightarrow S_{ox} + E_{red} \tag{6-1}$$

$$S'_{ox} + E_{red} \rightarrow S'_{red} + E_{ox} \tag{6-2}$$

In the case of glucose biosensor, if the enzyme is GO_x for catalytic reaction, then S_{red}, S_{ox}, S'_{ox}, and S'_{red} correspond to glucose, gluconic acid, oxygen, and hydrogen peroxide, respectively.

The nanosensors, nanoprobes, and other nanosystems are revolutionizing the fields of enzymatic biosensors to enable rapid analysis.[5] Applications of nanomaterials including (metals, metal–oxides, carbon, CP, etc.) have grown exponentially toward eventual translation into biomolecular recognition layers on their surfaces.[5] Nanostructured materials have been utilized to modify the surfaces of glass, silicone, metal, carbon, and so on integrating with electrochemical and optical transducers. They can be used to obtain the enhanced immobilization of biomolecules due to the ultrafine structure resulting in an enhanced surface area. Nanomaterials exhibit good catalytic properties that are different from their bulk due to their high surface area and high edge concentration. In addition, the nanoparticles on the surfaces of electrodes can be used to improve ET that improves the biosensor's efficacy. Wang et al. have described the advances in electrochemical glucose biosensors.[2] Nanomaterials can play a crucial role for attaching biomolecules (enzyme) on substrates as well as signal amplification to monitor thereby improving the characteristics of biosensors. Pumera et al. have described the nanoscale materials including gold nanoparticles, quantum dots as biomolecule tracers, novel nanobiolabels, and enzyme tag loaded carbon nanotubes (CNTs) for development of electrochemical biosensors.[6] In particular, the electrochemical biosensors with recent advances in nanotechnology offer simple, cost-effective, and efficient tools to measure the target biomolecules. Thus, the electrochemical detection plays an important role in development of enzymatic biosensors because of the low cost, ease of use, portability, label-free operation, simplicity, and higher signal-to-noise ratio. Umar et al. have fabricated a highly sensitive enzymatic biosensor based cholesterol oxidase immobilized zinc oxide nanoparticles.[7] Ali et al. have explored urease and glutamate dehydrogenase as a co-immobilization material on zinc oxide nanoparticles for detection of urea.[8] Kaushik et al. have chosen GO_x as a model enzyme to functionalized chitosan modified iron oxide (Fe_3O_4) nanoparticles for detection of glucose concentration.[9] Thus, enzymes coupled with nanomaterials (Figure 6.1) provide a prospect to develop a point-of-care (POC) device for enzymatic detection using the electrochemical technique.

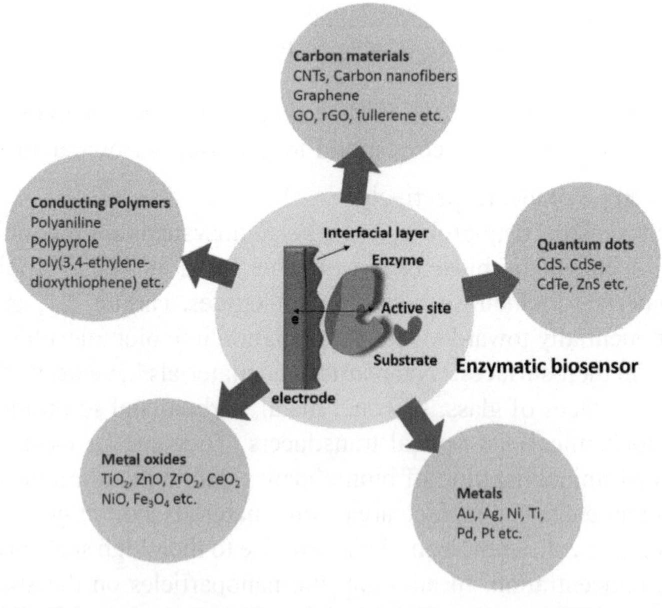

FIGURE 6.1 Schematic representation of enzymatic biosensor and different nanomaterials for enzymatic biosensor.

To use an enzymatic biosensor, outside the laboratory, it either has to be very simple or automated with regard to sample processing and reagent addition. Automation and miniaturization of analytical techniques, as well as the development of online and remote sensing devices can be achieved by microfluidics technology.[10] Use of enzyme-coupled nanoparticles, nanotubes, and nanowires in microfluidic assembly is currently of much interest for real-time monitoring of target analytes. Their integration with a microfluidic platform through a combination of microfabrication and nanomaterial synthesis allows the unique functionality of the enzymatic biosensor devices.[11] Recently, the advanced microfluidic technology has been explored in the field of enzymatic biochips for POC diagnostics.[10] The integrated microfluidic devices have capability to automatic sample analysis using minimum reagent and power requirements in a portable format (size: 1–100 mm). With integration of fluidic microchannels, reservoirs, and valves, these microfluidic devices are designed to perform multiple functions simultaneously like sample and reagents handling, preparation, and detection.[12] High sensitivity and selective detection of target analytes is the main aim, need to be achieved by these devices. A microfluidic biochip with a three-electrode configuration has been developed for the detection of cholesterol using bienzyme functionalized CNTs and nickel oxide nanoparticle composites.[13]

6.2 ENZYME BASED ELECTROCHEMICAL BIOSENSOR

The electrochemical detection technique is widely used for the development of an enzymatic sensor. These biosensors are economic, easy to fabricate, have high signal-to-noise ratio, and have potential to integrate with microfluidic and nanomaterials for better sensing performance. An electrochemical sensor measures current, voltage, impedance, or capacitance based on an enzymatic catalysis of a reaction that can produce electrons.[2] It consists of a reference electrode, a counter electrode, and a working electrode. In this biosensor, the target analyte encompasses an enzymatic reaction that takes place on the working electrode's surface. This catalytic reaction on the sensor's surface can cause the ET through the double layer that changes the measurable signal. The rate of flow of electrons or measured current or potential is directly proportional to target analyte concentration. In the electrochemical configuration, bioreceptor molecules or biorecognition elements are either coated physically onto or covalently bonded to a transducer surface. The bioreceptor molecules react specifically with target analytes to be detected. Based on their operating principle by which chemical information can be converted into a measurable signal, an electrochemical sensor can be classified into three categories namely: potentiometric, amperometric, and impedimetric biosensors. The immobilization of enzyme molecules on an electrochemical transducer surface is an important step for fabrication of a biosensor.[13,14] The selectivity, stability, sensitivity, and reproducibility of the biosensor are the dominant parameters that can be improved by selecting an effective method for immobilization. We now discuss the various methods that have been reported for the immobilization of desired biomolecules onto solid support materials. They include the following:

- Entrapment and encapsulation
- Covalent binding
- Cross-linking
- Adsorption

In biomolecules' functionalization, the noncovalent immobilization indicates a π–π stacking, electrostatic interaction, entrapment in polymers or metal–oxides, or van der Waals forces between the nanomaterials and the biological entities. Covalent functionalization relies on the covalent attachment of enzyme biomolecules to nanomaterials by classic amide coupling reactions, cross-linking, or conjugation chemistry and click chemistry. This approach can improve the biosensor stability and reproducibility.

It is known that the enzymatic biosensor provides a powerful tool for the clinical diagnostics. In an enzymatic biosensor, an enzyme is immobilized onto a suitable transducer surface and it produces a specific signal upon reaction with a detectable analyte.[15] The main purpose is to convert an enzymatic reaction into a suitable analytical signal (electrochemical, colorimetric, optical properties, etc.). The thrust is toward developing biosensors for the easy detection of the most important parameters like blood sugar, urea, and cholesterol that may serve as an indicator about the onset of a disease. Several devices based on electrochemical biosensors have been developed.[16] Most of these devices are based on enzyme functionalized screen printed disposable electrode strips that work on the principle of amperometric detection. These hand-held POC testing devices help in self-monitoring as well as getting quick, reliable, and accurate results using small volume of samples. Many approaches have been followed to obtain improved characteristics of glucose and other analytes sensing devices by coupling it with microfluidics. Hou et al. have reported glucose microfluidics chip based on polymethyl methacrylate using CO_2 laser and hot plate press bonding fabrication technique.[17] Compared to the conventional detection methods, this chip has the advantage of self-rotation effect resulting in improved mixing in a micromixer and promises rapid and low-cost detection of glucose. Recent developments in nanotechnology and engineering have enabled immobilization of enzymes on to nanoparticles and nanostructures, which in turn can be functionalized onto the electrodes. Yu et al. fabricated a microfluidic chip where enzymes were immobilized on a single-walled CNT onto the polydimethylsiloxane (PDMS) channel.[18] Ali et al. reported a bienzyme functionalized anatase-TiO_2 and chitosan nanocomposite integrated microfluidic device (Figure 6.2) for monitoring cholesterol concentrations.[16] The optical three-dimensional (3D) profiling image before (a) and after (b) incorporation of anatase-TiO_2 and chitosan onto an indium tin oxide (ITO) electrode, atomic force microscopic image (c) of anatase-TiO_2 nanoparticles, and a photographic image with sensor response curve (d) are shown in Figure 6-3.[16]

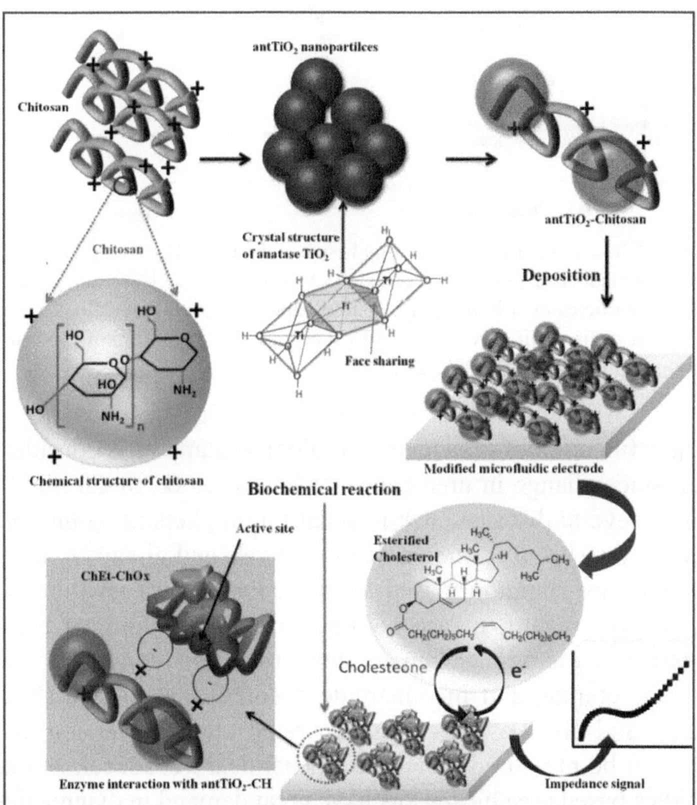

FIGURE 6.2 Enzyme immobilization onto the surface of antTiO2-CH for cholesterol detection.[16]

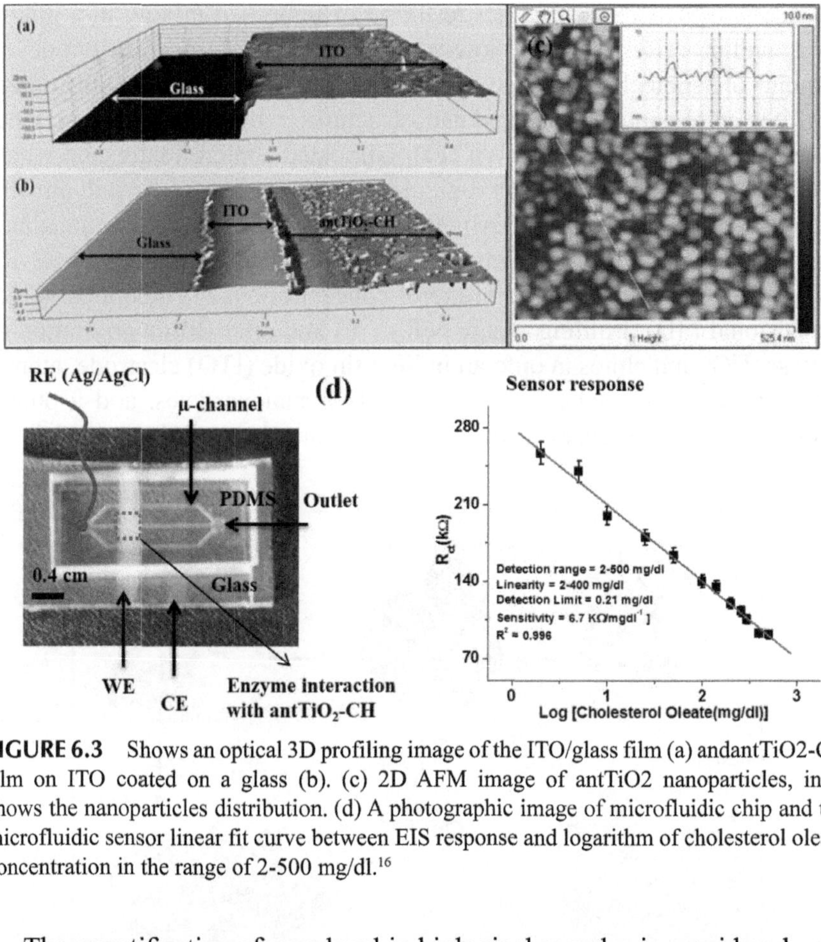

FIGURE 6.3 Shows an optical 3D profiling image of the ITO/glass film (a) andantTiO2-CH film on ITO coated on a glass (b). (c) 2D AFM image of antTiO2 nanoparticles, inset shows the nanoparticles distribution. (d) A photographic image of microfluidic chip and the microfluidic sensor linear fit curve between EIS response and logarithm of cholesterol oleate concentration in the range of 2-500 mg/dl.[16]

The quantification of urea level in biological samples is considered very important since change in urea concentration in blood (8–20 mg/dL) is an indicator of several diseases such as renal failure, hepatic failure, nephritic syndrome, and so on. It is known that the urease and glutamate dehydrogenase enzymes are commonly used for detection of urea level in biological samples. In this regard, urease converts urea into hydrogen bicarbonate and ammonia, whereas glutamate dehydrogenase converts ammonium ions, α-ketoglutarate, and nicotinamide adenine dinucleotide (NADH) to L-glutamate and NAD+.[8] Electrons generated after conversion of NADH to NAD+ can be measured by an electrochemical transducer. Miniaturized microfluidics based urea biosensors have great demand in diagnostics due to their higher sensitivity, low consumption of reagents, portability, and high accuracy along with reusability. Srivastava et al. covalently immobilized enzymes onto the gold surface through self-assembled monolayer with the help of 10-carboxy-1-decanthiol.[19] However, the sensitivity and detection range have been found to be improved when enzymes are immobilized on the titania–zirconia composite surface of the microfluidic biochip and a schematic representation of this sensor is shown in Figure 6.4.[20]

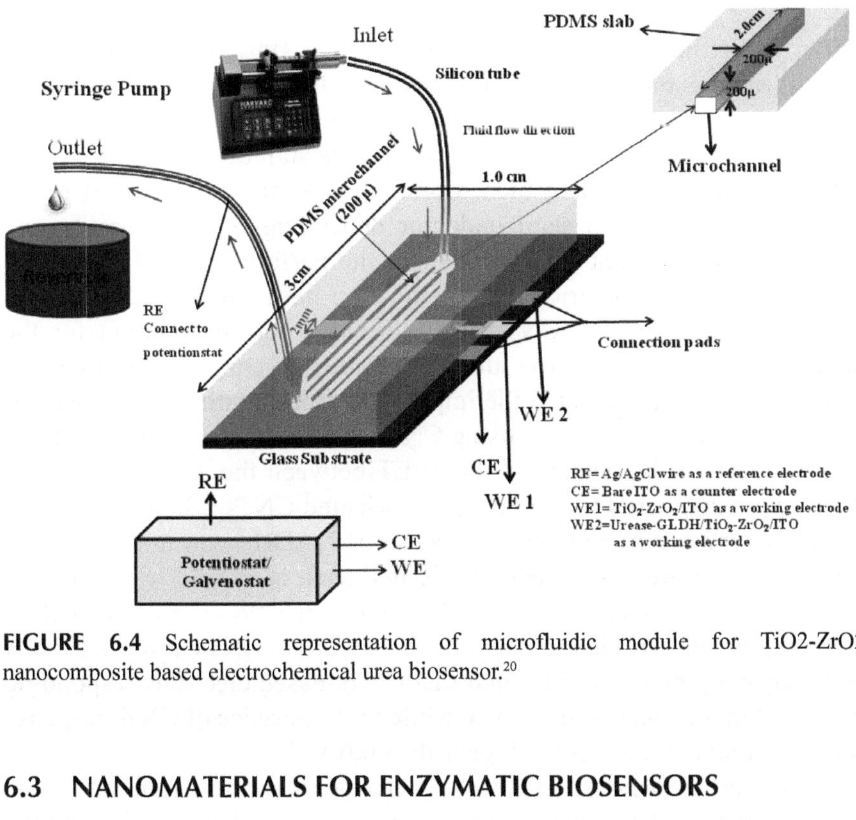

FIGURE 6.4 Schematic representation of microfluidic module for TiO2-ZrO2 nanocomposite based electrochemical urea biosensor.[20]

6.3 NANOMATERIALS FOR ENZYMATIC BIOSENSORS

The applications of nanostructured materials have revolutionized the concept of biosensors including enzymatic biosensors.[21] The utilization of nanomaterials has been known to improve the critical biosensing characteristics such as sensitivity, detection limit, and others.[22, 23] Irrespective of the type, all nanomaterials (metal nanoparticles, CNTs, graphene oxide (GO), metal–oxides, nanocomposites, etc.) offers higher surface-to-volume ratio resulting in the enhanced immobilization of bioreceptors on its surface. Different types of nanomaterials can provide the following advantages for the development of enzymatic biosensors such as (1) enhanced loading of biomolecules, (2) catalysis of electrochemical enzymatic reactions, (3) superior electron transport phenomenon, (4) labeling of the biomolecules/proteins, and (5) acting as a reactant. We now discuss the application of carbon materials, metal–oxides, and CPs for enzymatic biosensors.

6.3.1 CARBON MATERIALS BASED ENZYMATIC BIOSENSORS

For example, introduction of CNTs by Ijima in 1991 has created special interest among biological scientists across the globe due to their high electrical conductivity, high chemical stability, ease of functionalization, and unique electrochemical and electrocatalytic properties for the development of enzymatic biosensors.[24] Electronic properties of CNTs range from metallic to semiconductor depending on the chirality and the diameter of the nanotube influences the performances of enzymatic biosensors.[8] Direct electrical communication between the redox centers of the enzymes and the CNT electrode has received attention for the development of mediator-free third generation of enzymatic biosensors. Gooding et al. demonstrated the improvement in the direct electrochemistry of enzymatic biosensors using CNTs.[25] Yu et al. have concluded that aligned CNTs offers enhanced direct ET between the redox proteins of the enzyme compared to the randomly oriented CNTs.[26] In an enzymatic biosensor, the electrochemical oxidation of NADH is an important step for the development of amperometric detection as it work as a cofactor for NAD+/NADH-dependent dehydrogenases reaction.[27] Musameh et al. have explained the electro-catalytic behavior of CNTs toward NADH with varying potentials and found that CNTs-based electrode responding to NADH in the range of 0.1–1.0 V while in the absence of CNTs response was only found at a potential higher than 0.6 V.[27]

Other carbon nanomaterial such as GO or its reduced form due to its inherent properties including extraordinary electrical, thermal conductivity, low temperature quantum hall effect, and very high electron mobility at room temperature has proved to be an efficient platform for the fabrication of enzymatic biosensors.[28] In this context, the role of graphene as a conducting component between the redox site and electrode facilitates the direct monitoring of the ET and hence results in the efficient detection of glucose or other target analytes. In addition, the integration of metal nanoparticles into GO to achieve direct electrochemistry of H_2O_2 at lower potential offers a solution to the conventional problem of enzymatic biosensor.[29] Liu et al.

have developed a highly efficient glucose biosensor using covalent attachment between carboxylic group of graphene and amino group to obtain a high sensitive enzymatic biosensor.[30] Manjunatha et al. have utilized GO for an enzymatic biosensor which offers high ET rate constant (K_s = 0.78 s^{-1}) for routine quantification of free cholesterol.[31] Therefore, GO and its composite with metal, metal–oxide, and polymer with excellent electrocatalytic activity for H_2O_2 is a promising material for the fabrication of enzymatic biosensors.

6.3.2 METAL–OXIDE NANOPARTICLES FOR ENZYMATIC BIOSENSORS

Among the various types nanostructures (Figure 1) utilized for the fabrication of enzymatic biosensors, metal–oxides nanostructures (MNO_x) due to their exceptional electrical properties owing to electron and phonon confinement, enhanced surface reaction activity, modified surface work function, and high catalytic and strong adsorption ability of the enzyme on its surface has become a preferred transducer material.[32] In biological recognition events, active sites of enzymes are directly attached with MNOx surface which facilitate the direct ET between the enzyme and MNOx resulting in improved sensitivity and detection limit. Further, the MNO_x-based electrodes offer a biocompatible and electroactive surface for the immobilization of the enzyme offering favorable conformation, orientation, and preservation of the biological activity for a long period of time. In enzymatic biosensors, active sites of enzymes are deeply embedded with thick insulating protein shells preventing oxidation/reduction on the electrode's surface.[33] The MNO_x can overcome this limitation by activating the active sites of enzyme molecules and can prevent the biodegradation of the enzymes. It has been known that ET rate between the active site of the enzyme and electrode surface varies with the type of metal–oxide nanoparticles. Zinc oxide (ZnO), cerium oxide (CeO_2), titania (TiO_2), zirconia (ZrO_2), and so on have been widely used for the fabrication of enzymatic biosensors. An enzymatic biosensor based on ZnO nanocombs having 3D porous network with high specific surface area and good biocompatibility for GO_x provides high sensitivity (15.33 μA mM^{-1} cm^{-2}) for detection of glucose.[34] A schematic representation of enzymatic microfluidic device using nanostructured nickel oxide for detection of total cholesterol is shown in Figure 6-5.[35]

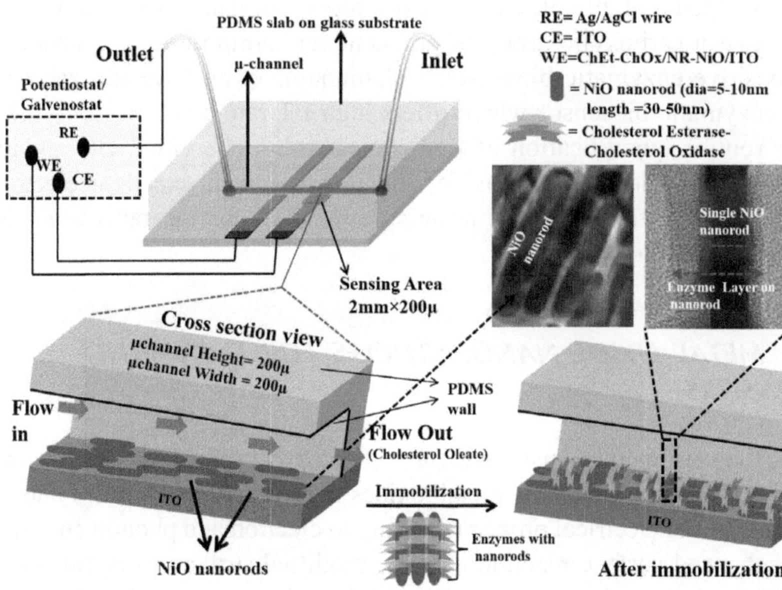

FIGURE 6.5 Nickel oxide nanorods based enzymatic microfluidic device for detection of total cholesterol.[35]

In addition to cholesterol and glucose, MNO_x have been utilized as highly suitable transducer materials for the development of enzymatic biosensors for the detection of urease, glutamate dehydrogenase, lipase, and other target analytes. Table 6-1 shows the utilization of various nanostructured materials for the development of enzymatic biosensors.

6.3.3 CPS FOR ENZYMATIC BIOSENSORS

CPs are found to play an important role in the development of enzymatic biosensors since they contain π-electrons cloud in their backbone offering excellent electrical conductivity.[36] Owing to the higher electrical conductivity, they are also referred as "synthetic metals". In addition to the electrical conductivity, low energy optical transition, low ionization potential, and high electron affinity are the other motivating factors to prove CPs as a suitable platform for enzymatic biosensors.[37] Further, CPs can directly be deposited on the surface of an electrode while simultaneous trapping of the biomolecules offers a control over the spatial distribution on its surface, and the film thickness and enzyme activity can be adjusted by modulating

TABLE 6.1 Utilization of nanostructured materials for the development of enzymatic biosensors.

Working electrodes	Detection range (mg/dL)	Sensitivity (μA/mg $dL^{-1}cm^{-2}$)	K_m Value (mg/dL)	Cost (based on element availability)	Response time (s)	Shelf-life (days)	Ref.
Nano CeO_2	10–400 mg/dL	2 μA/mgdL^{-1} cm^{-2}	76 mg/dL	Expensive (rare earth metal)	15 s	–	57
NanoFe_3O_4	50–200 mg/dL	–	17 mg/dL	Inexpensive	–	15 days	58
Graphene	36–252 mg/dL			Inexpensive		7 days	59
Nano NiO	10–400 mgdL^{-1}	0.808 μA/mgdL^{-1}	(25.52 mg/dL)	Inexpensive	15 s	70 days	60
Titania–zirconia	5–100 mg/dL	2.74 μA [log mM]$^{-1}$ cm^{-2}		Inexpensive	10 s	28 days	5
Nano-Tm_2O_3	8–400 mg/dL	0.9245 μA/ (mg/dL)/cm^2	–	Expensive (rare earth metal)	40 s	–	61
NanoNb_2O_5NR	25–500 mg/dL	0.267 μA/ (mg/dL)/ cm^2	0.07 mg/dL	Inexpensive	$<$10 s	$>$180 days	62

the state of the polymer. Further, CPs can provide 3D electrical conducting structures thereby originating the concept of "electrical wiring" that offers an efficient flow of electrons from the surface electrode to the active site of the enzymes.[38] The electrical conductivity of CPs can easily be modified by changing the pH and redox potential of the surrounding environment. CPs offers ion exchange and size exclusion properties resulting in the highly sensitive enzymatic device.[39] Several enzymatic biosensors including glucose and cholesterol with improved sensing parameters have been developed using CPs. Forzani *et al.* have utilized the aforementioned properties of polymer (polyaniline) to design a glucose biosensor with ultrafast response (<200 ms).[40] Singh *et al.* have designed a cholesterol biosensor by conjugating cholesterol esterase and cholesterol oxidase with conducting polypyrole for total cholesterol estimation.[41]

6.4 ENZYMES BASED GLUCOSE/CHOLESTEROL BIOSENSORS

Diabetes is considered as one of the leading public health problems across the globe.[42] It is one of the foremost causes of death and disability indicating the need for essential diagnosis and management of diabetes mellitus. It is still a challenge to provide an efficient and reliable glycemic control and remains a subject of significant amount of research.[43] Since the last two decades, electrochemical biosensors have played a critical role for the determination of glucose. Amperometric detection based on the GO_x immobilized on nanomaterial modified transducer electrode has been a matter of research.

In 1962, Clark and Lyons proposed the method for determination of glucose by monitoring the oxygen consumed using flowing enzymatic reaction [6-3 to 6-6].

$$\text{Glucose} + \text{Oxygen} \xrightarrow{GO_x} \text{Gluconic acid} + \text{Hydrogen peroxide} \quad (6\text{-}3)$$

This method of detection has further been modified in 1973 for the monitoring of blood glucose based on amperometric (anodic) monitoring of the released hydrogen peroxide:

$$H_2O_2 \longrightarrow O_2 + 2H^+ + 2e^- \quad (6\text{-}4)$$

Since then research toward the development of blood glucose monitoring become a hot topic among the research community. Further, the first generation of glucose biosensor was developed by use of an enzyme such as GO_x, depending on the oxygen consumed and hence the generation and detection of hydrogen peroxide (Figure 6-6). The flavin adenine dincleotide (FAD) of the GO_x results in the reduced form of $FADH_2$ by the following reaction:

$$GO_x\ (FAD) + Glucose \longrightarrow GO_x\ (FADH_2) + Gluconolactone \quad (6\text{-}5)$$

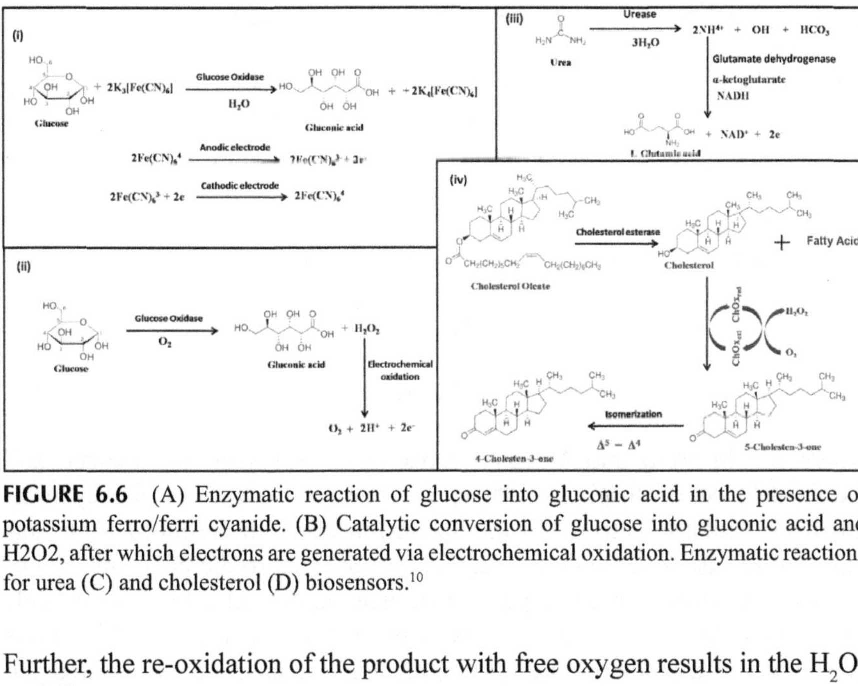

FIGURE 6.6 (A) Enzymatic reaction of glucose into gluconic acid in the presence of potassium ferro/ferri cyanide. (B) Catalytic conversion of glucose into gluconic acid and H2O2, after which electrons are generated via electrochemical oxidation. Enzymatic reactions for urea (C) and cholesterol (D) biosensors.[10]

Further, the re-oxidation of the product with free oxygen results in the H_2O_2 as follows:

$$GO_x\ (FADH_2) + O_2 \longrightarrow GO_x\ (FAD) + H_2O_2 \quad (6\text{-}6)$$

This simple measurement of H_2O_2 allows designing a device for glucose detection. Further, modification has been done to work on second generation of glucose biosensors.

GO_x lacks the ability to direct transfer of electrons to the electrodes, since the FAD redox center is surrounded by a thick insulating protein layer. To overcome this challenge nonphysiological shuttling electrons were applied to solve oxygen deficiency and this becomes the prominent feature of the second-generation biosensor. Initially, redox polymers such as poly(vinylpyridine) were covalently attached with osmium complex electron relays resulting in the reduced distance between polymer and the FAD leading to higher current and prompt response.[44] Different types of nanostructured materials including metal, metal–oxide, and carbon-based materials (GO and CNTs) were found to be very effective toward the development of glucose biosensors. Xiao et al. have utilized gold nanoparticles, linked to the Au electrode via dithiol bridge, and N-aminoethyl flavin adenine dinucleotide (amino-FAD), was linked to the particles; however, this procedure enhances the bioelectrocatalytic oxidation of glucose resulting in the improved ET rate (K_{et} = 5000 s^{-1}).[45] Patolsky et al. have utilized CNTs as an electrical connector between active sites of the enzyme and electrode and observed that the electrons were able to travel a distance of more than 150 nm[46] due to incorporation of CNTs, thereby resulting in an efficient biosensor for glucose detection via the following reactions (6-7–6-9):

$$\text{Glucose} + \text{GO}_x\,(\text{ox}) \longrightarrow \text{Gluconic acid} + \text{GO}_x\,(\text{red}) \qquad (6\text{-}7)$$

$$\text{GO}x(\text{red}) + 2\text{M}\,(\text{ox}) \longrightarrow \text{GO}x\,(\text{Ox}) + 2\text{M}\,(\text{red}) + 2\text{H}^+ \qquad (6\text{-}8)$$

$$2\text{M}\,(\text{red}) \longrightarrow 2\text{M}\,(\text{ox}) + 2\text{e}^- \qquad (6\text{-}9)$$

where the M (red) and M (ox) respectively represent the reduced and oxidized forms of different mediators (ferrocene derivatives, ferricyanide, artificial mediators, etc.).[47] In the above reactions, the reduced form gets oxidized resulting in the current signal proportional to the concentration of the glucose. However, since the last decade significant efforts were made to design a glucose biosensor that is independent of mediators. This can be considered as a landmark step if complicated mediators could be avoided from the glucose design and efforts are directed to design a third generation glucose biosensor (Figure 6-7).

Significant increase in clinical disorders such as heart disease, hypertension, arteriosclerosis, coronary artery disease, cerebral thrombosis, and so on due to abnormal levels of cholesterol concentrations in blood serum have made the public interested in measuring the cholesterol level. The atherosclerosis is a known to be a critical condition in which arteries becomes blocked due to the accumulation of cholesterol.[48] Deposition of cholesterol inside the walls of arteries can lead to formation of plaques resulting in blockages and interruption of the blood circulation that can causes heart attack, angina, and myocardial infarction. If the flow of oxygen-rich blood to the heart muscle is reduced or blocked, angina or a heart attack can occur. In this context, biosensors have shown a great potential to monitor the accurate amount of cholesterol in human blood. In a cholesterol biosensor, cholesterol oxidase

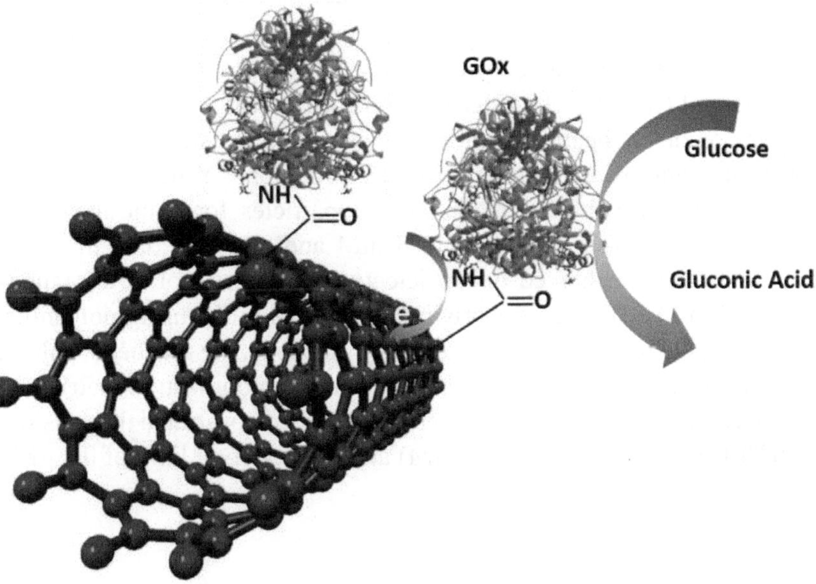

FIGURE 6.7 A schematic representation of 3rd generation of glucose biosensor using carbon nanotube.

(ChO_x), a FAD-containing enzyme is the most commonly used biosensing component. It catalyzes the dehydrogenation of C(3)-OH of the cholestan system.[48] The cholesterol further catalyzes the oxidation and isomerization 3β-hydroxysteroids with double bond D^5-D^6 of the steroid ring resulting in D^4-ketosteroid and hydrogen peroxide.[35] This oxidized FAD has a role of primary acceptor for hydride and alcohol, while the reduced FAD transfers the redox equivalent to dioxygen as the final acceptor.[49] Cholesterol esterase (ChEt) is another important enzyme used in the determination of the cholesterol level in human blood. It is a bile salt activated lipase that catalyzes the hydrolysis of dietary cholesterol esters, triacylglycerols, and phospholipids by the process of serine protease mechanism.[50] Utilizing these mentioned biomolecules for the fabrication of cholesterol biosensor needs a matrix for immobilization, methods for immobilization, and a transducer material. Different kinds of materials such as CPs, nanostructured materials including carbon-based materials (CNTs and graphene) have been utilized as the matrix for the immobilization due to their excellent electrical and mechanical properties. Many researchers have used CPs such as polypyrrole, polyaniline, and so on for highly selective enzymatic biosensor toward cholesterol monitoring in human blood. Vidal et al. have designed a cholesterol biosensor by entrapping ChO_x into the polypyrole film.[51] Brahim et al. have utilized the entrapment of ChO_x within a composite of poly(2-hydroxethyl methacrylate)-tertraethyleneglycol diacrylate/polypyrol membrane for entrapment of ChO_x for cholesterol biosensor fabrication.[52] Khan et al. have used zinc oxide nanoparticles uniformly dispersed in chitosan to fabricate a hybrid nanocomposite for cholesterol determination.[53] Yang et al. have used layer-by-layer assembly of platinum nanoparticles for cholesterol detection in the range from 0.01 mM to 3 mM and response time of <30 s.[54] Dhand et al. have achieved linear detection of cholesterol in the range of 1.29–12.93 mM with high sensitivity of 6800 nA mM^{-1} using polyaniline–CNT composite for cholesterol detection.[55] Recently, the graphene–paltinum nanocomposites have also been utilized as an immobilization matrix and a transducer surface for cholesterol detection and demonstrated the sensitivity of 2.07 ± 0.1 μA/μM/cm^2 and 0.2 μM) and the detection limit of 0.2 μM.[56]

6.5 CONCLUSIONS

To summarize, we have discussed in this chapter the applications and prospects of enzymatic biosensors. The basic principle of enzymatic biosensors especially electrochemical transduction and immobilization of enzymes has also been discussed in this chapter. The applications of various nanomaterials such as GO, CNTs, CPs, metal–oxides, and so on for the development of enzymatic biosensors toward clinical diagnostics have also been discussed in this chapter. A board spectrum of the development of enzymatic biosensors including cholesterol and glucose biosensors has been added with the reported literature. The nanostructured materials can improve the biosensor's sensitivity, limit of detection, selectivity, and reproducibility.

The enzymatic biosensors play an important role in the development of POC devices. Recent advancements in nanotechnology and microfabrication technology have assisted in achieving superior sensing characteristics of the enzymatic biosensor. However, it is still a challenge to develop enzymatic biosensor devices for real-time monitoring due to the instability of enzyme molecules on the transducer's surface. The structural innovation of nanotechnology with surface functionalization combined with miniaturization can realize the development of compact devices for health care. In future, the development of portable, hand-handled, and cost-effective biosensors with multiple analytes' detection is believed to play a critical role in the diagnostics of various diseases that pose a threat to humanity.

ACKNOWLEDGMENTS

We acknowledge the National Science Foundation under grant DBI-1331390, the United State Agriculture Department's National Institute of Food and Agriculture under grant 2013-68004-20374, the Iowa Department of Transportation, the Iowa Corn Promotion Board, and the Iowa State University's Plant Sciences Institute Faculty Scholar Program. We acknowledge Dr. Liang Dong, Department of Electrical and Computer Engineering, Iowa State University, Ames, USA for his support and the Department of Science & Technology Centre on Biomolecular Electronics, Biomedical Instrumentation Section, CSIR-National Physical Laboratory, New Delhi in the preparation of this book chapter.

KEYWORDS

- **Biosensors**
- **Nanomaterials**
- **Nanoprobes**
- **Biomolecules**
- **Microfabrication**
- **Graphene**

REFERENCES

1. Newman, J. D.; Setford, S. J. Enzymatic Biosensors. *Mol. Biotechnol.* **2006,** *32*, 249–68.
2. Wang, J. Electrochemical Glucose Biosensors. *Chem. Rev.* **2008,** *108*, 814–825.
3. Wang, J. Glucose Biosensors: 40 Years of Advances and Challenges. *Electroanalysis* **2001,** *13*, 983–988.
4. Wilson, G. S.; Hu, Y. Enzyme-Based Biosensors for In Vivo Measurements. *Chem. Rev.* **2000,** *100*, 2693–2704.
5. Wang, J. Nanomaterial-Based Electrochemical Biosensors. *Analyst* **2005,** *130*, 421–426.
6. Pumera, M.; Sanchez, S.; Ichinose, I.; Tang, J. Electrochemical Nanobiosensors. *Sensor. Actuat. B* **2007,** *123,* 1195–1205.
7. Umar, A.; Rahman, M. M.; Vaseem, M.; Hahn, Y.-B. Ultra-Sensitive Cholesterol Biosensor Based on Low-Temperature Grown ZnO Nanoparticles. *Electrochem. Commun.* **2009,** *11*, 118–121.
8. Ali, A.; Ansari, A. A.; Kaushik, A.; Solanki, P. R.; Barik, A.; Pandey, M. K.; Malhotra, B. D. Nanostructured Zinc Oxide Film for Urea Sensor. *Mater. Lett.* **2009,** *63*, 2473–2475.

9. Kaushik, A.; Khan, R.; Solanki, P. R.; Pandey, P.; Alam, J.; Ahmad, S.; Malhotra, B. D. Iron Oxide Nanoparticles–Chitosan Composite Based Glucose Biosensor. *Biosens. Bioelectron.* **2008,** *24*, 676–683.
10. Kumar, S.; Kumar, S.; Ali, Md. A.; Anand, P.; Agrawal, V. V.; John, R.; Maji, S.; Malhotra, B. D. Microfluidic-Integrated Biosensors: Prospects for Point-of-Care Diagnostics. *Biotechnol. J.* **2013,** *8*, 1267–1279.
11. Pumera, M. Nanomaterials Meet Microfluidics. *Chem. Comm.* **2011,** *47*, 5671–5680.
12. Chin, C. D.; Linder, V.; Sia, S. K. Commercialization of Microfluidic Point-of-Care Diagnostic Devices. *Lab on a Chip* **2012,** *12*, 2118–2134.
13. Ali, Md. A.; Srivastava, S.; Solanki, P. R.; Reddy, V.; Agrawal, V. V.; Kim, C.; John, R.; Malhotra, B. D. Highly Efficient Bienzyme Functionalized Nanocomposite-Based Microfluidics Biosensor Platform for Biomedical Application. *Sci. Rep.* **2013,** *3*, 1–9.
14. Gerard, M.; Chaubey, A.; Malhotra, B. D. Application of Conducting Polymers to Biosensors. *Biosens. Bioelectron.* **2002,** *17*, 345–359.
15. Scouten, W. H.; Luong, J. H. T.; Brown, R. S. Enzyme or Protein Immobilization Techniques for Applications in Biosensor Design. *Trends Biotechnol.* **1995,** *13*, 178–185.
16. Ali, Md. A.; Srivastava, S.; Mondal, K., Chavhan, P. M., Agrawal, V. V.; John, R.; Sharma, A.; Malhotra, B. D. A Surface Functionalized Nanoporous Titania Integrated Microfluidic Biochip. *Nanoscale* **2014,** *6*, 13958–13969.
17. Hou, H.-H.; Wang, Y.-N.; Chang, C.-L.; Yang, R.-J.; Fu, L.-M. Rapid Glucose Concentration Detection Utilizing Disposable Integrated Microfluidic Chip. *Microfluid. Nanofluid.* **2011,** *11*, 479–487.
18. Yu, J.; Roux, R. L.; Gu, Y.; Yunus, K.; Matthews, S.; Shapter, J. G.; Fisher, A. C. Integration of Enzyme Immobilised Single-Walled Carbon Nanotube Arrays into Microfluidic Devices for Glucose Detection. *ICONN* **2008,** 137–140.
19. Srivastava, S.; Solanki, P. R.; Kaushik, A.; Ali, M. A.; Srivastava, A.; Malhotra, B. D. A Self Assembled Monolayer Based Microfluidic Sensor for Urea Detection. *Nanoscale*, **2011,** *3*, 2971–2977.
20. Srivastava, S.; Ali, Md. A.; Solanki, P. R.; Chavhan, P. M.; Pandey, M. K.; Mulchandani, A.; Srivastava, A.; Malhotra, B. D. Mediator-Free Microfluidics Biosensor Based on Titania–Zirconia Nanocomposite for Urea Detection. *RSC Advances* **2013,** *3*, 228–235.
21. Penn, S. G.; He, L.; Natan, M. J. Nanoparticles for Bioanalysis. *Curr. Opin. Chem. Biol.* **2003,** *7*, 609–615.
22. Wang, J. Nanoparticle-Based Electrochemical DNA Detection. *Anal. Chim. Acta* **2003,** *500*, 247–257.
23. Wang, J. Nanomaterial-Based Amplified Transduction of Biomolecular Interactions. *Small* **2005,** *1*, 1036–1043.
24. Wang, J. Carbon-Nanotube Based Electrochemical Biosensors: A Review. *Electroanalysis* **2005,** 17, 7–14.
25. Gooding, J. J. Protein Electrochemistry Using Aligned Carbon Nanotube Arrays. *J. Am. Chem. Soc.* **2003,** 125, 9006–9007.
26. Yu, X.; Chattopadhyay, D.; Galeska, L.; Papadimitrakopoulos, F.; Rusling, J. F. Peroxidase Activity of Enzymes Bound to the Ends of Single-Wall Carbon Nanotube Forest Electrodes. *Electrochem. Commun.* **2003,** *5*, 408–411.
27. Musameh, M.; Wang, J.; Merkoci, A.; Lin, Y. Low-Potential Stable NADH Detection at Carbon-Nanotube-Modified Glassy Carbon Electrodes. *Electrochem. Commun.* **2002,** *4*, 743–746.

28. Geim, A. K.; Novoselov, K. S. The Rise of Graphene. *Nat. Mater.* **2007,** *6*, 183–191.
29. Rahman, M. M.; Ahammad, A. J.; Jin, J.-H.; Ahn, S. J.; Lee, J.-J. A Comprehensive Review of Glucose Biosensors Based on Nanostructured Metal-Oxides. *Sensors* **2010,** *10*, 4855–4886.
30. Liu, Y.; Yu, D.; Zeng, C.; Miao, Z.; Dai, L. Biocompatible Graphene Oxide-Based Glucose Biosensors. *Langmuir* **2010,** *26*, 6158–6160.
31. Manjunatha, R.; Suresh, G. S.; Melo, J. S.; D'Souza, S. F.; Venkatesha, T. V. An Amperometric Bienzymatic Cholesterol Biosensor Based on Functionalized Graphene Modified Electrode and Its Electrocatalytic Activity Towards Total Cholesterol Determination. *Talanta* **2012,** *99*, 302–309.
32. Pandey, P.; Datta, M.; Malhotra, B. D. Prospects of Nanomaterials in Biosensors. *Anal. Lett.* **2008,** *41*, 159–209.
33. Solanki, P. R., Kaushik, A.; Agrawal, V. V.; Malhotra, B. D. Nanostructured Metal Oxide-Based Biosensors. *NPG Asia Mater.* **2011,** *3*, 17–24.
34. Wang, J. X.; Sun, X. W.; Wei, A.; Lei, Y.; Cai, X. P.; Li, C. M.; Dong, Z. L. Zinc Oxide Nanocomb Biosensor for Glucose Detection. *Appl. Phys. Lett.* **2006,** *88*, 233106–233106.
35. Ali, Md. A.; Solanki, P. R.; Patel, M. K.; Dhayani, H.; Agrawal, V. V.; John, R.; Malhotra, B. D. Malhotra, A Highly Efficient Microfluidic Nano Biochip Based on Nanostructured Nickel Oxide. *Nanoscale* **2013,** *5*, 2883–2891.
36. Lu, W.; Zhao, H.; Wallace, G. G. Pulsed Electrochemical Detection of Proteins Using Conducting Polymer Based Sensors. *Anal. Chim. Acta* **1995,** *315*, 27–32.
37. Heller, A. Electrical Wiring of Redox Enzymes. *Acc. Chem. Res.* **1990,** *23*, 128–134.
38. Gregg, B. A.; Heller, A. Cross-Linked Redox Gels Containing Glucose Oxidase for Amperometric Biosensor Applications. *Anal. Chem.* **1990,** *62*, 258–263.
39. Teasdale, P. R.; Wallace, G. G. Molecular Recognition Using Conducting Polymers: Basis of an Electrochemical Sensing Technology—Plenary Lecture. *Analyst* **1993,** *118*, 329.
40. Forzani, E. S.; Zhang, H.; Nagahara, L. A.; Amlani, I.; Tsui, R.; Tao, N. A Conducting Polymer Nanojunction Sensor for Gglucose Detection. *Nano Lett.* **2004,** *4*, 1785–1788.
41. Singh, S.; Chaubey, A.; Malhotra, B. D. Amperometric Cholesterol Biosensor Based on Immobilized Cholesterol Esterase and Cholesterol Oxidase on Conducting Polypyrrole Films. *Anal. Chim. Acta* **2004,** *502*, 229–234.
42. Reach, G.; Wilson, G. S. Can Continuous Glucose Monitoring be used for the Treatment of Diabetes? *Anal. Chem.* **1992,** *64*, 381A–386A.
43. Turner, A. P.; Chen, B.; Piletsky, S. In Vitro Diagnostics in Diabetes: Meeting the Challenge. *Clin. Chem.* **1999,** *45*, 1596.
44. Pishko, M. V.; Katakis, I.; Lindquist, S. E.; Ye, L.; Gregg, B. A. Direct Electrical Communication Bbetween Graphite-Electrodes and Surface Adsorbed Glucose-Oxidase Redox Polymer Complexes. *Angew. Chem. Int. Edit.* **1990,** *29*, 82–89.
45. Xiao, Y.; Patolsky, F.; Katz, E.; Hainfeld, J. F.; Willner, I. Plugging into Eenzymes': Nanowiring of Redox Enzymes by a Gold Nanoparticle. *Science* **2003,** *299*, 1877–1881.
46. Patolsky, F.; Weizmann, Y.; Willner, I. Long-Range Electrical Contacting of Redox Enzymes by SWCNT Connectors. *Angew. Chem. Int. Ed.* **2004,** *43*, 2113–2117.
47. Shichiri, M.; Kawamori, R.; Yamasaki, Y.; Hakui, N.; Abe, H. Wearable Artificial Endocrine Pancreas with Needle-Type Glucose Sensor. *Lancet* **1982,** *2*, 1129–1131.
48. Arya, S. K.; Datta, M.; Malhotra, B. D. Recent Advances in Cholesterol Biosensor. *Biosen. Bioelectron.* **2008,** *23*, 1083–1100.

49. Pollegioni, L.; Wels, G.; Pilone, M. S.; Ghisla, S. Kinetic Mechanisms of Cholesterol Oxidase from Streptomyces Hygroscopicus and Brevibacterium Sterolicum. *Eur. J. Biochem.* **1999,** *264*, 140–151.
50. Lin, G., Chiou, S.-Y.; Hwu, B.-C.; Hsieh, C.-W. Probing Structure–Function Relationships of Serine Hydrolases and Proteases with Carbamate and Thiocarbamate Inhibitors. *Protein J.* **2006,** *25*, 33–43.
51. Vidal, J. C.; Garcia, E.; Castillo, J. R. In Situ Preparation of a Cholesterol Biosensor: Entrapment of Cholesterol Oxidase in an Overoxidized Polypyrrole Film Electrodeposited in a Flow System: Determination of Total Cholesterol in Serum. *Anal. Chim. Acta* **1999,** *385*, 213–222.
52. Brahim, S.; Narinesingh, D.; Guiseppi-Elie, A. Amperometric Determination of Cholesterol in Serum Using a Biosensor of Cholesterol Oxidase Contained within a Polypyrrole–Hydrogel Membrane. *Anal. Chim. Acta* **2001,** *448*, 27–36.
53. Khan, R.; Kaushik, A.; Solanki, P. R.; Ansari, A. A.; Pandey, M. K.; Malhotra, B. D. Zinc Oxide Nanoparticles-Chitosan Composite Film for Cholesterol Biosensor. *Anal. Chim. Acta* **2008,** *616*, 207–213.
54. Yang, M.; Yang, Y.; Yang, H.; Shen, G.; Yu, R. Layer-by-Layer Self-Assembled Multilayer Films of Carbon Nanotubes and Platinum Nanoparticles with Polyelectrolyte for the Fabrication of Biosensors. *Biomaterials* **2006,** *27*, 246–255.
55. Dhand, C.; Arya, S. K.; Datta, M.; Malhotra, B. D. Polyaniline–Carbon Nanotube Composite Film for Cholesterol Biosensor. *Anal. Biochem.* **2008,** *383*, 194–199.
56. Dey, R. S.; Raj, C. R. Development of an Amperometric Cholesterol Biosensor Based on Graphene−Pt Nanoparticle Hybrid Material. *J. Phys. Chem. C* **2010,** *114*, 21427–21433.
57. Ansari, A. A.; Kaushik, A.; Solanki, P. R.; Malhotra, B. D. Sol–Gel Derived Nanoporous Cerium Oxide Film for Application to Cholesterol Biosensor. *Electrochem. Commun.* **2008,** *10*, 1246.
58. Kouassi, G. K.; Irudayaraj, J.; McCarty, G. Examination of Cholesterol Oxidase Attachment to Magnetic Nanoparticles. *J. Nanotechnol.* **2005,** *3*, 1477.
59. Shan, C.; Yang, H.; Song, J.; Han, D.; Ivaska, A.; Niu, L. Direct Electrochemistry of Glucose Oxidase and Biosensing for Glucose Based on Graphene. *Anal. Chem.* **2009,** *81*, 2378–2382.
60. Singh, J.; Kalita, P.; Singh, M. K.; Malhotra, B. D. Nanostructured Nickel Oxide-Chitosan Film for Application to Cholesterol Sensor. *Appl. Phys. Lett.* **2011,** *98*, 123702.
61. Singh, J.; Roychoudhary, A.; Srivastava, M.; Solanki, P. R.; Lee, D. W.; Lee, S. H.; Malhotra, B. D. A Dual Enzyme Functionalized Nanostructured Thulium Oxide Based Interface for Biomedical Application. *Nanoscale* **2014,** *6*, 1195.
62. Singh, C.; Pandey, M. K.; Biradar, A. M.; Srivastava, A. K.; Sumana, G. A Bienzyme-Immobilized Highly Efficient Niobium Oxide Nanorod Platform for Biomedical Application. *RSC Adv.* **2014,** *4*, 15458.

CHAPTER 7

NANOBIOSESORS FOR GENOSENSING

CHANDRA MOULI PANDEY

Biomedical Instrumentation Section, CSIR-National Physical Laboratory, New Delhi-110012

Department of Biotechnology, Delhi Technological University, Delhi, India

Email: cmp.npl@gmail.com

CONTENTS

DNA biosensors (genosensors), based on nucleic acid recognition processes, have witnessed a burgeoning interest in the commercial as well as research fields. Recent times have perceive an upsurge in the development of genosensors for sensitive, rapid, specific, and inexpensive testing of genetic and infectious diseases, and for the detection of DNA interactions. The main principle governing the detection by genosensors relies on specific DNA hybridization, directly on the surface of a physical transducer. This chapter deals with the main DNA immobilization techniques, new micro- and nanotechnological platforms for biosensing and the transduction mechanisms in genosensors. Further, the applications of nanomaterials, particularly conducting polymers (CPs), noble metal nanoparticles (MNPs), carbonaceous material, quantum dots (QDs), and metal–oxide for the signal amplification of these genosensors have also been discussed.

7.1 INTRODUCTION

The determination of genes is a current demand in fields like gene expression monitoring, pharmacogenomic research, clinical diagnostics, viral and bacterial identification, and genetic identification.[1,2] Genome sequencing has allowed detection of inherited disease causing point mutations and human pathogens through their peculiar, specific nucleic acid (NA) sequences.[3,4] In this context, the growing demand for rapid, simple, inexpensive, and portable testing methods has encouraged research in the field of DNA sensors or genosensors.[5,6] The current research on genosensors is prominently increasingly in the literature as they offer better diagnosis, prevention, and treatment for many infectious disease.[7] Conventional methods for the analysis of specific gene sequences are usually based on either direct sequencing or DNA hybridization.[8] The latter method is more preferred due to its simplicity to be used in the diagnostic laboratory than the direct sequencing method.[8] The DNA hybridization principle relies on the identification of the target gene sequence by a DNA probe resulting in the formation of a double-stranded hybrid with high efficiency and high specificity in the presence of other, noncomplementary DNA.[9] To be a high-performance biosensor, the immobilized DNA probe should be very specific and able to discriminate even a single base-pair mismatch between different target DNA (tDNA) strains. Conventional DNA microarrays also make use of sequence-specific DNA detection, but their low efficiency is usually due to large size of biological samples and complex treatment, making it difficult to obtain real-time outputs.[10] Moreover, the technology is still too expensive to turn them valuable in point-of-care (POC) diagnosis. In general, DNA biosensors are able to overcome these demerits thus allowing easier, quicker, and cost-effective traditional hybridizing assays, while keeping high sensitivity and specificity of detection.[11]

Electrochemical (EC) biosensors represent a leading approach for fast, low-cost, portable, and sensitive determination of the genetic disorder.[2] These biosensors offer exciting breakthrough for various clinical applications, ranging from "alternative site" testing emergency room screening, bedside monitoring, or POC diagnostics.[12] For the fabrication of such powerful devices there is an urgent need for innovative efforts in the design of new material and novel fabrication processes.[13] A range of conventional macromolecular material matrices have been proposed for the development of EC DNA biosensing devices. In this context, nanostructured materials (NMs) are being widely explored in DNA biosensors' application, as they provide high sensitivity, selectivity, and stability which are faced by other EC biosensors.[14] Further, these NMs modified electrodes not only help in improving the catalytic activity of the transducer but it also support effective biomolecular reaction on the electrode's surface.[12] The NMs facilitate the performance of the biosensor thus allowing miniaturization and speed, reducing the use of reagent, and sample consumption.[15] A number of NMs including CPs, metals and their oxides, semiconductors, carbon species, has been widely investigated to enhance the response of the genosensors.[16] This chapter mainly focuses on the development and validation of genosensors that incorporate nanomaterials as either a signal transducer or as an electro-active species for direct detection of an analyte. It also covers some recent developments in nanotechnology, namely CPs, MNPs, carbon nanotubes (CNTs), and hierarchical hybrid materials, which are responsible for improved sensitivity and selectivity in genosensor development.

7.2 DNA HYBRIDIZATION MECHANISM

EC DNA biosensors are of major interest due to their tremendous application in obtaining sequence-specific information in a faster, simpler, and cheaper manner.[17] EC detection of DNA hybridization typically involves the monitoring of a current response, arising due to Watson–Crick base-pair recognition event (optimized conditions) into a measurable analytical signal.[18] The first step toward designing of a DNA biosensor is the immobilization of single stranded oligonucleotides probe (ODN) on a suitable transducer surface which can distinguish its complementary (target) DNA sequence via hybridization.[19] For the hybridization process the probe-coated electrode is immersed into a solution of a target DNA (tDNA) whose nucleotide sequence is to be tested. The hybridization of DNA usually occurs between a known probe DNA (pDNA) sequence and the other counterpart, that is, tDNA and in some cases DNA–ribonucleic acid (RNA) and RNA–RNA hybridizations are also observed.[20] The designing of the DNA probes is either through chemical methods or by molecular biology where a probe is obtained by reverse transcription of a previously isolated and specific messenger RNA, or inferring its nucleotide sequence based on the amino acid sequence of the protein expressed by that DNA.[21] The duplex formation is detected by the change in the configuration of the DNA or by using an appropriate hybridization indicator (Figure 7-1).[22] Conventional NA hybridization methods, like gel electrophoresis and Southern blotting, are usually lengthy and labor intensive, and so is also the intrinsic biomolecular recognizing event of most genosensors. Both in vivo and in vitro, that is, as onto a transducer surface (solid support), NA hybridization is stronger and specific when the complimentarity degree between two DNA chains increases. Further, with the increase in the stability and specificity the linkage may reach a maximum (100%).[4] The mechanism of DNA hybridization over solid supports is still greatly unknown and unpredictable due to the

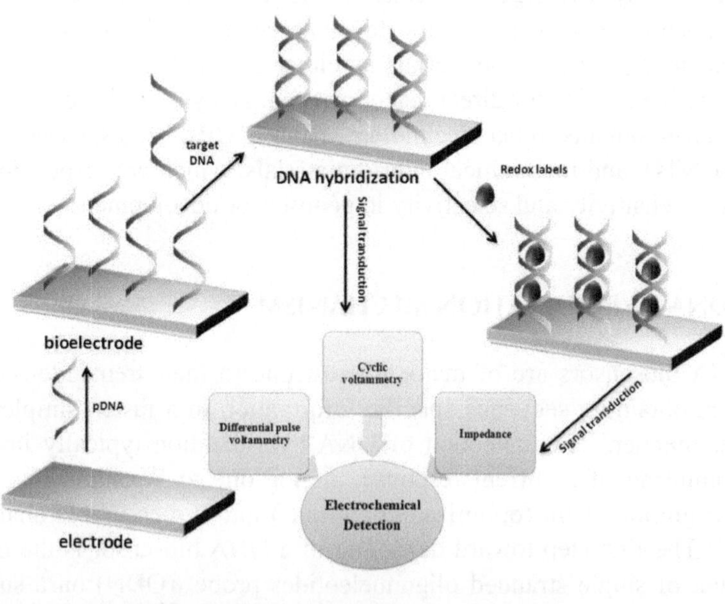

FIGURE 7-1 Scheme showing the DNA hybridization mechanism.

difficulty in accurately determining the concentration of the immobilized NA.[2] The most common accepted hypothesis at the solid/liquid interface is diffusion of an analyte toward the surface of the sensor (bi-dimensional diffusion, adsorption, and desorption).[15] The hybridization process in solution and at the interface are quite similar, but in the former the hybridization rate is typically very higher, assuming identical DNA sequences and conditions.[23] This may be due to the partial unavailability of many linking groups in the immobilized chain, eventually involved in that immobilization process.[24] The rate of hybridization also decreases with the secondary structure level of one or both chains which can be easily overcome by selecting a proper probe sequence.[25]

The stratagem to immobilize a pDNA onto a suitable transducer surface for the recognition of specific disease (cancer, bioterrorism agent, bacterial pathogens, etc.) play a fundamental role in rapid detection of a genetic disorder.[26] The employed immobilization strategy on the electrode surface effects the sensing properties of the biosensor in terms of sensitivity, specificity, operational stability, long-term use, and detection of long linear concentration range of the analyte.[15] Further, the selection of a suitable immobilization method depends on the nature of the solid surface, immobilized pDNA, and the transduction mechanism.

7.3 IMMOBILIZATION TECHNIQUES

The crucial factor for designing a biosensor is the development of immobilization techniques which strongly stabilizes the DNA on the transducer's surface.[27] Mostly, the integration of the DNA with the signal transducer is achieved by immobilizing the pDNA on the modified electrode surface.[28] Many different immobilization methods such as adsorption, cross-linking, covalent binding, entrapment, encapsulation, and others have been reported for the binding of DNA with the transducers surface (Figure 7-2).[29] The selection of a suitable immobilization method primarily relies on the nature of the biological element, type of transducer used, physicochemical properties of the analyte and the operating conditions for the biosensor.[11] The simplest and the traditional method used is physical adsorption of the biocomponent is based on van der Waals attractive forces where the biomolecule is directly adsorbed onto bare surfaces of bulk materials.[30] Hydrophilic surfaces are more preferred for effective NA hybridization as they facilitate exposure of hybridizing bases, despite being also prone to DNA detachment with increasing ionic strength.[31] However, the introduction of functional groups is still the main strategy to functionalize solid surfaces with biomolecules.

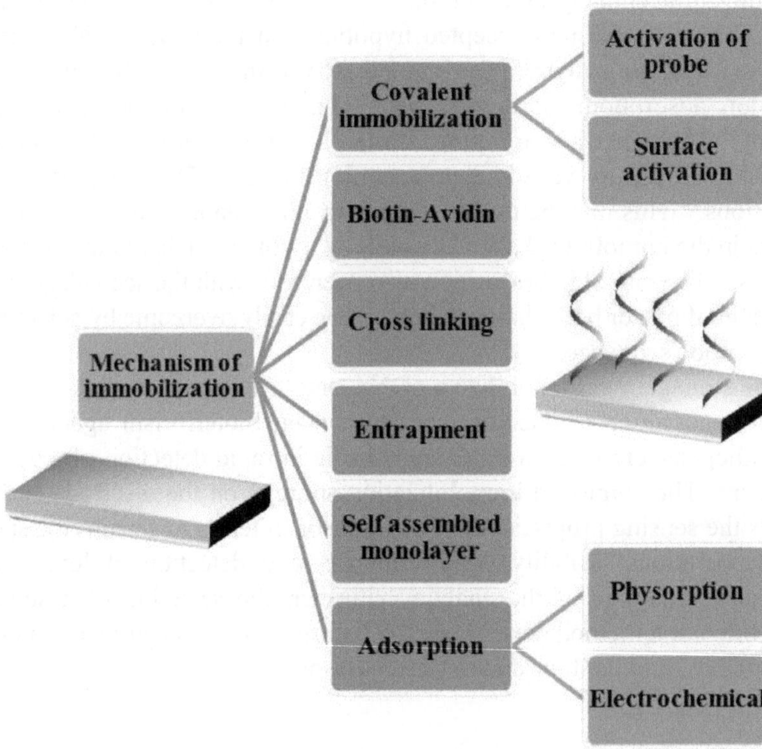

FIGURE 7-2 Method of DNA immobilization in a genosensor.

Several polymers are being used simultaneously like a polymeric gel of polyvinyl alcohol cross-linked to polyallylamin chloride and poly(L-lysine) in polystyrene-modified surfaces.[32] Despite DNA attachment to solid surfaces being usually stronger via covalent linkages, however, adsorption, may be preferred due to its slighter adverse effect over DNA structure, that is, adsorption process avoids breakdown of DNA structure.[26] Further, to increase the detection sensitivity range one-point covalent immobilization of DNA with minimum steric hindrance is favored. Technique like self-assembled monolayers (SAMs) are being spontaneously used to generate an ultrathin and highly ordered layer, similar to the cell's microenvironment.[10] The versatility of SAMs for several applications arises from the fact that they are thermodynamically stable and there is possibility of controlling the hydrophilicity degree and the chain length of the polymer.[33] The utilization of SAMs avoids conformational changes over the immobilized biocomponent, capable of affecting its activity which assures higher homogeneity and reproducibility of the electrode's surface.[34]

7.4 GENOSENSOR AND NANOSCALE MATERIALS

Fabrication of an efficient genosensor requires the selection of a suitable substrate for immobilizing the sensing materials. With the advancement of nanotechnology powerful DNA biosensors based on nanosized labels and amplification platforms have been developed.[15] Nanoparticles can be used to modify the oligonucleotides or the electrode surfaces for the effective acceleration of electron transfer between the electrode and the pDNA, thus leading to rapid current responses for target DNA. By this way the resulting transducers enhance the quantity and activity of the immobilized redox active biomacromolecules which result in enhancement in the sensitivity and the stability of the electrode.[13] Considering their unique physicochemical properties, in particular the high surface nominal area, nanomaterials provide interesting opportunities for development of novel design of genosensors.[35] Various classes of nanomaterials, such as noble MNPs (Au, Ag, Pt, Pd, etc.), metal–oxide nanoparticles, polymeric and inorganic–organic nanocomposites, CNTs, and QDs, are being frequently used in genosensor fabrication (Figure 7-3). These nanomaterials provide high surface-to-volume ratio, excellent electro-catalytic activity as well as good biocompatibility and novel electron transport properties.[16] Further, the morphology (shape, size, diameter, surface condition, crystal structure, etc.) and the quality of

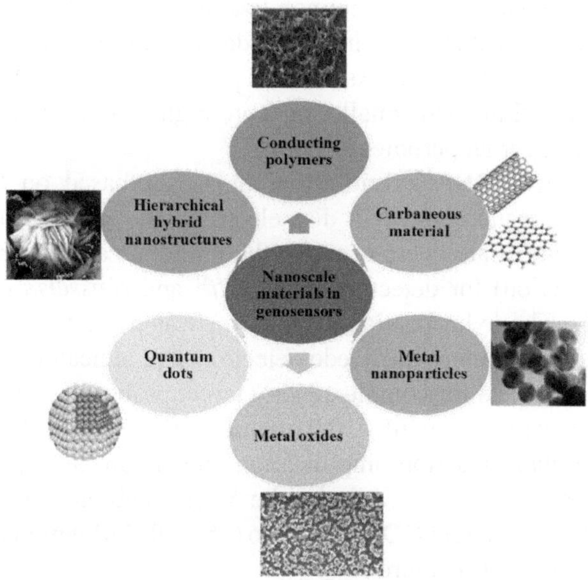

FIGURE 7.3 Application of nanoscale materials in genosensor fabrication.

nanomaterials provides a remarkable electron transport properties. However, the immobilization of the ODN) onto the surfaces of nanomaterials can retain their bioactivity because of the biocompatibility of nanoparticles. The difference in the charge of the nanomaterials and the DNA help in electrostatically adsorbing the biomolecule onto the nanomaterial surface.[35] In the following sections, we discuss the nanobiosensing platform involved in the construction of portable EC genosensor devices.

7.4.1 GENOSENSORS BASED ON CP

Polymeric nanomaterials used for the immobilization of DNA include CPs like polyaniline (PANI), poly(phenylenevinylene), polypyrrole (PPy), polythiophene, polyacetylene, and polyindol.[36] The advantageous properties of these CPs are their remarkable high electrical conductivity, ease of processability, low ionization potentials, and high electron affinity which make them relevant materials for designing novel biosensors.[37] Further, the conductivity of these polymers strongly depends on solution pH and the oxidation state. Whereas by the modification of the deposition method the thickness and shape of the polymeric film (from nanometer to micrometer range) can be easily controlled.[38] These unmatched properties of the polymeric nanomaterials provide enhanced sensitivity, selectivity, stability, better signal transduction, and flexibility for the immobilization of DNA.[36] Short DNA probes have been entrapped in PPy film as the dopant during film growth to maintain the affinity for target DNAs.[39] It was proposed that in a DNA-doped PPy sensor, thinner films with smaller or more highly concentrated dopant ions produced stronger amperometric signals.[40]

Ultrasensitive DNA hybridization biosensor based on PANI electrochemically deposited onto a Pt disc electrode and indium–tin oxide (ITO) coated glass substrate, were used for the immobilization of pDNA (biotin/avidin interaction) for detection of *E. coli*[41] and *Neisseria gonorrhoeae*,[42] respectively. The hybridization mechanism was detected using both direct EC oxidation of guanine and a redox electroactive indicator, methylene blue. The detection limits for complementary target probe, *E. coli* genomic DNA and *E. coli* were 0.009 ng μL^{-1}, 0.01 ng μL^{-1} and *E. coli* cells mL^{-1} without polymerase chain reaction amplification and it can be used 5–7 times at temperatures of 30–45 °C while for the *N. gonorrhoeae* the detection limit of complementary target ODN was up to 0.5×10^{-15} M within 60 s of hybridization time at 25 °C (Figure 7-4).

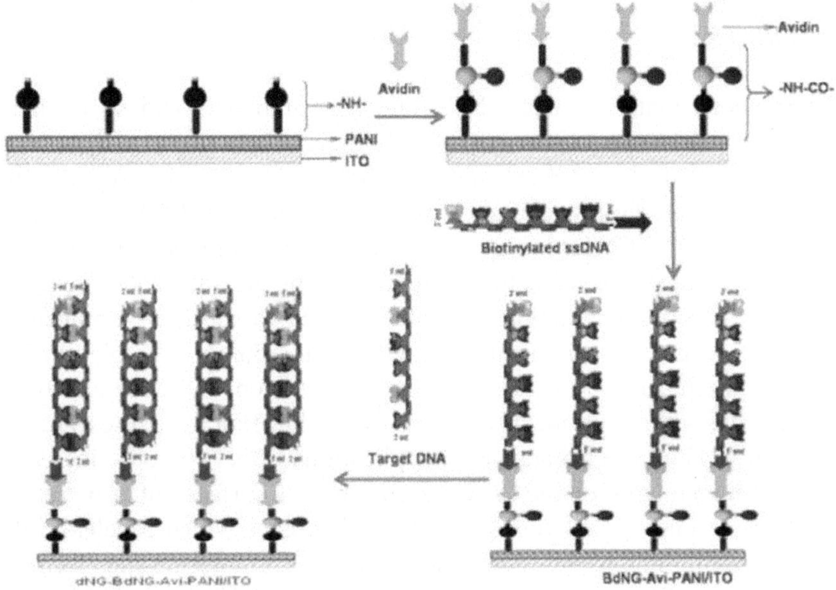

FIGURE 7-4 DNA hybridization detection using PANI bioelectrode.[41]

Until now, CPs with different morphologies (nanoparticles, nanowires, and nanotubes) have been prepared by chemical or EC methods.[37] It is proposed that single-molecule detection can be achieved by adjusting the nanowire's conductivity to a value closer to the lower end of the semiconductor.[43] Further, the orientation and location control of the nanotubes of CPs on a graphite electrode facilitate the transfer of electron from the redox active site of DNA molecule to the electrode's surface. The EC results show that the conducting PANI nanotube arrays have signal enhancement capability, and the limit of detection obtained at varying tDNA concentration was 1.0 fM. The fabricated biosensors can easily differentiate the perfect matched target DNA from one-nucleotide mismatched DNAs.[44]

Several advantageous properties like the ease of preparation and modification of the physical properties of CPs make them attractive for the construction of DNA biosensing devices. However, these polymeric nanomaterials are usually less used in the construction of genosensors because of their relative low conductivity as well as their phase morphology, leading to low detection sensitivity. Much research will be needed to improve the stability and processability of these polymers if they are to be used for POC diagnostics.

7.4.2 GENOSENSORS BASED ON CARBONACEOUS MATERIALS

Carbonaceous materials like CNT, fullerene, graphene, and so on have attracted considerable attention because of their unique structure-dependent electrical, chemical, and mechanical properties.[45] CNTs, first reported by Iijima (1991), are hollow cylinders made of single-walled CNTs or multi-walled CNTs (MWCNTs) layers of graphene.[46] The introduction of oxygen functional groups on CNT helps in the improvement of the adhesive properties, or selectively functionalizes the surface to meet the application demands.[47] Other properties like good biocompatibility, flexible surface chemistry, high surface area, ease of functionalization, enhanced electronic conductivity, and a high mechanical resistance have motivated researchers to exploit CNT in electro-analytical applications.[4] CNT not only increases the immobilized DNA on the CNTs based substrate surface but also provide huge surface energy and flexible surface chemistry which accelerate the electron transfer rate between the redox active DNA molecule and electrode resulting in amplification of the EC signal of DNA hybridization.[45] Hembram *et al.* studied the electrical and optical properties of MWCNTs/DNA nanocomposite by covalently immobilizing DNA at the ends of defective sites and elucidate the wrapping of DNA on the CNTs is due to van der Waals force.[47] Thus, alteration of the electronic properties of nanotubes is done either by chemically functionalizing them with a moiety or by varying the structure whose intrinsic properties are electrically configurable.[15] MWCNT modified screen printed electrodes were used for the fast and sensitive detection of DNA and RNA from the electro-oxidation of guanine and adenine residues and the resulting transducer could detect calf thymus DNA concentration ranging from 17.0 $\mu g \cdot mL^{-1}$ to 345 $\mu g \cdot mL^{-1}$ with a detection limit of 2.0 $\mu g \cdot mL^{-1}$ and yeast tRNA ranging from 8.2 $\mu g \cdot mL^{-1}$ to 4.1 $mg \cdot mL^{-1}$.[48] Most of the CNT based DNA biosensors published in the literature have focused on the ability of surface-confined CNTs to promote electron transfer reactions.[45]

With the discovery of graphene, there was a marked revolution in the field of DNA-based sensors. Due to their astounding properties (high carrier mobility, thermal conductivity, and biocompatibility), it has been predicted that the nanomaterials of graphene will be among the candidate materials for post-silicon electronics.[49] Graphene is a semimetal zero gap nanomaterial with demonstrated ability to be employed as an excellent candidate for DNA attachment and detection due to its large surface area (up to 2630 m^2/g) and unique sp^2 (sp^2/sp^3)-bonded network.[50] Graphene oxide GO has been successfully adopted as a platform to discriminate DNA sequences between pDNA and double-stranded DNA (dsDNA). It was also revealed that the oxidation signals of DNA stacked graphene nanofibers were almost two to four times higher than those on CNT-based electrode, edge-plane pyrolytic graphite and glassy carbon, or graphite microparticle based electrodes.[51] The high EC activity could be assigned to better electron exchange between DNA bases and the edge-plane-like defective sites of the sheets as active sites for oxidation of DNA bases. Varying intensity GO reduction signals of GO nanoplatelets (37 nm) were used for DNA analysis where different binding ability of GO nanoplatelets to fully matched dsDNA, one-base-mismatched dsDNA, and unhybridized sequences, with varying intensities (GO reduction signals) were observed for discrimination of single nucleotide polymorphism related to Alzheimer's disease. Though the EC reduction mechanism of GO is not yet thoroughly investigated, one might be certain that hydrogen ions participate in the reduction process.[52] Thus, it can be said that the electrode fabrication techniques using carbon-based materials and the hybridization indication techniques may help in developing ultrasensitive, selective, and miniaturized EC DNA biosensor which can be used for DNA sequence analysis in practical applications, such as early cancer detection and POC use.

7.4.3 GENOSENSORS BASED ON MNPS

Noble MNPs (Au, Ag, Pt, and Pd) have played a pivotal role in the development of DNA biosensors by improving the existing biosensing techniques leading to more specific and highly sensitive biomolecular diagnostics.[53] MNPs not only help in improving the sensing performance of the genosensor but it also enhance the transfer of electron rate between the electrode surface and the redox active DNA.[54] However, majority of these MNPs have been used as labels in EC DNA sensing to enhance the loading of electrochemical active species for signal amplification.[55] The combined properties of MWCNTs and Pt and Pd nanoparticles dispersed in nafion-modified glassy carbon electrode glassy carbon electrode (GCE) for construction of sensitivity-enhancing EC DNA biosensing. These nanoparticles accelerate the electron transfer rate of redox molecule and showed the detection limit where the sensitivities are in the pico- and femtomolar range.[56]

MNPs like gold nanoparticles (AuNPs) have been widely explored for DNA hybridization detection.[57] The selective and sensitive recognition of target DNA by chemically attaching probe DNA onto functionalized AuNPs was reported by Glynou et al. where the AuNPs amplify DNA recognition and transduction events, which has been used as an ultrasensitive method for electrical biosensing of DNA.[58] Self-assembly was used for the co-adsorption of probe DNA-functionalized AuNPs because these SAMs require very small amount of ODN which results in the binding of desired DNA in the near vicinity to the transducer surface. This self-assembly provides biocompatible microenvironment to undergo direct electron transfer reactions and also amplify the EC signal by increasing the binding sites for the DNA immobilization.[34] Multilayer AuNPs assembled on electropolymerized poly-2,6-pyridinedicarboxylic acid film on GCE electrode was used for sequence-specific DNA sensor, related to phosphinothricin acetyltransferase trans gene (PAT, transgenic plants). The high efficiency of the biosensor arises from the amalgamation of the electrocatalytic properties of AuNPs with the flexibility and biocompatibility of the polymeric materials.[26] GCE modified with functionalized AuNPs (cysteamine/polyglutamic acid) was applied for the immobilization of the pDNA onto the electrode surface for development of NA sensors. The same group reported that the peak current increases linearly with increase in the complementary target DNA concentration (9.0×10^{-11} M to 4.8×10^{-9} M) and with a detection limit of 4.2×10^{-11} M.[59] A novel and sensitive sandwich EC DNA biosensor was prepared using magnetic microbeads and AuNPs modified with bar code and lead sulfide nanoparticles.[60] To increase the number of carboxylic groups on the surface, the magnetic microspheres were coated with four layers of polyelectrolytes, which enhanced the amount of the captured DNA resulting in enhanced sensitivity, selectivity, and detection limit. Silver nanoparticles (AgNPs) were also used for detection of short DNA on a gold electrode surface in connection with the use of AgNPs as a label conjugate.[61] The observed Ag/AgCl redox process signal of the AgNP labels was subsequently used to quantify the amount of DNA with a detection limit up to 1 pM. Ultrasensitive NA detection based on interdigited microelectrodes was also reported by attaching hematin molecules with hybridized DNA to act as a catalyst for accelerating the reduction of ammonical silver ions to form AgNPs.[62] This alteration in conductance of the AgNPs directly correlated with the number of the hybridized DNA molecules and under optimized conditions the biosensor was sensitive up to 1 fM. Functionalized AgNPs (3–5 nm) were also used as an electroactive label on the surface of gold electrode modified with thiolated natural probe peptide NA and 6-mercapto-1-hexanol as linker for detection of ODN from the H5N1 bird flu virus. The fabricated DNA sensor has good response to target DNA with a detection limit up to 10 fM.

To overcome the limitations of conventional molecular probes for the detection of DNA and RNA, DNA-conjugated MNPs have been used as promising biological tags.[63] A number of highly selective and sensitive DNA detection methods have been developed based on DNA-conjugated MNPs. The remarkable sensitivity and selectivity achieved in these methods are results of the excellent optical and catalytic properties of MNPs and the enhanced binding properties of the DNA strands immobilized on the MNPs.[53] Significant advances in developing new methodologies (surface-enhanced Raman spectroscopy and fluorescent quenching/enhancement approaches) by using noble MNPs will unravel the current paradigms and emerging challenges in genosensor fabrication.[64]

7.4.4 GENOSENSORS BASED ON METAL OXIDES

Nanostructured metal–oxide (NMO) based biosensors have been widely used to convert a hybridization event into an analytical signal in NA sensors.[35] These NMOs facilitates the analytical capacities of sensor devices by increasing the surface-to-volume ratio which increases the detection sensitivity of the constructed transducers to a single molecular detection level.[65] Ansari et al. exploited sol-gel derived nanostructured zinc oxide (ZnO) film deposited onto ITO glass substrate to immobilization of 20-mer thiolated ODN probe for detection of target DNA (*N. gonorrhoeae*) using a hybridization technique. The EC response of the proposed genosensor was linear in the concentration range of target DNA from 0.000524 fmol to 0.524 nmol, with a detection limit of 0.000704 fmol and hybridization time of 60 s.[66] Considerable efforts have been taken on the fabrication of an EC DNA biosensor, by immobilizing DNA onto nanocrystalline transparent metal oxide (ZrO_2). This metal oxide has drawn considerable attention due to its unique physical, chemical, and optical properties, thus making it a promising matrix for sensing applications. Moreover, the nanostructures provide increased surface area for DNA immobilization that may lead to the improved limit of detection. Besides this, it was observed that ZrO_2 is thermally stable, chemically inert, nontoxic, and has good affinity for groups containing oxygen thus facilitating covalent immobilization of biomolecules without using any cross-linker.[67,68] In spite of being an interesting matrix for biomolecule immobilization, nanostructured zirconium oxide (ZrO_2) based electrode suffers a major drawback of cracking and aggregation leading to limited application of ZrO_2 nanoparticles to biosensing which can be overcome by modifying ZrO_2 nanoparticles with chitosan. Chitosan not only provides excellent film-forming and adhesion abilities but is also nontoxic and biocompatible.[65] Similarly, Feng et al. utilized cerium oxide–chitosan composite matrix for immobilization of pDNA for development of DNA biosensor for colorectal cancer gene detection. This chitosan introduced CeO_2 nanocomposite matrix demonstrated enhanced loading of DNA probe on the surface of electrode yielding good biocompatibility, nontoxicity, and excellent electronic conductivity.[69] Doping of MWCNTs, ZrO_2 nanoparticles, and chitosan modified onto GCE provides a sensitive detection of DNA hybridization using electroactive daunomycin as an indicator.[70] The coupling of MWCNTs with chitosan and ZrO_2 nanoparticles provides enhanced electroactive surface area for higher amount loading of pDNA and excellent electron transfer ability between the ODNs and the electrode surface.[71] In another work, chitosan-doped ZnO nanoparticles were used for voltammetric detection of DNA hybridization where the nanostructure ZnO provides enhanced active surface area for the immobilization of DNA. The fabricated nanobiocomposite provides a conducive environment for the DNA to retain its bioactivity under considerably extreme conditions and the ZnO nanoparticles in the biocomposite offer excellent affinity to pDNA. The reported biosensor have a detection limit of 1.09×10^{-11} mol·L^{-1} and can easily discriminate the complementary target and two-base-mismatched sequences.[72] Iron oxide microspheres and self-doped PANI nanofibers modified carbon ionic liquid electrode was used for immobilization of pDNA for sensitive impedomatric detection of sequence-specific DNA of phosphoenolpyruvate carboxylase gene where $Fe(CN)_6]^{3-/4-}$ was employed as an internal indicator.[73] The strong adsorption ability of Fe_2O_3 microspheres and excellent conductivity of self-doped PANI nanofibers enhanced the sensitivity of DNA hybridization recognition. In the same study, DNA hybridization events were monitored with a label-free electrochemical impedance spectroscopy EIS strategy. The EC response of the genosensor was measured in the concentration range from 1.0×10^{-13} mol/L to 1.0×10^{-7} mol/L, with a detection limit 2.1×10^{-14} mol/L.[74]

Rare earth semiconductor metals (praseodymium oxide) were also used for the immobilization of single-stranded ODN for label free rapid detection of DNA hybridization and the change in electrical impedance was measured. In a similar report, thiol-modified ODN was immobilized on the surface of praseodymium oxide for impedomatric detection of unlabeled DNA hybridization.[75,76] The proposed EC AC impedomatric biosensor showed ultrasensitivity for the detection of complementary ODNs in solution without the use of a label reagent. Pandey et al. electrophoretically deposited cationic poly(lactic-co-glycolic acid) (PLGA) microspheres onto ITO glass substrate for detection of probe sequence specific for *E. coli*. Under the optimal conditions, this biosensor shows a detection limit of 8.7×10^{-14} M and is found to retain about 81% of the initial activity after nine cycles of use. These prepared electrodes are readily available as sensors for environmental preservation because PLGA is a biodegradable and biocompatible polymer (Figure 7-5).[32]

In spite of their unique tunable porosity, biocompatibility, optical transparency, excellent thermal stability, chemical inertness, and negligible swelling in both aqueous and nonaqueous solutions, the film-forming ability, thickness, and brittleness of these NMOs require further research for the construction of suitable transducer surface for NA detection.[35] Efforts need to be taken to seek a new method of immobilization which could overcome the disadvantages of biomolecule immobilization onto these matrices.

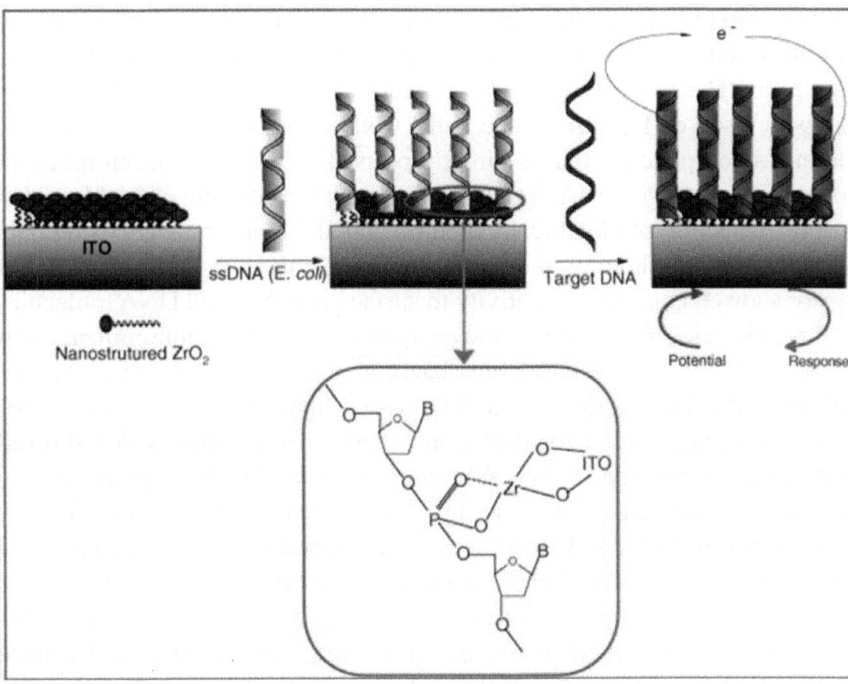

FIGURE 7.5 Scheme showing the immobilization of pDNA onto sol-gel derived ZrO2 deposited on ITO electrode.[68]

7.4.5 GENOSENSORS BASED ON QDS

Semiconductor QDs have currently been used as the basis for signal development in a variety of biomedical applications.[77] Some of these applications include amplifiers, biosensors, high resolution cellular imaging, tumor targeting, and diagnostics.[78] The use of QDs as transducers for the development of biosensors and bioprobes has attracted considerable interest as they provide unique optical properties as donors in fluorescence resonance energy transfer to be well suited for optical multiplexing.[77] Majority of QD based biosensing technologies reported in the literature operate in bulk solution environments, and to reach a steady-state signal the selective binding events at the surface of QDs are often associated with relatively long periods.[77] QDs not only provide high surface-area-to-volume ratio but also they have a tunable band gap making them more favorable to be used in DNA sensors.[79] Recently, a number of research article reported on the application of semiconductor QDs for genosensor application. In this context, EC assays based on QD nanocrystals as tracers were used. These QDs exhibit sharp and well resolved EC signals due to the well defined oxidation potentials of the metal components, that is directly proportional to the concentration of corresponding target DNA.[80] GaN nanowires have been developed for label free EC detection of target DNA (anthrax lethal factor sequence) using dual routes—EIS and photoluminescence measurements.[81] This fabricated DNA sensor showed enhanced sensitivity to surface immobilized DNA molecules as it provided high surface binding energies for enhanced immobilization of probe DNA and surface enhanced charge transfer capability to the analyte. Self-assembled CdSe QDs onto ITO coated glass substrate was used for genosensing application for detection of cancer in humans which showed that the QDs provide a favorable conformation for DNA probe immobilization.[82] The same author group, [82] has prepared Langmuir–Blodgett monolayers of tri-n-octylphosphinevoxide-capped cadmium selenide QDs (QCdSe) onto ITO coated glass substrate. The behavior of the monolayer has been studied at the air–water interface under various subphase conditions. This nanopatterned QCdSe based platform has been used to fabricate an EC DNA biosensor for detection of chronic myelogenous leukemia by covalently immobilizing the thiol terminated oligonucleotide probe sequence via a displacement reaction (Figure 7-6). The results of EC response studies reveal that this biosensor can detect target DNA in the range of 10^{-6} M to 10^{-14} M within 120 s.[79]

A number of different NA sensing modalities are expected to benefit from using QD based detection methods, the most common of which still

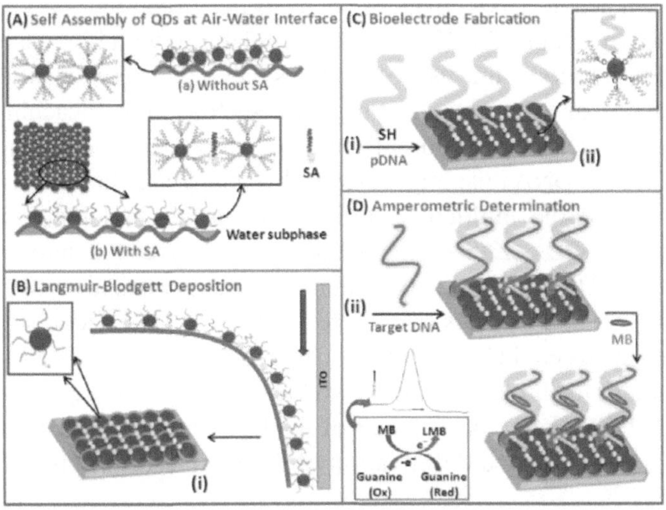

FIGURE 7-6 Steps involved in fabrication of pDNA/QCdSe-SA-LB/ITO bioelectrode.[79]

remains to be fluorescence.[83,84] QDs of varied colors have been tagged to different molecular and cellular target DNA molecules and their concentrations are ascertained by the amount of light transmitted in a particular color. Gerion et al. employed QD conjugated DNA oligonucleotides as hybridization targets for the identification of single nucleotide polymorphism and single base deletion of tumor suppressor gene *P53* on a complementary DNA microarray.[85] The hybridization and detection studies were performed at room temperature and signal-to-noise ratio was monitored to be >10. On the same platform, microarray-based multiple allele detection was studied, using hybridization of multicolor nanocrystals conjugated to two sequences specific for the hepatitis B and hepatitis C virus. Similarly, Liang et al. employed streptavidin coated QD probes conjugated to capture biotinylated microRNA targets derived from rice.[86] With the microarray assays the sensitivity was calculated down to sub-femtomolar concentrations, which was several orders of magnitude better than those reported by dye-based methods. In addition, the use of QDs amplified the signals and thus precluded the use of any external agent. In order to extend the spectral windows, Shepard et al. combined QDs with standard cyanine dyes. This was, however, hard to achieve with conventional fluorophores as multiple excitation lasers coupled to multiple spectral detection windows separated by appropriate filters were the pre-requisites.[87]

Single-molecule DNA imaging has also been reported via QDs where the two extreme ends of a DNA primer were labeled with biotin and digoxigenin.[88] Furthermore, the use of QDs as DNA labeling agent exceeded the limitations of using intercalating dyes, such as photobleaching, photo-induced cleavage, modification of the DNA properties, and so on, conventionally used for DNA imaging. A report by Mingyong Han and co-workers discusses about the multicolor optical coding for biological assays by different-sized QDs (ZnS–CdSe nanocrystals) embedded into polymeric microbeads.[89] DNA hybridization studies demonstrated that by using 10 intensity levels and 6 colors to code for one million NA sequences, coding and target signals could be simultaneously read at the single-bead level. The mechanism for the binding of the NA to the QDs was investigated indicating that NPs bind to the helix structure of the DNA in a nonintercalative way, resulting in the observed deactivation of the luminescence emission of the QDs.

Developing multiplex detection is an ever-expanding research tool for QDs in optical arrays. However, a majority of applications occur in bulk solutions which limit the development of reusable sensors, resulting in the hindrance of routine analytical applications due to high toxicity of QDs.[90] In the near future the major challenges tackled will be the development of new synthetic and immobilization techniques that will enhance the stability, sensitivity, and binding specificity of DNA with the QD nanoassemblies.

7.4.6 GENOSENSOR BASED ON HIERARCHICAL HYBRID NANOSTRUCTURES

To understand the biomolecule–nanomaterial interactions for genosensor application there has been an increasing interest in interfacing biological molecules with hybrid materials.[91] In the design and the realization of bio/nonbiointerfaces with specific properties, such as chemical stability, wettability, and biomolecules' immobilization ability are key features in the miniaturization and optimization processes of genosensors.[92] Hierarchical structures based on organic–inorganic hybrids materials have enthralled researchers as they coalesce the potential distinct properties of organic and inorganic components within a single molecular framework.[93] Kelley and co-workers reported that hierarchical Pd-nanostructured microelectrodes (NMEs) tethered by peptide NAs allow rapid, ultrasensitive, and label-free detection of complementary DNA and micro RNA (miRNA) by using an electrocatalytic reporter system. The authors realized precise detection of fewer than 100 target DNA molecules by using these hierarchical Pd NMEs.[94] The same NMEs was also used for specific and sensitive detection of miRNA and it was found that the fractal Pd NMEs were able to distinguish not only two closely related sequences, miR-26a and miR26-b, but also a precursor miR-21 sequence (full length and double stranded) and a mature miR-21 sequence (shorter and single stranded).[95] Pandey et al. developed a hierarchical cystine (Cys) based impedimetric sensor that exhibited linear response to a clinical DNA sample of *E. coli* in the concentration range from 10^{-6} M to 10^{-14} M with a response time of 30 min.[96] The synthesis of these Cys based hierarchical materials requires mild reaction conditions (neutral pH, room temperature, and aqueous solution) (Figure 7-7). The other advantage of using biomolecules for materials' synthesis is the elegant control on the shape, size, chemistry, and crystal structure which plays a crucial role in determining the properties of the synthesized material, making them excellent material for genosensor application.[97] Although significant progress has been achieved in this field, the study of hierarchical hybrid nanostructures is at an early stage and several fundamental and practical problems remain unsettled. Still there is a need to understand how the material properties affect the structure, activity, and stability of conjugated bioanalytes and identify optimal conditions to preserve functionality following biomolecule immobilization.

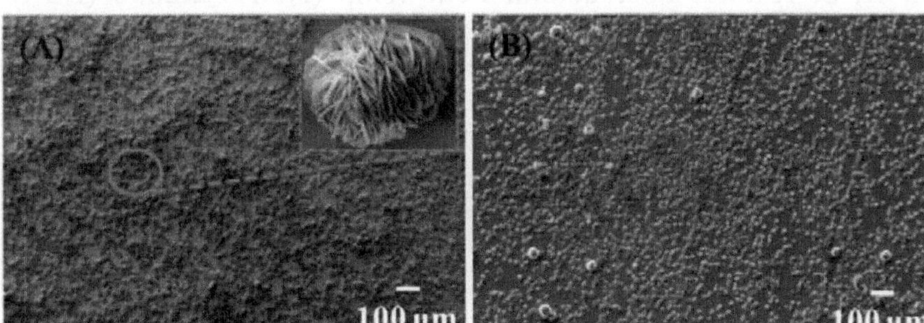

FIGURE 7-7 Scanning electron microscopic image of (A) cystine flower self assembled on Au electrode, and (B) pDNA/CysFl/Au bioelectrode.[96]

7.5 CONCLUSIONS

With the development of microfabrication technology toward chips and arrays, there is an increased improvement in DNA sensing strategies or gene detection. It is hoped that continued combined efforts in microelectronics, surface/interface chemistry, molecular biology, and analytical chemistry will lead to the establishment of genosensor technology.[98] In this regard, the contribution of nanoscale materials in genosensor development is a timely area of activity, which may be further be used as sensing platform, in addition to amplification stages. These nanostructured materials based EC DNA devices have a number of key features, including high sensitivity, exquisite selectivity, fast response time, rapid recovery (reversibility), and potential for miniaturization on a large scale, which make them superior from other sensor technologies available in the market. In the near future, development of advanced techniques aiding in molecule-level detection in simple nanosensor devices is expected. To fully grasp the potential applicability of nanostructures in EC sensors, several issues related to their fabrication methods need to be addressed. With further increase in the sensitivity, flexibility, and miniaturization capabilities, these genosensors have the potential to become the next generation of field deployable analytical instruments.

ACKNOWLEDGMENTS

CMP acknowledges Prof. BD Malhotra (DTU,Delhi) for interesting discussions and University Grant Commission, India for providing Dr. D. S. Kothari Postdoctoral Fellowship (No.F.4-2/2006 (BSR)/CH/14-15/0164).

KEYWORDS

- **Genosensors**
- **Quantum dots**
- **DNA**
- **Conducting polymers**
- **Self-assembled monolayers**
- **Nanoparticles**

REFERENCES

1. Pividori, M. I.; Merkoçi, A.; Alegret, S. Electrochemical genosensor design: immobilisation of oligonucleotides onto transducer surfaces and detection methods. *Biosens. Bioelectron.* **2000,** *15*, 291-303.
2. Hashimoto, K.; Ito, K.; Ishimori, Y. Novel DNA sensor for electrochemical gene detection. *Anal. Chim. Acta* **1994,** *286*, 219-224.
3. Tuppen, H. A. L.; Blakely, E. L.; Turnbull, D. M.; Taylor, R. W. Mitochondrial DNA mutations and human disease. *Biochim. Biophys. Acta* **2010,** *1797*, 113-128.
4. Drummond, T. G.; Hill, M. G.; Barton, J. K. Electrochemical DNA sensors. *Nat. Biotech.* **2003,** *21*, 1192-1199.
5. Martín-Fernández, B.; Miranda-Ordieres, A. J.; Lobo-Castañón, M. J.; Frutos-Cabanillas, G.; de-los-Santos-Álvarez, N.; López-Ruiz, B. Strongly structured DNA sequences as targets for genosensing: Sensing phase design and coupling to PCR amplification for a highly specific 33-mer gliadin DNA fragment. *Biosens. Bioelectron.* **2014,** *60*, 244-251.
6. Beattie, K. L.; Beattie, W. G.; Meng, L.; Turner, S. L.; Coral-Vazquez, R.; Smith, D. D.; McIntyre, P. M.; Dao, D. D. Advances in genosensor research. *Clin. Chem.* **1995,** *41*, 700-6.
7. Wei, F.; Lillehoj, P. B.; Ho, C.-M. Nanotechnology-enhanced electrochemical detection of nucleic acids. *Pediatr. Res.* **2010,** *67*, 458-468.
8. Marrazza, G.; Chiti, G.; Mascini, M.; Anichini, M. Detection of human apolipoprotein E genotypes by dna electrochemical biosensor coupled with PCR. *Clin. Chem.* **2000,** *46*, 31-37.
9. Siddiquee, S.; Yusof, N. A.; Salleh, A. B.; Bakar, F. A.; Heng, L. Y. Electrochemical DNA biosensor for the detection of specific gene related to Trichoderma harzianum species. *Bioelectrochemistry* **2010,** *79*, 31-36.
10. Nimse, S.; Song, K.; Sonawane, M.; Sayyed, D.; Kim, T. Immobilization techniques for microarray: challenges and applications. *Sensors* **2014,** *14*, 22208-22229.
11. Privett, B. J.; Shin, J. H.; Schoenfisch, M. H. Electrochemical Sensors. *Anal. Chem.* **2008,** *80*, 4499-4517.
12. Wang, J. Electrochemical biosensors: Towards point-of-care cancer diagnostics. *Biosens. Bioelectron.* **2006,** *21*, 1887-1892.
13. Bonanni, A.; del Valle, M. Use of nanomaterials for impedimetric DNA sensors: A review. *Anal. Chim. Acta* **2010,** *678*, 7-17.
14. Ivnitski, D.; Abdel-Hamid, I.; Atanasov, P.; Wilkins, E. Biosensors for detection of pathogenic bacteria. *Biosens. Bioelectron.* **1999,** *14*, 599-624.
15. Abu-Salah, K.; Alrokyan, S. A.; Khan, M. N.; Ansari, A. A. Nanomaterials as analytical tools for genosensors. *Sensors* **2010,** *10*, 963-993.
16. Mikołaj, D.; Zbigniew, S. Nanoparticles and nanostructured materials used in modification of electrode surfaces. In *Functional Nanoparticles for Bioanalysis, Nanomedicine, and Bioelectronic Devices Volume 1*, American Chemical Society: 2012; Vol. 1112, pp 313-325.
17. Wang, J. From DNA biosensors to gene chips. *Nucleic Acids Res.* **2000,** *28*, 3011-3016.
18. Paleček, E.; Bartošík, M. Electrochemistry of nucleic acids. *Chem. Rev.* **2012,** *112*, 3427-3481.

19. Tichoniuk, M.; Ligaj, M.; Filipiak, M. Application of DNA hybridization biosensor as a screening method for the detection of genetically modified food components. *Sensors,* **2008,** *8*, 2118-2135.
20. Moore, R. L.; McCarthy, B. J. Comparative study of ribosomal ribonucleic acid cistrons in enterobacteria and myxobacteria. *J. Bacteriol.* **1967,** *94*, 1066-1074.
21. Amann, R.; Ludwig, W. Ribosomal RNA-targeted nucleic acid probes for studies in microbial ecology. *FEMS Microbiol. Rev.* **2000,** *24*, 555-565.
22. Millan, K. M.; Mikkelsen, S. R. Sequence-selective biosensor for DNA based on electroactive hybridization indicators. *Anal. Chem.* **1993,** *65*, 2317-2323.
23. Goda, T.; Singi, A. B.; Maeda, Y.; Matsumoto, A.; Torimura, M.; Aoki, H.; Miyahara, Y. Label-free potentiometry for detecting dna hybridization using peptide nucleic acid and dna probes. *Sensors,* **2013,** *13*, 2267-2278.
24. Datta, S.; Christena, L. R.; Rajaram, Y. R. S. Enzyme immobilization: an overview on techniques and support materials. *3 Biotech* **2013,** *3*, 1-9.
25. Shendure, J.; Ji, H. Next-generation DNA sequencing. *Nat. Biotech.* **2008,** *26*, 1135-1145.
26. Hu, K.; Lan, D.; Li, X.; Zhang, S. Electrochemical DNA biosensor based on nanoporous gold electrode and multifunctional encoded DNA−Au bio bar codes. *Anal. Chem.* **2008,** *80*, 9124-9130.
27. Grieshaber, D.; MacKenzie, R.; Vörös, J.; Reimhult, E. Electrochemical biosensors - sensor principles and architectures. *Sensors,* **2008,** *8*, 1400-1458.
28. Mikkelsen, S. R. Electrochecmical biosensors for DNA sequence detection. *Electroanalysis* **1996,** *8*, 15-19.
29. Carrara, S.; Ghoreishizadeh, S.; Olivo, J.; Taurino, I.; Baj-Rossi, C.; Cavallini, A.; Op de Beeck, M.; Dehollain, C.; Burleson, W.; Moussy, F. G.; Guiseppi-Elie, A.; De Micheli, G. Integrated biochip platforms for advanced healthcare. *Sensors,* **2012,** *12*, 11013-11060.
30. Kim, D.; Herr, A. E. Protein immobilization techniques for microfluidic assays. *Biomicrofluidics* **2013,** *7*, 041501.
31. Kastantin, M.; Schwartz, D. K. DNA hairpin stabilization on a hydrophobic surface. *Small* **2013,** *9*, 933-941.
32. Pandey, C. M.; Sharma, A.; Sumana, G.; Tiwari, I.; Malhotra, B. D. Cationic poly(lactic-co-glycolic acid) iron oxide microspheres for nucleic acid detection. *Nanoscale* **2013,** *5*, 3800-3807.
33. Heimel, G.; Romaner, L.; Zojer, E.; Bredas, J.-L. The interface energetics of self-assembled monolayers on metals. *Acc. Chem. Res.* **2008,** *41*, 721-729.
34. Pandey, C. M.; Singh, R.; Sumana, G.; Pandey, M. K.; Malhotra, B. D. Electrochemical genosensor based on modified octadecanethiol self-assembled monolayer for *Escherichia coli* detection. *Sens. Actuators, B* **2011,** *151*, 333-340.
35. Solanki, P. R.; Kaushik, A.; Agrawal, V. V.; Malhotra, B. D. Nanostructured metal oxide-based biosensors. *NPG Asia Mater.* **2011,** *3*, 17-24.
36. Gerard, M.; Chaubey, A.; Malhotra, B. D. Application of conducting polymers to biosensors. *Biosens. Bioelectron.* **2002,** *17*, 345-359.
37. Pan, L.; Qiu, H.; Dou, C.; Li, Y.; Pu, L.; Xu, J.; Shi, Y. Conducting polymer nanostructures: template synthesis and applications in energy storage. *Int. J. Mol. Sci.* **2010,** *11*, 2636-2657.
38. Xia, L.; Wei, Z.; Wan, M. Conducting polymer nanostructures and their application in biosensors. *J. Colloid Interface Sci.* **2010,** *341*, 1-11.

39. Wang, J.; Jiang, M.; Fortes, A.; Mukherjee, B. New label-free DNA recognition based on doping nucleic-acid probes within conducting polymer films. *Anal. Chim. Acta* **1999,** *402*, 7-12.
40. Rodriguez, M. I.; Alocilja, E. C. Embedded DNA-polypyrrole biosensor for rapid detection of *Escherichia coli. Sensors Journal, IEEE* **2005,** *5*, 733-736.
41. Arora, K.; Prabhakar, N.; Chand, S.; Malhotra, B. D. *Escherichia coli* Genosensor Based on Polyaniline. *Anal. Chem.* **2007,** *79*, 6152-6158.
42. Singh, R.; Prasad, R.; Sumana, G.; Arora, K.; Sood, S.; Gupta, R. K.; Malhotra, B. D. STD sensor based on nucleic acid functionalized nanostructured polyaniline. *Biosens. Bioelectron.* **2009,** *24*, 2232-2238.
43. Ramanathan, K.; Bangar, M. A.; Yun, M.; Chen, W.; Myung, N. V.; Mulchandani, A. bioaffinity sensing using biologically functionalized conducting-polymer nanowire. *J. Am. Chem. Soc.* **2004,** *127*, 496-497.
44. Chang, H.; Yuan, Y.; Shi, N.; Guan, Y. Electrochemical DNA biosensor based on conducting polyaniline nanotube array. *Anal. Chem.* **2007,** *79*, 5111-5115.
45. Ebbesen, T. W.; Ajayan, P. M. Large-scale synthesis of carbon nanotubes. *Nature* **1992,** *358*, 220-222.
46. Kushwaha, S. K. S.; Ghoshal, S.; Rai, A. K.; Singh, S. Carbon nanotubes as a novel drug delivery system for anticancer therapy: a review. *Braz. J. Med. Biol. Res.* **2013,** *49*, 629-643.
47. Hembram, K. P. S. S.; Rao, G. M. Studies on CNTs/DNA composite. *Mater. Sci. Eng. C* **2009,** *29*, 1093-1097.
48. Ye, Y.; Ju, H. Rapid detection of ssDNA and RNA using multi-walled carbon nanotubes modified screen-printed carbon electrode. *Biosens. Bioelectron.* **2005,** *21*, 735-741.
49. Karimi, H.; Yusof, R.; Rahmani, R.; Ahmadi, M. T. Optimization of DNA sensor model based nanostructured graphene using particle swarm optimization technique. *J. Nanomater.* **2013.**
50. Liu, B.; Sun, Z.; Zhang, X.; Liu, J. Mechanisms of DNA sensing on graphene oxide. *Anal. Chem.* **2013,** *85*, 7987-7993.
51. Pumera, M. Graphene in biosensing. *Mater. Today* **2011,** *14*, 308-315.
52. Feng, L.; Wu, L.; Qu, X. New Horizons for diagnostics and therapeutic applications of graphene and graphene oxide. *Adv. Mater.* **2013,** *25*, 168-186.
53. Nath, N.; Chilkoti, A. Noble metal nanoparticle biosensors. In *Radiative Decay Engineering*, Geddes, C.; Lakowicz, J., Eds. Springer US: 2005; Vol. 8, pp 353-380.
54. Castañeda, M. T.; Merkoçi, A.; Pumera, M.; Alegret, S. Electrochemical genosensors for biomedical applications based on gold nanoparticles. *Biosens. Bioelectron.* **2007,** *22*, 1961-1967.
55. Peng, H.-I.; Miller, B. L. Recent advancements in optical DNA biosensors: Exploiting the plasmonic effects of metal nanoparticles. *Analyst* **2011,** *136*, 436-447.
56. Chang, Z.; Fan, H.; Zhao, K.; Chen, M.; He, P.; Fang, Y. Electrochemical DNA biosensors based on palladium nanoparticles combined with carbon nanotubes. *Electroanalysis* **2008,** *20*, 131-136.
57. Conde, J.; Dias, J. T.; Grazú, V.; Moros, M.; Baptista, P. V.; de la Fuente, J. M. Revisiting 30 years of biofunctionalization and surface chemistry of inorganic nanoparticles for nanomedicine. *F.Chem.* **2014,** *2*, 48.
58. Glynou, K.; Ioannou, P. C.; Christopoulos, T. K.; Syriopoulou, V. Oligonucleotide-functionalized gold nanoparticles as probes in a dry-reagent strip biosensor for dna analysis by hybridization. *Anal. Chem.* **2003,** *75*, 4155-4160.

59. Yang, J.; Yang, T.; Feng, Y.; Jiao, K. A DNA electrochemical sensor based on nanogold-modified poly-2,6-pyridinedicarboxylic acid film and detection of PAT gene fragment. *Anal. Biochem.* **2007,** *365*, 24-30.
60. Ding, C.; Zhang, Q.; Lin, J.-M.; Zhang, S.-s. Electrochemical detection of DNA hybridization based on bio-bar code method. *Biosens. Bioelectron.* **2009,** *24*, 3140-3143.
61. Ting, B. P.; Zhang, J.; Gao, Z.; Ying, J. Y. A DNA biosensor based on the detection of doxorubicin-conjugated Ag nanoparticle labels using solid-state voltammetry. *Biosens. Bioelectron.* **2009,** *25*, 282-287.
62. Kong, J. M.; Zhang, H.; Chen, X. T.; Balasubramanian, N.; Kwong, D. L. Ultrasensitive electrical detection of nucleic acids by hematin catalysed silver nanoparticle formation in sub-microgapped biosensors. *Biosens. Bioelectron.* **2008,** *24*, 787-791.
63. Doria, G.; Conde, J.; Veigas, B.; Giestas, L.; Almeida, C.; Assunção, M.; Rosa, J.; Baptista, P. V. Noble metal nanoparticles for biosensing applications. *Sensors* **2012,** *12*, 1657-1687.
64. Ngo, H. T.; Wang, H.-N.; Fales, A. M.; Vo-Dinh, T. Label-Free DNA biosensor based on sers molecular sentinel on nanowave chip. *Anal. Chem.* **2013,** *85*, 6378-6383.
65. Malhotra, B. D.; Maumita, D.; Pratima, R. S. Opportunities in nano-structured metal oxides based biosensors. *J. Phys.* **2012,** *358*, 012007.
66. Ansari, A. A.; Singh, R.; Sumana, G.; Malhotra, B. D. Sol-gel derived nano-structured zinc oxide film for sexually transmitted disease sensor. *Analyst* **2009,** *134*, 997-1002.
67. Das, M.; Sumana, G.; Nagarajan, R.; Malhotra, B. D. Zirconia based nucleic acid sensor for Mycobacterium tuberculosis detection. *Appl. Phys. Lett.* **2010,** *96*, 133703.
68. Solanki, P. R.; Kaushik, A.; Chavhan, P. M.; Maheshwari, S. N.; Malhotra, B. D. Nanostructured zirconium oxide based genosensor for *Escherichia coli* detection. *Electrochem. Commun.* **2009,** *11*, 2272-2277.
69. Feng, K.-J.; Yang, Y.-H.; Wang, Z.-J.; Jiang, J.-H.; Shen, G.-L.; Yu, R.-Q. A nano-porous CeO2/Chitosan composite film as the immobilization matrix for colorectal cancer DNA sequence-selective electrochemical biosensor. *Talanta* **2006,** *70*, 561-565.
70. Zhu, N.; Zhang, A.; Wang, Q.; He, P.; Fang, Y. Electrochemical detection of DNA hybridization using methylene blue and electro-deposited zirconia thin films on gold electrodes. *Anal. Chim. Acta* **2004,** *510*, 163-168.
71. Yang, Y.; Wang, Z.; Yang, M.; Li, J.; Zheng, F.; Shen, G.; Yu, R. Electrical detection of deoxyribonucleic acid hybridization based on carbon-nanotubes/nano zirconium dioxide/chitosan-modified electrodes. *Anal. Chim. Acta* **2007,** *584*, 268-274.
72. Liu, Z.-M.; Liu, Y.-L.; Shen, G.-L.; Yu, R.-Q. Nano-ZnO/Chitosan composite film modified electrode for voltammetric detection of DNA hybridization. *Anal. Lett.* **2008,** *41*, 1083-1095.
73. Cheng, G.-F.; Huang, C.-H.; Zhao, J.; Tan, X.-L.; He, P.-G.; Fang, Y.-Z. A novel electrochemical biosensor for deoxyribonucleic acid detection based on magnetite nanoparticles. *Chinese Journal of Anal. Chem.* **2009,** *37*, 169-173.
74. Zhang, W.; Yang, T.; Li, X.; Wang, D.; Jiao, K. Conductive architecture of Fe_2O_3 microspheres/self-doped polyaniline nanofibers on carbon ionic liquid electrode for impedance sensing of DNA hybridization. *Biosens. Bioelectron.* **2009,** *25*, 428-434.
75. Shrestha, S.; Mills, C. E.; Lewington, J.; Tsang, S. C. *J. Phys. Chem. B* **2006,** *110*, 25633-25637.
76. Shrestha, S.; Yeung, C. M. Y.; Mills, C. E.; Lewington, J.; Tsang, S. C. Modified rare earth semiconductor oxide as a new nucleotide probe. *Angew. Chem. Int. Ed.* **2007,** *46*, 3855-3859.

77. Vannoy, C. H.; Tavares, A. J.; Noor, M. O.; Uddayasankar, U.; Krull, U. J. Biosensing with quantum dots: a microfluidic approach. *Sensors,* **2011,** *11*, 9732-9763.
78. Sapsford, K.; Pons, T.; Medintz, I.; Mattoussi, H. Biosensing with luminescent semiconductor quantum dots. *Sensors* **2006,** *6*, 925-953.
79. Sharma, A.; Pandey, C. M.; Matharu, Z.; Soni, U.; Sapra, S.; Sumana, G.; Pandey, M. K.; Chatterjee, T.; Malhotra, B. D. Nanopatterned cadmium selenide langmuir–blodgett platform for leukemia detection. *Anal. Chem.* **2012,** *84*, 3082-3089.
80. Huang, H.; Li, J.; Tan, Y.; Zhou, J.; Zhu, J.-J. Quantum dot-based DNA hybridization by electrochemiluminescence and anodic stripping voltammetry. *Analyst* **2010,** *135*, 1773-1778.
81. Chen, R.-S.; Chen, H.-Y.; Lu, C.-Y.; Chen, K.-H.; Chen, C.-P.; Chen, L.-C.; Yang, Y.-J. Ultrahigh photocurrent gain in m-axial GaN nanowires. *Appl. Phys. Lett.* **2007,** *91*, 223106-223106-3.
82. Sharma, A.; Sumana, G.; Sapra, S.; Malhotra, B. D. Quantum dots self assembly based interface for blood cancer detection. *Langmuir* **2013,** *29*, 8753-8762.
83. Martín-Palma, R.; Manso, M.; Torres-Costa, V. Optical biosensors based on semiconductor nanostructures. *Sensors* **2009,** *9*, 5149-5172.
84. Ma, Q.; Su, X. Recent advances and applications in QDs-based sensors. *Analyst* **2011,** *136*, 4883-4893.
85. Gerion, D.; Chen, F.; Kannan, B.; Fu, A.; Parak, W. J.; Chen, D. J.; Majumdar, A.; Alivisatos, A. P. Room-temperature single-nucleotide polymorphism and multiallele DNA detection using fluorescent nanocrystals and microarrays. *Anal. Chem.* **2003,** *75*, 4766-4772.
86. Liang, R.-Q.; Li, W.; Li, Y.; Tan, C.-y.; Li, J.-X.; Jin, Y.-X.; Ruan, K.-C. An oligonucleotide microarray for microRNA expression analysis based on labeling RNA with quantum dot and nanogold probe. *Nucleic Acids Res.* **2005,** *33*, e17.
87. Shepard, J. R. E. Polychromatic Microarrays: Simultaneous multicolor array hybridization of eight samples. *Anal. Chem.* **2006,** *78*, 2478-2486.
88. Crut, A.; Géron-Landre, B.; Bonnet, I.; Bonneau, S.; Desbiolles, P.; Escudé, C. Detection of single DNA molecules by multicolor quantum-dot end-labeling. *Nucleic Acids Res.* **2005,** *33*, e98.
89. Han, M.; Gao, X.; Su, J. Z.; Nie, S. Quantum-dot-tagged microbeads for multiplexed optical coding of biomolecules. *Nat. Biotech.* **2001,** *19*, 631-635.
90. Frasco, M. F.; Chaniotakis, N. Semiconductor quantum dots in chemical sensors and biosensors. *Sensors (Basel, Switzerland)* **2009,** *9*, 7266-7286.
91. Martinez-Manez, R.; Sancenon, F.; Biyikal, M.; Hecht, M.; Rurack, K. Mimicking tricks from nature with sensory organic-inorganic hybrid materials. *J. Mater. Chem.* **2011,** *21*, 12588-12604.
92. Stefano, L. D.; Rea, I.; Rendina, I.; Giocondo, M.; Houmadi, S.; Longobardi, S.; Giardina, P. *Organic-inorganic interfaces for a new generation of hybrid biosensors.* 2011.
93. Holzinger, M.; Le Goff, A.; Cosnier, S. Nanomaterials for biosensing applications: a review. Front. Chem. 2014, 2, 63.
94. Soleymani, L.; Fang, Z.; Sargent, E. H.; Kelley, S. O. Programming the detection limits of biosensors through controlled nanostructuring. *Nat. Nano.* **2009,** *4*, 844-848.

95. Yang, H.; Hui, A.; Pampalakis, G.; Soleymani, L.; Liu, F.-F.; Sargent, E. H.; Kelley, S. O. Direct, Electronic microRNA detection for the rapid determination of differential expression profiles. *Angew. Chem. Int. Ed.* **2009,** *48*, 8461-8464.
96. Pandey, C. M.; Tiwari, I.; Sumana, G. Hierarchical cystine flower based electrochemical genosensor for detection of *Escherichia coli* O157:H7. *RSC Advances* **2014,** *4*, 31047-31055.
97. Pandey, C. M.; Sumana, G.; Malhotra, B. D. Microstructured cystine dendrites-based impedimetric sensor for nucleic acid detection. *Biomacromolecules* **2011,** *12*, 2925-2932.
98. Erdem, A. Nanomaterial-based electrochemical DNA sensing strategies. *Talanta* **2007,** *74*, 318-325.

CHAPTER 8

NANOMATERIAL-BASED IMMUNOSENSORS FOR CLINICAL DIAGNOSTICS

MANOJ KUMAR PATEL[1] and PRATIMA SOLANKI[2*]

[1]*Department of Chemistry, College of Arts and Sciences, Oklahoma State University, Stillwater, Oklahoma, USA*

[2]*Special Centre for Nanosciences, Jawaharlal Nehru University, New Delhi, India*

**E-mail: partima@mail.jnu.ac.in*

CONTENTS

Nanotechnology has played a significant role in the development of biosensing devices over the past few decade. The development, testing, optimization, and validation of new biosensors have become highly interdisciplinary efforts involving experts in chemistry, biology, physics, engineering, and medicine. Various types of nanomaterials are being utilized for the fabrication of biosensing electrode to monitor their respective response studies. Nanomaterials based electrochemical immunosensors amplify the sensitivity by facilitating greater loading of the larger sensing surface with biorecognition molecules as well as improving the electrochemical properties of the transducer. Nanomaterials have similar dimension and provide high surface area to loading of biomolecules like proteins , enzymes, antigens (Ags), antibodies (Abs) and DNA. In this chapter, we have discussed about recent progress in the fields of nanotechnology, nanomaterial-based immunosensors, and their applications in the field of clinical diagnostics.

8.1 INTRODUCTION

Nanostructured materials (NMs) have recently been considered as suitable materials for the development of various biosensors such as enzymatic sensors, genosensors, immunosensors, and so on.[1,2] Their unique surface properties such as stability, electrical conductivity, and various favorable structural and catalytic properties have enabled enhanced interaction of biomolecules and hence provided improved biosensing properties.[3] These NMs could be synthesized in different size and shape such as rods, wires, tubes, pores, particles, and arrays of these shapes.[4,5] Various biocompatible nanomaterials (e.g., nobel metal or metal–oxides, carbon nanoparticles, etc.) with unique physical and chemical properties have been successfully applied in immunosensing interface fabrication to achieve improved immobilization of Abs or Ags as well as to increase the transduction efficiency.[6,7] The enormous signal enhancement associated with nanoparticles as signal tags and forming nanoparticles–biomolecules assemblies also provides a great probability for ultrasensitive and selective immunosensors.[8]

In recent decades, biosensors/biochips have been envisaged to compensate and complement conventional diagnostic methods due to their easy operation and transport, require no expensive reagents, and provide results in a few minutes.[9,10] Among them, immunosensors based on electrochemical detection have the advantage of being extremely sensitive, quick, low-cost, and highly amenable to microfabrication and it is also easy to measure the changes in electrical/electrochemical properties resulting from the Ag–Ab reaction on electrode's surface.[11] The development of immunoassay technology is a success story especially for the clinical laboratory and still continues to be a vibrant area of research.[12] In the clinical laboratory, a future substitution of immunoassays by immunosensors simply depends on the superiority and versatility of the new technology.[12,13] For application in point-of-care diagnostics or when they are temporarily implanted into the patient additionally depends on the reliable and accurate analysis of the target analyte, without any matrix interferences.[14] Immunosensors generally have high sensitivity and specificity, cost-effective, require small sample, and exhibit wide applications in biomedical and environmental fields.[15,16]

Nanotechnology is a multidisciplinary area which covers various disciplines such as biological sciences, physical sciences, chemical sciences, material science, and engineering.[17,18] It provides excitingly new possibilities for advanced development of new analytical methods and instruments for biomedical applications.[19] The enormous signal enhancement associated with the use of nanomaterials' amplifying labels and with the formation of nanoparticle–Ab–Ag assemblies provides the basis for ultrasensitive immunosensings and immunoassays.[20]

Present chapter is focused on the development and application of nanomaterials based electrochemical immunosensor for clinical diagnostics applications. Here, we introduce the main principles of immunosensor, possible interaction between the Abs–Ag and different transducer based immunosensor.

8.2 PRINCIPLE OF IMMUNOSENSORS

Abs are biological molecules, proteins that exhibit very specific binding capabilities for specific Ags. Abs are made up of hundreds of individual amino acids arranged in a highly ordered sequence to form a complex structure. Abs can be produced against a particular foreign particle, inhaled by an organism, and their identification and fight against them is the principal function of the immune system. The identification and communication take place in the blood and lymph; thereafter, certain proteins are synthesized to identify the invader and to prohibit its harmful effects.

Abs are divided into five classes (immunoglobulins), namely, IgG, IgM, IgE, IgM, and IgD based on their structures and biological functions. However, IgM and IgG are the first and second Abs produced in response to invading non-self substances. Abs are highly specific and bind constantly with their corresponding Ags. Ags are also molecules of proteins and polysaccharides and may originate within the body ("self") or from the external environment ("non-self") and have the ability to bind with the variable fragment Ag-binding (Fab) region of an Ab. The different Abs have a potential to discriminate between specific epitopes present on the surface of the Ag. Haptens are the molecules that are recognized by the Abs but do not elicit an immune response.

Abs has been classified into two major categories such as monoclonal Abs (McAbs) and polyclonal Abs according to their functional domain. McAb is a homogeneous Ab generally produced in the laboratory and consists of a single Ag binding site, produced by a single B cell clone. A McAb can bind to only one epitope on the Ag and Abs does not form lattices with monomeric and homogeneous proteins. However, polyclonal Abs prepared from immunized animals consist of complex mixtures of different Abs produced by many different B cell clones. The polyclonal Abs can form lattices with homogeneous, monomeric protein Ags because the Ab can interact with a different epitope on the Ag.

8.3 INTERACTION BETWEEN AG AND AB

The interaction between the Ag and specific Ab may be understood as analogous to a lock and key phenomena, which is highly specific for geometrical configurations. The Ab is highly specific toward the Ag and thus this unique property of Abs is the key to their usefulness in immunosensors where only the specific Ag fits into the Ab binding site. The interaction of Ag and Abs take place as either of the two ways: competitive and noncompetitive. The competitive interaction takes place between free and bound Ag for a limited amount of labeled Abs or between Ag present in the sample and labeled Ag for a limited amount of Ab. The detection can be achieved directly and indirectly (Figure 8-1a–c).

In a competitive format, unlabeled analyte (usually the Ag) in the test sample is measured by its ability to compete with the labeled Ag in the sample. The response decreases with the increase of analyte concentration, which requests a high background signal toward zero analyte. In contrast, noncompetitive immunoassay formats (usually the sandwich type formats) give the highest level of sensitivity and specificity because of the use of a couple of match Abs (Figure 8-1b). The measurement of the labeled analyte (usually the Ab) is directly proportional to the amount of Ag present in the sample, thus resulting that the detectable signal increases with the increasing target analyte. Therefore, the sandwich type assay is one of the most popular schemes in the immunosensings and immunoassays. Although the Ag–Ab reaction can cause the change of detectable signal to some extent, the change is comparatively little. High affinity Abs and appropriate labels are usually employed for the amplification of detectable signal.

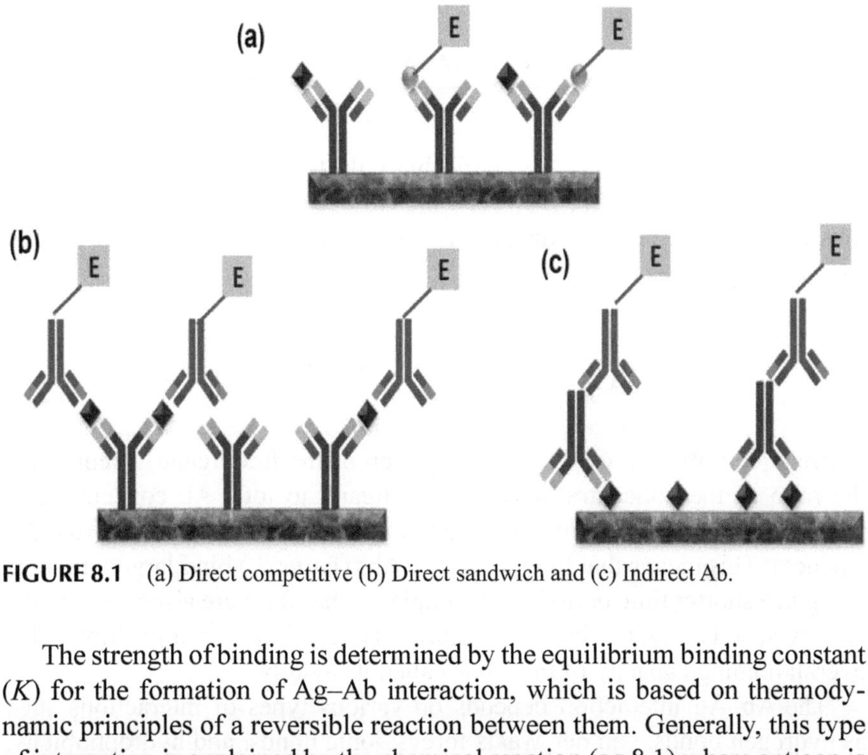

FIGURE 8.1 (a) Direct competitive (b) Direct sandwich and (c) Indirect Ab.

The strength of binding is determined by the equilibrium binding constant (*K*) for the formation of Ag–Ab interaction, which is based on thermodynamic principles of a reversible reaction between them. Generally, this type of interaction is explained by the chemical reaction (eq 8.1) where antigen is the Ag, Ab is the antibody, and Ag–Ab is the Ag–Ab complex either in free ([Ab],[Ag]) or bound ([Ab–Ag]) state. In eqs. 8.2–8.4, *K* is the equilibrium constant:

$$A_g + A_b \Leftrightarrow A_gA_b \tag{8.1}$$

The equilibrium association (a) constant can be represented as

$$K_a = \frac{K_{on}}{K_{off}} = \frac{[A_bA_g]}{[A_b][A_g]} \tag{8.2}$$

Reciprocally, the dissociation (d) constant will be

$$K_d = \frac{K_{off}}{K_{on}} = \frac{[A_b][A_g]}{[A_bA_g]} \tag{8.3}$$

where the ratio of K_a and K_d describes the binding affinity

$$K = \frac{K_a}{K_d} = \frac{[A_bA_g]}{[A_b][A_g]} \tag{8.4}$$

However, these equations are applicable only to a single epitope binding, that is one Ag on one Ab. However, the Ab necessarily has two paratopes, thus complex interactions occurr; therefore, multiple binding equilibrium can be explained as

$$K_a = \frac{K_{on}}{K_{off}} = \frac{[A_bA_g]}{[A_b][A_g]} = r/c(n-r) \tag{8.5}$$

where at equilibrium, *c* is the concentration of the free ligand, *r* represents the ratio of the concentration of bound ligand to total Ab concentration, and *n* is the maximum number of binding sites per Ab molecule (the Ab valence). This means that higher affinity Abs (larger *K*) bind larger amounts of Ag in a shorter time period. The complexes that form are also more stable. However, *K* is affected by temperature, pH, and buffer composition. This overall binding capacity of the Ab is called its avidity.

The Ab–Ag interaction depends on various types of interactions such as hydrogen bonds, van der Waals forces, ionic bonds, and hydrophobicity. The reaction is reversible and favors complex formation under physiological conditions. Binding is very specific and requires the correct three-dimensional structure of an Ag. The amount of complex formed depends on the concentration of Ab and Ag. Both Ab and Ag (if large enough) have multiple sites for binding to occur. Therefore, extensive cross-linking can occur when both are present in the solution. When Ab and Ag reach equivalence, then large immune complexes form which can precipitate out of solution.

The interaction of Abs and Ag when occurred on solid state devices that couple immunochemical reactions to appropriate transducers called immunosensors, affinity ligand based biosensing. Generally, an immunosensor consists of a biorecognition element and a transducer. The biorecognition element Ag or Ab immobilize onto a solid electrode surface and the subsequent binding of event as analytes (Ag or Ab) is transformed into a measurable signal by the transducer. The specificity of the molecular recognition of Ag by Ab to form a stable complex is the basis of both the analytical immunoassay in solution and the immunosensor on solid state interfaces. The merits of immunosensors are obviously related to the selectivity and affinity of the Ab–Ag binding reaction.

8.4 DIFFERENT TRANSDUCERS BASED IMMUNOSENSOR

There are many transducer based immunosensor detection system has been developed such as electrochemical (amperometric potentiometric, impedimetric, conductometric, and capacitative), optical and waveguide. Electrochemical transducer can either be used as direct (nonlabeled) or as indirect (labeled) immunosensor. The direct sensors are able to detect the physical changes during the immune complex formation, whereas the indirect sensors use signal-generating labels which allow more sensitive and versatile detection modes when incorporated into the complex.

8.4.1 ELECTROCHEMICAL IMMUNOSENSORS

Electrochemical immunosensors are a type of chemical sensors which provide high sensitivity, low detection limits, have ability to be miniaturized, low cost of electrode, and mass production of electrochemical transducers. The high specificity of biological recognition processes and label free detection have made these sensors attractive when compared to other approaches based on quartz crystal microbalance (QCM) and surface plasmon resonance (SPR).[21,22] In this section, we have summarized the fabrication of an immunosensor using different transducers, with more emphasis on electrochemical-based immunosensors.

8.4.2 AMPEROMETRIC IMMUNOSENSOR

Amperometric immunosensors are designed to measure a current flow generated by an electrochemical reaction at the electrode's surface at constant voltage. The aim of the test is to detect the presence of Abs in serum via the formation of Ag–Ab complexes. Generally, Ag or Abs are immobilized onto the nanomaterials' surface, conductive electrode (metal or metal–oxide), and so on through ample molecular linkers. This principle is similar to the enzyme-linked immunosorbent assay (ELISA) tests with electrochemical detection, where redox species generated by a redox enzyme (enzymatic label) are converted into a measurable current. Different types of labels including enzymes, radioisotopes, and metal compounds have been explored for monitoring the Ab–Ag interactions.[23,24] The nanogold enwrapped grapheme nanolabels platform has been fabricated for the conjugation of horse radish peroxidase (HRP) anticarcinoembryonic Ag (anti-CEA) for CEA detection.[25] The sandwich type immunoassay has the advantages of high specificity and sensitivity because of increased interaction of Ab with Ag. Figure 8-2 illustrates a proposed scheme for fabrication of sandwich assay based immunosensor.

The sandwich assay strategies can be applied for the development of an amperometric immunosensor. It depends on two stages with the use of two types of Abs: in the first stage the Abs (primary) are immobilized on the electrode or modified with nanomaterials that could bound with the analyte and in the second stage the "sandwich" structure is formed with other types of Abs (secondary) being additionally introduced. These secondary Abs could be capped with the enzyme and fluorescence materials, that is, quantum dots as shown in Figure 8-2. These enzymes labeled secondary Abs provide further confirmation of specific binding between the primary Abs and Ags.

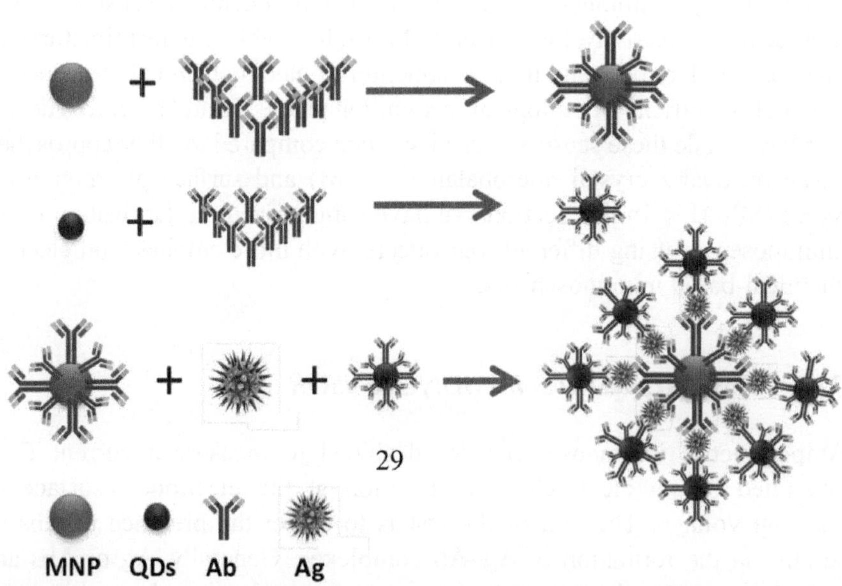

FIGURE 8.2 Shows the different immobilization mechanisms.

Mostly, HRP, glucose oxidase, alkaline phosphatase (AP), and *p*-aminophenyl phosphate (*p*-APP) are adopted as enzyme substrates and have been explored as fast electro-catalytic reactions for sensor application. It has been found that HRP and AP-based immunosensor allows a steady reaction rate for a prolonged time and shows long-term stability even in nonsterile environments. To accomplish higher signal amplification for electrochemical immunosensors, the enzymatic amplification is usually combined with an additional amplification step such as redox cycling of enzymatically produced electroactive species.[26]

8.4.3 POTENTIOMETRIC IMMUNOSENSOR

The Nernst equation provides the fundamental principle of all potentiometric transducers and according to this equation, potential changes are logarithmically proportional to the specific ion activity. Moreover, the change in the steady-state potential of the potentiometric sensor can be provoked by changes in ion, pH, or redox state at the surface. These changes can be caused by electrochemical, chemical, or biological interactions. The potentiometric sensors are the solid state ion selective field effect transistors and pH electrode-based glass ion selective electrodes. Moreover, it can be used for pH, ion, and chemical or gas sensing and can be found in blood gas analyzers such as those marketed by iStat Corp, Diametrics and others. Potentiometric sensors are being utilized in clinical diagnostics to identify high and low molecular compounds, to monitor the effectiveness of drug action and their metabolism, to identify the causes of the intoxication of the body, and to detect drug and doping agents in biological fluids.[27]

In the case of an immunosensor, the detection is based on the change in the potentiometric response (ΔE) before and after Ag–Ab reaction. Either Abs or Ags in aqueous solution has a net electrical charge polarity, which is correlated to the isoelectric points of the species and the ionic composition of the solution. Therefore, the potentiometric responses increase after the Ag–Ab reaction. The potentiometric responses of the immunosensors can be evaluated as followings:

$$\Delta E = E_2 - E_1 \tag{8.6}$$

where $E1$ is the value of the steady-state potentiometric response *vs* saturated calomel reference electrode (SCE) in a phosphate buffer solution (pH 7.0) before the Ag–Ab reactionand $E2$ represents the value of the steady-state potentiometric response (vs. SCE) after the Ag–Ab reaction under the same conditions.

Tang et al. have fabricated a potentiomatric immunosensor based on a layer of plasma-polymerized Nafion film deposited on the platinum electrode surface; thereafter, positively charged tris(2,2-bipyridyl) cobalt(III) ($Co(bpy)_3^{3+}$) and negatively charged gold nanoparticles (AuNPs) were assembled on the PPF-modified Pt electrode by the layer-by-layer technique.[28] Finally, hepatitis B surface antibody was electrostatically adsorbed on the surface of AuNPs. The electrochemical behavior of the $\{Au/Co(bpy)_3^{3+}\}n$ multilayer film modified electrodes was studied.

8.4.4 ELECTROCHEMICAL IMPEDANCE SPECTROSCOPY

It is well known that Newman and Martelet are the pioneers in formulating the concept of impedimetric-based immunosensors and a lot of work has been reported in this area utilizing different types of immunosensor. Electrochemical impedance spectroscopy (EIS) is particularly interesting and is a rapid approach for monitoring interfacial reaction mechanisms between Ab–Ag. It is well proved that the Ag–Ab complex acts as a layer disturbing ion diffusion and changing electrical capacitance which significantly affects the electrode response. Consequently, the EIS technique can be used for label-free detection of affinity interactions, including Ab–Ag, DNA–protein, and protein–protein interactions in real time.

Further, EIS has various advantages, including sensitive experimental measurements that may be made because the signal response is indefinitely steady and can be averaged over a long time frame. The resulting response can be treated theoretically by linearized current–potential characteristics and measurements are over a wide time (10^4–10^{-6} s) or frequency (10^{-4}–10^6 Hz) range. Since this phenomenon typically works close to equilibrium, detailed knowledge of the current versus potential curve over wide ranges of over potential is not required.

The nanomaterials based EIS immunosensors are of high significance due to improved electrical conductivity of sensing interface based on unique conducting properties of the nanoparticles and improved connectivity, easier chemical access to the chemical analyte, and significantly increased electrode's surface area.[29] Recently, Solanki et al. have fabricated a nickel oxide (NiO) and a nanocomposite of reduced graphene oxide (RGO) and anatase titania (ant-TiO_2) based impedematric immunosensor for the detection of *Vibrio cholerae* as shown in Figure 8-3.[30,31]

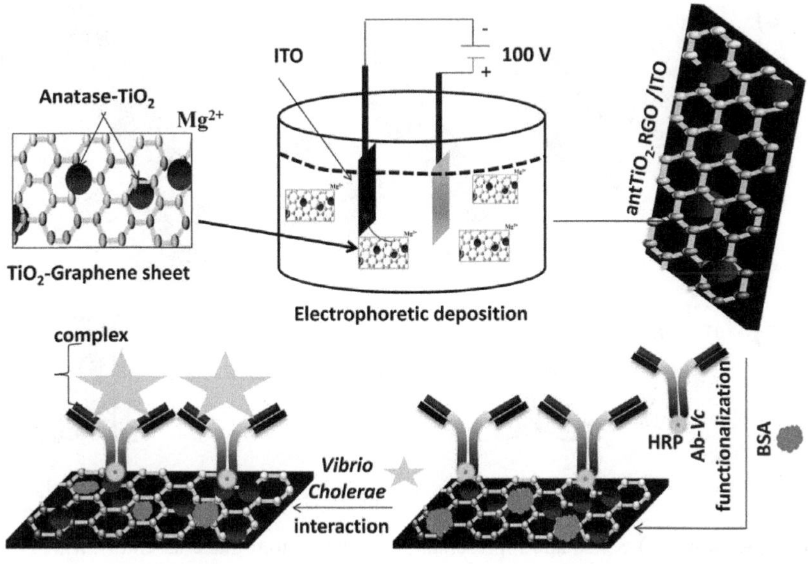

FIGURE 8.3 Fabrication of the immunosensor for *Vibrio cholerae* detection.[31]

8.4.5 CAPACITIVE AND CONDUCTOMETRIC IMMUNOSENSOR

These immunosensor transducers measure the alteration of the electrical conductivity in a solution at constant voltage, caused by biochemical (enzymatic) reactions which specifically generate or consume ions in the electrolyte. These changes can be measured using electrochemical techniques, in which the biorecognition element is available onto nanostructure materials including Au and Pt electrodes. There is some limitation for clinical applications such as high ionic strength of biological matrices makes it difficult to record the relatively small net conductivity changes caused by the biochemical reaction. To circumvent this problem, recently, an ion channel conductance immunosensor, mimicking biological sensory functions, was described. The basis of this technique is the fact that the conductance of a population of molecular ion channels, built of tethered gramicidin A and aligned across a lipid bilayer membrane, is changed by the Ab–Ag binding event. Bandodkar et al. have developed a nanostructured nanoporous polyaniline based parallel plate capacitor immunosensor to the detection of human IgG (HIgG) as shown in Figure 8-4.[32]

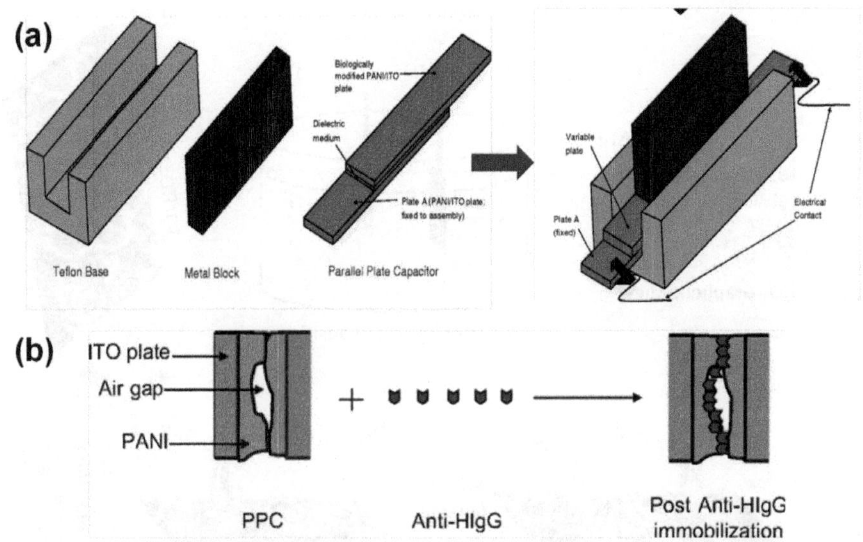

FIGURE 8.4 (a) Schematic representation of the custom made assembly with PPC used for the capacitance measurements and (b) An infinitesimal part of PPC clearly showing decrement in air gap after Anti-HIgG immobilization.[32]

Conductometric immunosensors monitor the changes in dielectric properties and thickness of the dielectric layer at the electrolyte/electrode interface, due to the Ag–Abs interaction. This allows the detection of a receptor-specific anatyte that has been immobilized on the insulating dielectric layer. Conductometric immunosensors have a number of merits, such as suitable for miniaturization and large-scale production, without reference electrode and with low driving voltage. Liang et al. have introduced a conductometric immunosensor based on immobilizing HRP-labeled interleukin-6 Ab onto dendrimer G4 and AuNPs modified composite film.[33] The conductometric immunosensor has been fabricated utilizing a streptavidin functioned magnetite nanoparticles ant their interaction was studied with biotinylated Abs.[34]

8.4.6 OPTICAL IMMUNOSENSORS

Optical immunosensors are most popular for bioanalysis and extensively explored in various groups of transducers. This is mainly because of the advantages of applying visible radiation compared to other transducer techniques. Additional benefits of using these sensors are the nondestructive operation mode and the rapid signal generation and reading. In particular, the introduction of fiber bundle optics ("optodes") as optical waveguides and sophisticated optoelectronics offers increased versatility of these analytical devices for clinical applications. Changes in refractive index (RI) fluorescence, adsorption, and luminescence, occur when light is reflected at sensing surfaces. This information is the physical basis for optical sensing techniques including photodiodes or photomultipliers. A reusable fiber optic probe based immunosensor fabricated by covalently immobilizing a microcystin-LR-ovalbumin (MC-LR-OVA) Abs conjugate onto a self-assembled thiol-silane monolayer through a hitherto bifunctional reagent.[35]

8.5 IMMOBILIZATION OF AB ONTO NANOMATERIALS' SURFACE

Immobilization of biomolecules (Abs or Ags) is a process in which an Ab or Ag is attached to several carriers or nanomaterial for fabrication of an immunosensor. Various methods have been reported for immobilization such as physical adsorption or ionic binding, covalent binding, and cross-linking. Physical adsorption or ionic bonding is the simplest and easiest method based on the nonspecific adsorption based on noncovalent interaction, such as hydrogen bonding, hydrophobic interactions, and van der Waals forces for Abs or Ags immobilized onto the electrode surface. For the development of immunosensor binding of Abs and Ags should be proper onto the electrode surface. The optimum density and adjusted orientation of the Abs are crucial factors because the binding of Abs onto sensing electrode surfaces can affect the reaction kinetic parameters of the immunosensor. The binding of Abs is based on its orientation that is, binding to fragment crystallizable (Fc) receptors (such as protein A or G or recombinant A/G fusion protein) on the surface and binding of other binding partners to structures, covalently linked to the Fc region of the Ab. For example, the biotin residue on the Fc binds to the surface-coated streptavidin and coupling to the solid support takes place via an oxidized carbohydrate moiety on the CH_2 Fc domain and the binding of Fab or side chain variable fragments to the surface of the device via a sulfhydryl group in its C terminal region. Various types of linkers can be applied for immobilization of Abs onto solid surfaces including glutaraldehyde, carbodiimide, succinimide ester, maleinimide, periodate, or galactose oxidase. For optical immunosensors, a noble metal surface (gold) or self-assembling monolayer technique appears to be exceptional.

8.6 APPLICATION OF IMMUNOSESNORS

Nanomaterials are being utilized for various kinds of immunosensors including detection of pathogens and biomarkers. Besides these clinical diagnostic applications, the evaluation of potential application areas in bacteriology or virology is a fast-growing area in the field of immunosensors. This includes serology, monitoring of food (fermentation) and hygiene processes, and detection of biological warfare products. There are also reports of environmental analyses of pesticides, herbicides, and hormone-active compounds. In this chapter, we discuss only the immunosensors based on nanomaterials for pathogen detection.

8.6.1 IMMUNOSENSORS FOR BACTERIAL PATHOGEN

We have recently discussed about the nanostructured material based immunosensors for the detection of a various virulent pathogen such as a bacteria and viruses for clinical diagnosis. These pathogens causes harmful diseases to the humans, animals and plants, leading to huge economic losses. The detection of these pathogens are very essential for proper diagnostics and therapeutics. Conventional diagnostic techniques such as radio immune assay (RIA) immunoassay, Western blot test, and ELISA is currently used for the detection of pathogenic microorganisms. However, these methods have various limitations such as tedious procedure, require highly qualified personnel, and consume a lot of time to yield the desired results. Using these techniques for detection of pathogens present in biological or food matrices requires an enrichment step for concentrating the pathogens from the complex media and dispensing them in buffer for their detection. Rapid and accurate detection of causative pathogen is essential in determining the choice of treatment in acute care settings.

8.6.1.1 V. CHOLERAE

Cholera is an acute infectious disease characterized by rapid onset of severe secretory diarrhea with the production of "rice water" stools. Without immediate rehydration treatment, death by fluid loss can occur within hours or days. *V. cholerae* is a causative agent for diarrhea, acidosis in humans, and is known to cause high lethality. There are limited number of immunosenosrs have been reported using different nanomaterials for detection of *V.*

cholerae. Recently, Solanki et al. have reported an immunosensor for detection of *V. cholerae* using nanostructured nickle oxide (NiO) and a composite of reduced graphene oxide (RGO) and titanium oxide (TiO_2) by the EIS technique and found detection limit as 10450 ng/mL as shown in Figure 8-5.[31]

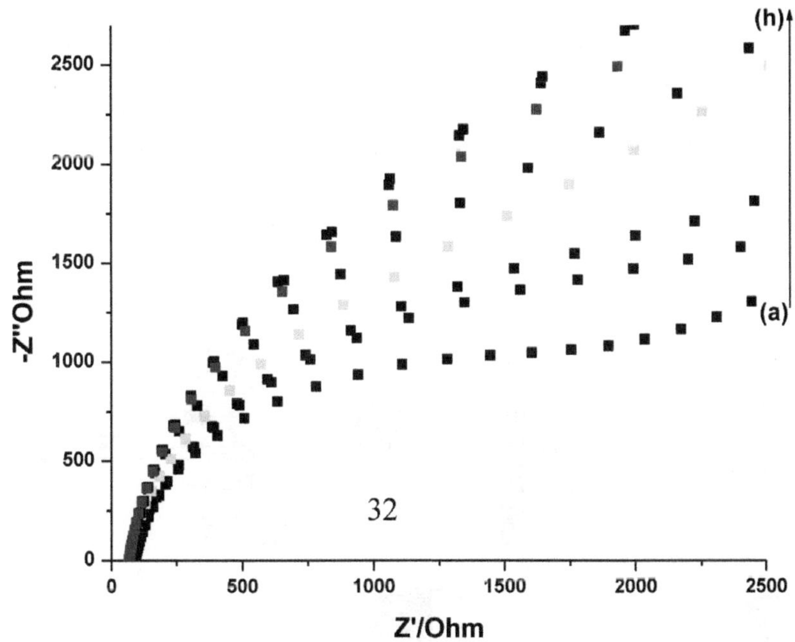

FIGURE 8.5 Electrochemical impedance response studies of BSA/Ab-*Vc*/RGO-antTiO2/ITO immunoelectrode as a function of *Vibrio cholerae* concentration [10-450 ng/mL].[31]

A chemiluminescence biosensor has been developed for the detection of cholera toxin (CT) using a supported lipid membrane as sensing surface and the HRP/ganglioside GM1-functionalized liposome as the detection probe. A disposable amperometric immunosensor for *V. cholerae* has been fabricated using the screen-printed electrode of homemade carbon inks consisting of a mixture of polystyrene and graphite particles. In this sensor, 1-naphthyl phosphate has been used as a substrate with the amperometric detection of hydrolysis product 1-naphthol and exhibits detection limit of 10^5 cells/mL in 55 min. Yu et al. have developed a direct one-step lateral flow biosensor for the simultaneous detection of both *V. cholerae* O1 and O139 serogroups using serogroup specific McAbs raised against lipopolysaccharides that were used to functionalize colloidal AuNPs by applying an immune chromatographic principle.[36]

A disposable amperometric immunosensor for *V. cholerae* has been fabricated using a screen-printed electrode based on carbon ink consisting of a mixture of polystyrene and graphite particles.[37] A gold-coated microcantilever based immunosensor was used for *V. cholerae* detection via resonance frequency shift followed by dynamic force microscopy and detection range from 1×10^3 CFU per mL to 1×10^7 CFU per mL.[38] Chen et al. have fabricated a novel chemilumincence biosensor for the detection of CT based on a lipid-supported membrane as sensing surface and the HRP/GM1-functionalized liposome as the detection probe as shown in Figure 8-6.[39] Loyprasert et al. have developed a capacitive immunosensor based on AuNPs with nonconducting film of polytyramine on a gold electrode, with a detection range of 0.1 aM–10 pM.[40]

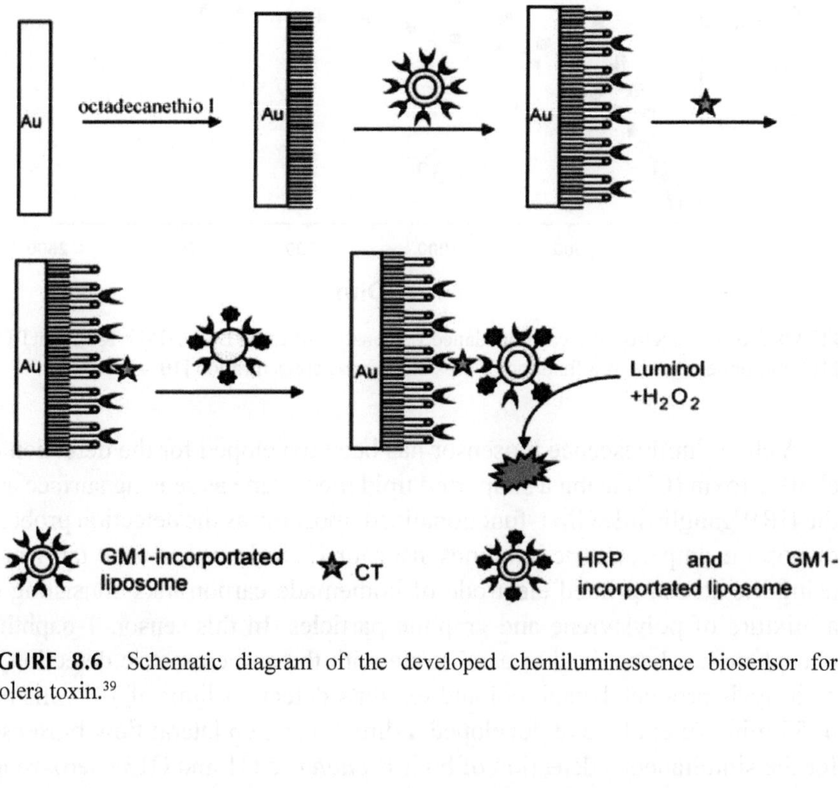

FIGURE 8.6 Schematic diagram of the developed chemiluminescence biosensor for cholera toxin.[39]

8.6.1.2 SALMONELLA TYPHI

Salmonella typhi is a bacterium, commonly found in the intestines of chickens and could affect the ovaries of healthy looking hens which subsequently leads to occurrence of this bacterium in raw eggs.[41] These infected eggs may cause illness, salmonellosis and manifests as fever, diarrhea, and abdominal cramps. Salmonellosis can spread from the intestines to the blood stream and then to other parts of the body, resulting in life-threatening infections in patients who are in poor health or weakened immune systems, and has been known to be fatal.[41] Among the over 2000 serovars that have been identified and characterized, *S. enteric* and *S. typhimurium* are the epidemiologically the most important ones because they are the causative agents in 80% of all human infections reported worldwide.[41] A sensitive and stable label free electrochemical impedance immunosensor for the detection of *S. typhimurium* in milk samples was developed by immobilizing anti-*Salmonella* Abs onto the AuNPs and poly (amidoamine)-multiwalled carbon nanotubes-chitosan nanocomposite film modified glassy carbon electrode has been used for immobilization of McAbs of *Salmonella* spp. onto AuNPs. The interaction of McAbs with *Salmonella* spp. was evaluated using the EIS technique as an efficient method for fabricating a capacitive immunosensor for the detection of *Salmonella* spp. in pork samples. The linear relationship between the relative change in capacitance and logarithm of *Salmonella* concentration was obtained in the range from 1.0×10^2 CFU mL^{-1} to 1.0×10^5 CFU mL^{-1}, represented as a linear regression equation: ΔC (%) = 17.18–14.92 log C (CFU mL^{-1}), and a correlation coefficient of 0.991. The lower detection limit of *Salmonella* Ag concentration was 1.0×10^2 CFU mL^{-1}. This capacitive immunosensor has the advantages of detection of high nonspecific interactions and the short analysis time (40 min) when compared with the polymerase chain reaction (PCR).

Dungchaia et al. have reported a highly sensitive detecting system for *S. typhi* based on AuNPs.[42] They immobilized McAbs on polystyrene microwells and captured *S. typhi*. After that a polyclonal Ab–colloidal gold conjugate was added to bind *S. typhi;* this was followed by a copper-enhanced solution with ascorbic acid and copper (II) sulfate added to the polystyrene microwells. The ascorbic acid reduced the copper (II) ions to copper (0), which was deposited onto AuNPs' tags. In this way, the released copper ions were detected by using anodic stripping voltammetry and the amount of deposit copper was related to the amount of AuNPs' tag. In this way the amount of *S. typhi* attached to the polyclonal Ab–colloidal gold conjugate was controlled. Therefore, the anodic stripping peak current was linearly dependent on *S. typhi*'s concentration, thus providing a detection limit of 98.9 CFU/mL.

8.6.1.3 LISTERIA MONOCYTOGENES

Listeria monocytogenes is a developing bacterial food-borne pathogen responsible for listeriosis, an illness characterized by meningitis, encephalitis, and septicaemia. Less commonly, infection can result in cutaneous lesions and flu-like symptoms. In pregnant women, the pathogen can cause bacteremia, and stillbirth or premature birth of the fetus. Sim et al. have reported a biosensor system integrated with microfluidic channel, enabling label free live cell detection of *L. monocytogenes*.[43] It consists of a waveguide sensor platform to determine the immune reaction and a birefringence measurement system that measures the phase difference between two orthogonal polarizations, arising from the RI change of the immune complex in the channel of the sensor's platform. Fabrication of the waveguide sensor for this pathogen detection system is compatible with conventional integrated circuit processes. The waveguide platform is overlaid with TiO_2 films of different thicknesses to optimize sensitivity. They experimentally confirmed a method for controlling the length of the TiO_2 film to efficiently increase the detection sensitivity in waveguide composition. Moreover, Park et al. have reported a dithiobis-succinimidyl propionate (DSP) modified immunosensor platform to detect *L. monocytogenes* in chicken skin.[44] Moreover, the authors of this study demonstrated the potential feasibility of a surface modified gold-coated immunosensor platform combined with a light microscopic imaging system to detect *L. monocytogenes* in chicken. The reactivity of custom-prepared Abs was significantly higher than the reactivity of commercial Abs and the custom-prepared Abs were highly specific with strains of *L. monocytogenes*. A gold-coated sensor platform modified with DSP exhibited a 40% increase in binding efficiency, compared to the gold-coated sensor platform.

8.6.1.4 CAMPYLOBACTER JEJUNI

Campylobacter jejuni is a Gram-negative, spiral, microaerophilic bacterium. It has now been identified as one of the main causes of bacterial food-borne diseases.[41] *C. jejuni* causes fever, diarrhea, abdominal cramps, and gastroenteritis in humans.[41] This bacterium can lead to the development of Guillian–Barré syndrome, a disorder of the peripheral nervous system often leading to partial paralysis. The food matrices that act as carriers for *C. jejuni* are poultry, meat, and milk.[41] *C. jejuni* is thermo-tolerant and can survive and grow at 40 °C. This bacterium lives in the intestines of chicken, and is often carried over with raw poultry. Eating undercooked chicken is the most frequent source of infection.[41] Wei et al. have developed a biosensor based on SPR for the rapid identification of *C. jejuni* in broiler samples.[45]

8.6.1.5 ESCHERICHIA COLI

Escherichia coli O157:H7 is a bacterial pathogen that is commonly found in the intestinal tracts in cattle and is carried over to the consumers via ground beef. Ingestion of the bacteria causes severe and bloody diarrhea and painful abdominal cramps. In a small number of cases, a complication called hemolytic uremic syndrome can occur which causes profuse bleeding and renal failure.[41] Five types of lectins from *Triticum vulgaris*, *Canavalia ensiformis*, *Ulex europaeus*, *Arachis hypogaea*, and *Maackia amurensis* were employed to evaluate the selectivity of the approach for binding *E. coli* O157:H7 effectively. Moreover, a gold interdigitated microelectrode impedance biosensor was fabricated for the detection of viable *E. coli* O157:H7. This sensor was fabricated using lithographic techniques. The surface of the electrode was immobilized with anti-*E.coli* IgG Abs. This approach is different from other studies where the change in impedance is measured in terms of growth of bacteria on the electrode, rather than the Ab/Ag bonding. The impedance values were recorded for frequency ranges between 100 Hz and 10 MHz. EIS was used to test the sensitivity and effectiveness of the sensor electrode by measuring the change in impedance values of electrodes before and after incubation with different concentrations of the bacteria. An equivalent electrical circuit model has also been proposed to explain the sensing mechanism. Also, Waswa et al. have reported a SPR-based biosensor for *E. coli* O157:H7 determination in spiked samples such as milk, apple juice, and ground beef extract using specific Abs.[46] In this biosensor light from an light-emitting diode is reflected off a gold surface, and the angle and intensity corresponding to the SPR minimum is measured and represented as a RI change corresponding to the Ag–Ab coupling at the sensor's surface.

8.6.1.6 MYCOBACTERIUM TUBERCULOSIS

Tuberculosis (TB) is a deadly infectious disease caused by *Mycobacterium tuberculosis*, and its early diagnosisis is essential for the proper treatment of patients and also for the prevention of further spread of the pathogen. Early diagnosis of active TB remains an elusive challenge, especially in individuals with disseminated TB and human immunodeficiency virus (HIV) coinfection. Recent studies have shown a promise for the direct detection of pathogen-specific biomarkers such as lipoarabinomannan (LAM) for the diagnosis of TB in HIV-positive individuals. Currently, traditional immunoassay platforms that suffer from poor sensitivity and high nonspecific interactions are used for the detection of such biomarkers for the direct detection of three TB specific biomarkers, namely LAM, early secretory Agic target 6 (ESAT6), and Ag 85 complex (Ag85), using a waveguide based optical biosensor platform. This platform combines detection within the evanescent field of a planar optical waveguide with functional surfaces that reduce nonspecific interactions in complex patient samples (urine, serum, etc.) within a short time. In addition, a QCM immunosensor was employed to screen for both whole Mtb bacilli and aMtb surface Ag, LAM. Microfluidic platform applied to the DNA detection of Mtb makes use of an optical colorimetric detection method based on AuNPs. The platform was fabricated using replica moulding technology in poly (dimethylsiloxane) (PDMS) patterned by high-aspect-ratio SU-8 moulds.

8.6.1.7 HELICOBACTER PYLORI

Helicobacter pylori is a Gram-negative bacterium, and is the most important etiological agent of chronic active type B gastritis and peptic ulcer diseases. The infection due to this microorganism is a risk factor in the development of gastric mucosa associated with lymphoid tissue lymphoma and adenocarcinoma. A portable microfluidic immunosensor has been fabricated which, coupled to laser-induced fluorescence (LIF) detection system for the determination of IgG Abs against *H. pylori* in human serum samples. The device has been fabricated and consists of a central channel (CC) with packed *H. pylori* Ag immobilized on 3-aminopropyl-modified controlled pore glass. Abs in serum samples reacted with the immobilized Ag and then, they were determined using alkaline phosphatase (AP) enzyme labeled second Abs specific to human IgG. In this regard, 4-methylumbelliferyl phosphate was employed as the enzymatic substrate.[47] Further, a microfluidic magnetic immunosensor was fabricated on a gold electrode and was used for detection of Abs of *H. pylori* in serum sample. During this process, Ags immobilized on magnetic microspheres were injected into microchannel devices and manipulated with the aid of an external removable magnet using a noncompetitive immunoassay.[48] The IgG Abs in human serum sample were quantified by AP enzyme labeled second Abs specific to human IgG. The *p*-APP was converted to *p*-aminophenol (*p*-AP) by AP and the electroactive product was detected on gold layer electrode at 0.250 V. Stege et al. developed an online and entirely automatized immune affinity assay CE to determine the concentration of anti-*H. pylori* IgG using magnetic nanobeads as a support for the immunological affinity ligands and an LIF as a detector as shown in Figure 8-7.[49] The separation was performed in 0.1 M glycine1–HCl, at pH 2.0, as the background electrolyte. In addition, Molina et al. reported a human serum IgG Abs to *H. pylori* quantization procedure based on the multiple use of an immobilized *H. pylori* Ag on an immuno-column incorporated into an a flow-injection analytical system.[50] The immuno-adsorbent column was prepared by packing 3-aminopropyl-modified controlled-pore glass (APCPG) covalently linking *H. pylori* Ags in a 3 cm of teflon tubing (0.5 i.d.). Abs in the serum sample were quantified by AP enzyme labeled second Abs specific to human IgG.

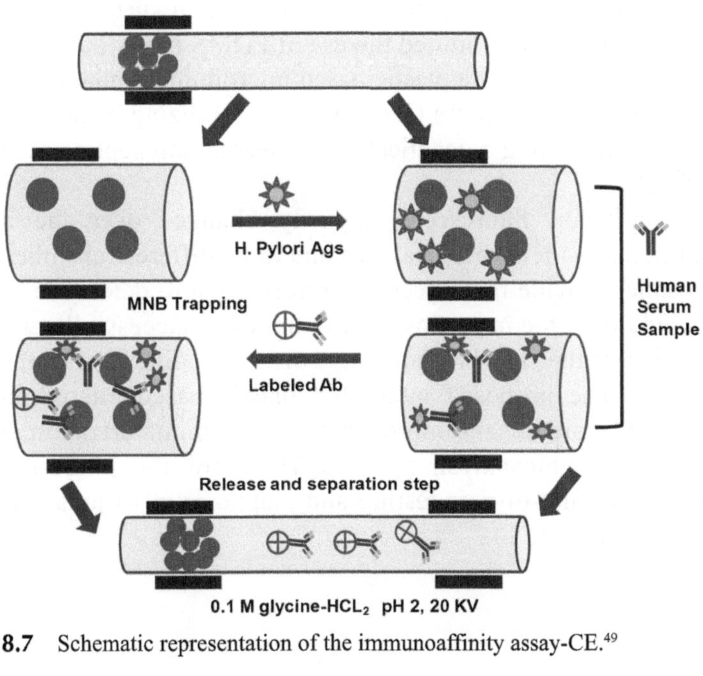

FIGURE 8.7 Schematic representation of the immunoaffinity assay-CE.[49]

8.6.2 IMMUNOSENSOR FOR VIRAL PATHOGEN DETECTION

Swine-origin influenza virus, a high-risk human influenza a virus (H1N1), is a serious health threat and potential leading cause of death all around the world.[51] The World Health Organization has reported that more than 16,000 cumulative deaths were reported from 213 countries due to H1N1 in February 2010.[52] Several laboratories have developed diagnostic methods that have been used to monitor the outbreaks of the virus. Such methods are as follows: (1) specific real-time PCR based detection method, (2) isolation of H1N1 influenza virus, and (3) detection of fourfold rise of neutralized Abs to the virus.

However, these methods require highly skilled personnel and expensive laboratory. In addition, they are not suitable for undeveloped countries because of the limited access to central laboratories and expensive costs. To overcome these issues, microfluidic immunoassay systems have been introduced because of their various advantages, including high throughput, high efficiency, low-cost, and minimized consumption of samples and reagents.[53] After the development of soft lithography techniques using PDMS, PDMS has become the most popular microfluidic device material and offers several advantages such as easy handling, good sealing properties, and high optical transparency.[54] However, the poor chemical stability in different types of organic solvents, difficulty in surface modification, and mass production has limited the use of PDMS in various applications. Lee et al. have developed a plastic based microfluidic immunosensor chip for the diagnosis of swine flu (H1N1) by immobilizing hemagglutinin Ag on a gold surface using a genetically engineered polypeptide as shown in Figure 8-8.[55]

The significant fluorescent intensity changes over the different concentrations to the serological Abs and three different chambers in one microchannel provide more accurate information to detect the H1N1 flu virus. In addition, the immunosensor chips were successfully applied for the detection of the H1N1without any surface modification of the microfluidic chip. The proposed integrated plastic based microfluidic chip could provide a significant improvement in the miniaturization and cost-effective way for bioanalysis systems. Therefore, this platform offers a perspective of point-of-care testing and diagnosis in various study areas of infectious disease.

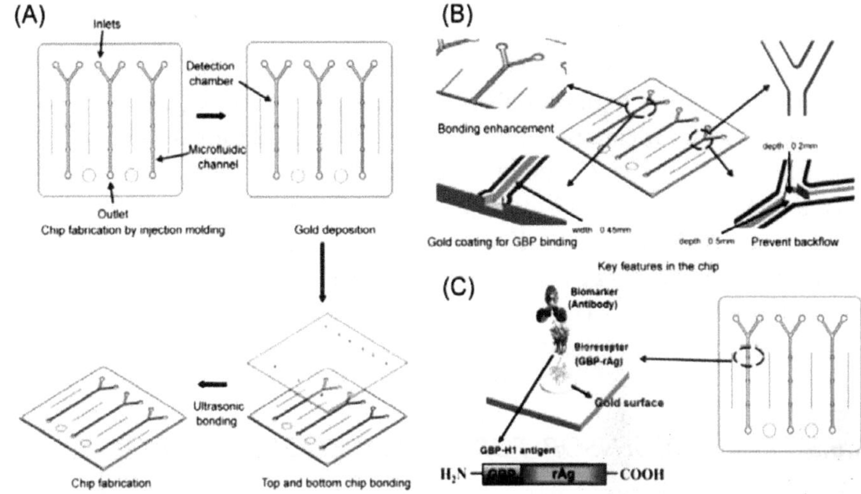

FIGURE 8.8 Schematic representation of the fabrication of (A) COC-based microfluidic chip, (B) detailed features of the chip, (C) GBP-H1a fusion protein immobilization step and Cy3-labeled Ab reaction in the chip.

8.6.2.1 HEPATITIS C VIRUS

Hepatitis C virus (HCV) infection, which presents as a persistent infection in up to 85% of all infected individuals, is a global health problem. Once HCV-infected patients develop cirrhosis or hepatocellular carcinoma, low cure rates and serious side effects shall be expected. That is why accurate and sensitive diagnosis of HCV in blood samples during the early stages of the infection is so crucial.[56] Resonant microcantilever arrays are developed for the purpose of label-free and real time analyte monitoring and bimolecular detection. Microelectromechanical systems' cantilevers made of electroplated nickel were functionalized with Hepatitis Abs.[57] An electrochemical immunosensor was developed for the detection of HCV core Ag.[58] The immunosensor consisted of graphitized mesoporous carbon–methylene blue nanocomposite as an electrode-modified material with AuNPs electrodeposited onto the electrode to immobilize the captured Abs and a HRP–DNA coated carboxyl multi-wall carbon nanotubes (CMWNTs) as a secondary Ab layer. The bridging probe and secondary Abs linked to the CMWNTs, and DNA concatemers were obtained by hybridization of the biotin-tagged signal and auxiliary probes. Finally, streptavidin–HRP were labeled on the secondary Ab layer via the biotin–streptavidin system.

8.6.2.2 JAPANESE ENCEPHALITIS VIRUS

Japanese encephalitis virus (JEV) is a leading cause of childhood encephalitis in Asia. It has a high mortality and high risk for subsequent infections. It is the most significant mosquito-borne viral encephalitis in several Asian countries. As many as 25 countries are at risk from the Japanese encephalitis, and approximately 3 billion people, including more than 700 million children under the age of 15 live in risk areas. Several conventional diagnostic methods have been developed for detection of JEV infection such as immunoglobulin M (IgM) assay, plaque reduction neutralization test, reverse transcription PCR (RT-PCR), or virus isolation.[59]

8.6.3 IMMUNOSENSOR FOR BIOMARKERS' DETECTION

Various metabolic biomarkers (substances that appear in biological fluids and tissues of the human body during the development of many somatic and infectious pathological processes) can be used in medical and biological research. The determination of these substances allows for diagnosis of diseases at early stages. An urgent problem in the clinical analysis is the need to detect single cells and microorganisms. In this connection, along with speed, high sensitivity and selectivity, one of the main requirements in performing a large number of routine analyses is the simplicity in preparation of samples for the analysis. Organic and inorganic NMs such as nanoparticles or nanowires are often used in biosensors to amplify the binding event by using some measurable change in a property such as electrical conductivity of a nanowire or change in the way the nanoparticle complex interacts with light. These nanomaterials are often used to modify or interact with the transducer which is responsible for measuring and transmitting the signal generated in the presence of the target analyte.

Biomarkers can be divided into three categories including (1) diagnostic biomarkers, which assist in early detection of a disease, (2) prognostic biomarkers, which help assess the malignant potential of tumors, and (3) predictive biomarkers, which can be used to differentiate between various cancers and help in designing therapy plans for the patient.[60] The analytical tools for biomarkers must be capable of operating at the level of differential diagnosis and to be specific enough not to produce numerous false-positive results.[61] The use of biomarkers in cancer diagnosis, staging, and monitoring response to cancer therapy is already a well established diagnostic procedure. An ideal biomarker for cancer is a protein or a protein fragment that can be detected very early in the patient's blood or urine, but not detected in healthy individuals.[62] Although cancer biomarkers are of great interest due to the incidence of cancer increasing dramatically over the past few decades, especially in the developed countries, biomarkers are also used in other areas of medicine such as psychiatry and endocrinology. Chen et al. have developed a electrochemical immunosensor for simultaneous detection of multiplex cancer biomarkers based on graphene nanocomposites.[63] The preparation of biofunctional carboxyl graphene nanosheets (CGS) nanocomposites and the multiplexed electrochemical immunoassay protocol are shown in Figure 8-9.

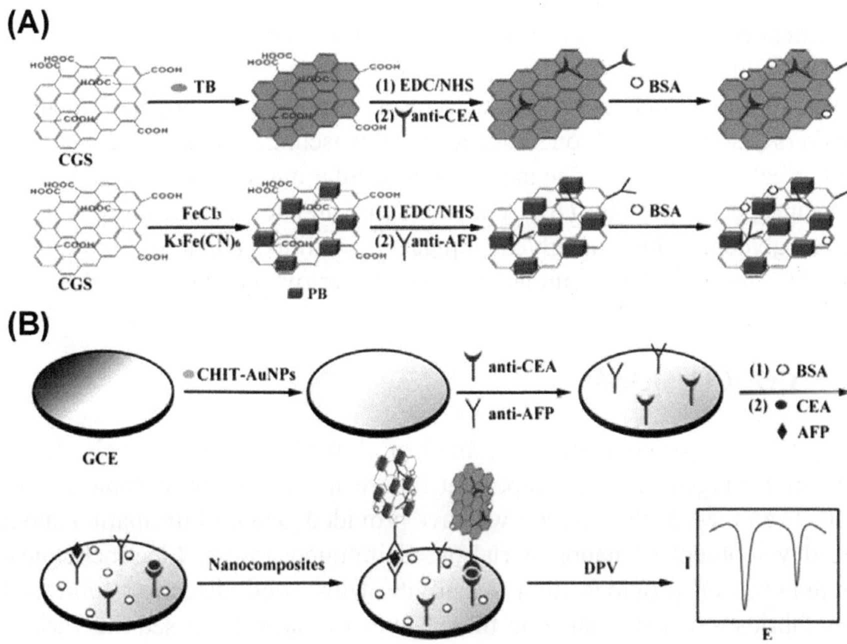

FIGURE 8.9 Schematic representation of the (A) preparation of biofunctional CGS nanocomposites and (B) multiplexed electrochemical immunoassay protocol.[63]

Biomarker detection has also made possible the development of personalized treatment plans for certain diseases. The treatment plan can take into consideration factors such as a patient's gender, age, height, weight, diet, and environment. It has been predicted that the use of biomarkers in the detection and treatment of a wide range of diseases will continue to grow in the future.[64] Since biomarkers have key role in preventive medicine, the need for accurate, reproducible, efficient, and easy detection and quantification procedures by nonspecialists has become highly important. Further, affordable detectors, fast sample processing times, minimal labor, small sample volumes, and ability to detect multiple biomarkers simultaneously are also very important. Furthermore, the storage stability and resistance to degradation of reagents in biosensors are important factors to consider. Also, as with any analysis method, automated or semiautomated, easy to use devices are preferred. Small, mass produced, and relatively inexpensive enzyme based electrochemical biosensors have been used by the public and the medical personnel, for example, in the management of diabetes by glucose monitoring as well as in the measurement of lactate and cholesterol for decades. However, biosensors which detect biomarkers of clinical significance, such as tumor markers or hormones are relatively new analytical tools in medicine.

Various electrochemical immunosensors are being developed to identify markers associated with oncological, cardiovascular, autoimmune, allergic, and infectious diseases. The nature of molecular markers is quite diverse as they are specific proteins (Abs, enzymes, etc.) and nonspecific protein molecules, glycoproteins and glycoconjugates, modified alkaloids, hormones, steroids, drugs, and other molecules of high- or low molecular weight.

8.7 CONCLUSIONS

As per the major concern in the field of clinical diagnostics, early detection of pathogens is very important before it reaches the chronic condition. Therefore, in this chapter we have provided general information about the development of nanomaterial-based immunosensors. Electrochemical immunosensor provides high sensitivity with good detection limit and low fabrication cost. Hence, in this chapter we have discussed the various types of nanomaterials including metal nanoparticles, carbon nanomaterials, semiconductor nanoparticles, metal–oxide nanostructures, and hybrid nanostructures. It has been observed that most of the studies are reported for communicable and noncommunicable diseases. However, such types of electrochemical immunosensors based on nanometarials can be used in point of care diagnostics.

ACKNOWLEDGMENT

PS is grateful to Jawaharlal Nehru University, University Grants Commission, and the Department of Science and Technology for financial support.

KEYWORDS

- **Immunosensors**
- **Nanostructured materials**
- **Immunoglobulins**
- **Ag–Ab reaction**
- **Diseases**

REFERENCES

1. Parolo, C.; de la Escosura-Muniz, A.; Merkoci, A. Electrochemical DNA Sensors Based on Nanoparticles. In *Electrochemical Biosensors*; Cosnier S.,Ed.; Pan Stanford: Singapore, 2015, p 195.
2. Freeman, M. H.; Hall, J. R.; Leopold, M. C. Monolayer-Protected Nanoparticle Doped Xerogels as Functional Components of Amperometric Glucose Biosensors. *Anal. Chem.* **2013,** *85*, 4057–4065.
3. Teles, F.; Fonseca, L. Applications of Polymers for Biomolecule Immobilization in Electrochemical Biosensors. *Mater. Sci. Eng. C.* **2008,** *28*, 1530–1543.
4. Wang, Z.; Qian, X.F.; Yin, J.; Zhu, Z. K. Large-Scale Fabrication of Tower-like, Flower-like, and Tube-like ZnO Arrays by a Simple Chemical Solution Route. *Langmuir* **2004,** *20*, 3441–3448.
5. Xia, Y.; Yang, P.; Sun, Y.; Wu, Y.; Mayers, B.; Gates, B.; Yin, Y.; Kim, F.; Yan, H. One-Dimensional Nanostructures: Synthesis, Characterization, and Applications. *Adv. Mater.* **2003,** *15*, 353–389.
6. Kimmel, D. W.; LeBlanc, G.; Meschievitz, M. E.; Cliffel, D. E. Electrochemical Sensors and Biosensors. *Anal. Chem.* **2011,** *84*, 685–707.
7. Festag, G.; Klenz, U.; Henkel, T.; Csáki, A.; Fritzsche, W. Biofunctionalization of Metallic Nanoparticles and Microarrays for Biomolecular Detection. Nanotechnologies for the Life Sciences, **2005**.
8. Liu, G.; Lin, Y.-Y.; Wang, J.; Wu, H.; Wai, C. M.; Lin, Y. Disposable Electrochemical Immunosensor Diagnosis Device Based on Nanoparticle Probe and Immunochromatographic Strip. *Anal. Chem.* **2007,** *79*, 7644–7653.
9. Tran, Q. H.; Nguyen, T. H. H.; Mai, A. T.; Nguyen, T. T.; Vu, Q. K.; Phan, T. N. Development of Electrochemical Immunosensors Based on Different Serum Antibody

Immobilization Methods for Detection of Japanese Encephalitis Virus. *Adv. Nat. Sci.: Nanosci. Nanotechnol.* **2012,** *3*, 015012.
10. Schlücker, S. SERS Microscopy: Nanoparticle Probes and Biomedical Applications. *Chem.Phys.Chem.* **2009,** *10*, 1344–1354.
11. Wang, J. Electrochemical Biosensors: Towards Point-of-Care Cancer Diagnostics. *Biosens. Bioelectron.* **2006,** *21*, 1887–1892.
12. Luppa, P. B.; Sokoll, L. J.; Chan, D. W. Immunosensors-Principles and Applications to Clinical Chemistry. *Clin. Chim. Acta* **2001,** *314*, 1–26.
13. Bezbaruah, A. N.; Kalita, H. *Sensors and Biosensors for Endocrine Disrupting Chemicals: State-of-The-Art and Future Trends*; IWA Publishing: London, 2010, pp 93–127.
14. Kost, G. J. *Point-of-Care Testing*; Wiley Online Library, 2006.
15. Morgan, C. L.; Newman, D. J.; Price, C. Immunosensors: Technology and Opportunities in Laboratory Medicine. *Clin. Chem.* **1996,** *42*, 193–209.
16. Van E. , J.; Gerlach, C.; Bowman, K. Bioseparation and Bioanalytical Techniques in Environmental Monitoring. J. *Chromatogr. B: Biomed. Sci. Appl.* **1998,** *715*, 211–228.
17. Porter, A. L.; Youtie, J. How Interdisciplinary is Nanotechnology? *J. Nanopart. Res.* **2009,** *11*, 1023–1041.
18. Sahoo, S.; Parveen, S.; Panda, The Present and Future of Nanotechnology in Human Health Care. *J. Nanomed. Nanotechnol. Biol. Med.* **2007,** *3*, 20–31.
19. Englebienne, P.; Hoonacker, A. V.; Verhas, M. Surface Plasmon Resonance: Principles, Methods and Applications in Biomedical Sciences. *J. Spectro.* **2003,** *17*, 255–273.
20. Pei, X.; Zhang, B.; Tang, J.; Liu, B.; Lai, W.; Tang, D. Sandwich-Type Immunosensors and Immunoassays Exploiting Nanostructure Labels: A Review. *Anal. Chim. Acta* **2013,** *758*, 1–18.
21. Solanki, P. R.; Dhand, C.; Kaushik, A.; Ansari, A. A.; Sood, K.; Malhotra, B. D. Nanostructured Cerium Oxide Film for Triglyceride Sensor. *Sens. Actuators, B: Chem.* **2009,** *141*, 551–556.
22. Ronkainen, N. J.; Halsall, H. B.; Heineman, W. R. Electrochemical Biosensors. *Chem. Soc. Rev.* **2010,** *39*, 1747–1763.
23. Viswanathan, S.; Rani, C.; Ho, J. A. Electrochemical Immunosensor for Multiplexed Detection of Food-Borne Pathogens Using Nanocrystal Bioconjugates and MWCNT Screen-Printed Electrode. *Talanta* **2012,** *94*, 315–319.
24. Liu, G.; Lin, Y. Nanomaterial Labels in Electrochemical Immunosensors and Immunoassays. *Talanta* **2007,** *74*, 308–317.
25. Zhong, Z.; Wu, W.; Wang, D.; Wang, D.; Shan, J.; Qing, Y.; Zhang, Z. Nanogold-Enwrapped Graphene Nanocomposites as Trace Labels for Sensitivity Enhancement of Electrochemical Immunosensors in Clinical Immunoassays: Carcinoembryonic Antigen as a Model. *Biosens. Bioelectron.* **2010,** *25*, 2379–2383.
26. Han, D.; Kim, Y.-R.; Kang, C. M.; Chung, T. D. Electrochemical Signal Amplification for Immunosensor Based on 3D Interdigitated Array Electrodes. *Anal. Chem.* **2014,** *86*, 5991–5998.
27. Ermolaeva, T.; Kalmykova, E. *Capabilities of Piezoelectric Immunosensors for Detecting Infections and for Early Clinical Diagnostics*. Advances In Immunoassay Technology, **2012,** 81.
28. Tang, D.; Yuan, R.; Chai, Y.; Fu, Y.; Dai, J.; Liu, Y.; Zhong, X. New amperometric and Potentiometric Immunosensors Based on Gold nanoparticles/tris(2,2′-bipyridyl)

cobalt (III) Multilayer Films for Hepatitis B Surface Antigen Determinations. *Biosens. Bioelectron.* **2005,** *21*, 539–548.
29. Suni, I. I. Impedance Methods for Electrochemical Sensors Using Nanomaterials. *Trends Anal. Chem.* **2008,** *27*, 604–611.
30. Solanki, P. R.; Ali, M. A.; Agrawal, V. V.; Srivastava, A.; Kotnala, R.; Malhotra, B. D. Highly Sensitive Biofunctionalized Nickel Oxide Nanowires for Nanobiosensing Applications. *RSC Adv.* **2013,** *3*, 16060–16067.
31. Solanki, P. R.; Srivastava, S.; Ali, M. A.; Srivastava, R. K.; Srivastava, A.; Malhotra, B. D. Reduced Graphene Oxide–Titania Based Platform for Label-Free Biosensor. *RSC Adv.* **2014,** *4*, 60386–60396.
32. Bandodkar, A. J.; Dhand, C.; Arya, S. K.; Pandey, M.; Malhotra, B. D. Nanostructured Conducting Polymer Based Reagentless Capacitive Immunosensor. *Biomed. Microdevices* **2010,** *12*, 63–70.
33. Liang, K.; Mu, W.; Huang, M.; Yu, Z.; Lai, Q. Interdigitated Conductometric Immunosensor for Determination of Interleukin-6 in Humans Based on Dendrimer G4 and Colloidal Gold Modified Composite Film. *Electroanalysis* **2006,** *18*, 1505–1510.
34. Hnaiein, M.; Hassen, W.; Abdelghani, A.; Fournier-Wirth, C.; Coste, J.; Bessueille, F.; Leonard, D.; Jaffrezic-Renault, N. A Conductometric Immunosensor Based on Functionalized Magnetite Nanoparticles For *E. coli* Detection. *Electrochem. Commun.* **2008,** *10*, 1152–1154.
35. Long, F.; He, M.; Zhu, A.; Shi, H. Portable Optical Immunosensor For Highly Sensitive Detection of Microcystin-LR in Water Samples. *Biosens. Bioelectron.* **2009,** *24*, 2346–2351.
36. Yu, C. Y.; Ang, G. Y.; Chua, A. L.; Tan, E. H.; Lee, S. Y.; Falero-Diaz, G.; Otero, O.; Rodríguez, I.; Reyes, F.; Acosta, A. Dry-Reagent Gold Nanoparticle-Based Lateral Flow Biosensor for the Simultaneous Detection of Vibrio Cholerae Serogroups O1 and O139. *J. Microbiol. Methods* **2011,** *86*, 277–282.
37. Mistry, K. K.; Layek, K.; Mahapatra, A.; RoyChaudhuri, C.; Saha, H. A Review on Amperometric-Type Immunosensors Based on Screen-Printed Electrodes. *Analyst* **2014,** *139*, 2289–2311.
38. Sungkanak, U.; Sappat, A.; Wisitsoraat, A.; Promptmas, C.; Tuantranont, A. Ultrasensitive Detection of Vibrio Cholerae O1 Using Microcantilever-Based Biosensor with Dynamic Force Microscopy. *Biosens. Bioelectron.* **2010,** *26*, 784–789.
39. Chen, H.; Zheng, Y.; Jiang, J.-H.; Wu, H.-L.; Shen, G.-L.; Yu, R.-Q. An Ultrasensitive Chemiluminescence Biosensor for Cholera Toxin Based on Ganglioside-Functionalized Supported Lipid Membrane and Liposome. *Biosens. Bioelectron.* **2008,** *24*, 684–689.
40. Loyprasert, S.; Hedström, M.; Thavarungkul, P.; Kanatharana, P.; Mattiasson, B. Sub-Attomolar Detection of Cholera Toxin Using a Label-Free Capacitive Immunosensor. *Biosens. Bioelectron.* **2010,** *25*, 1977–1983.
41. Sharma, H.; Mutharasan, R. Review of Biosensors for Foodborne Pathogens and Toxins. *Sens. Actuators, B: Chem.* **2013,** *183*, 535–549.
42. Dungchai, W.; Siangproh, W.; Chaicumpa, W.; Tongtawe, P.; Chailapakul, O. Salmonella Typhi Determination Using Voltammetric Amplification of Nanoparticles: A Highly Sensitive Strategy for Metalloimmunoassay Based on a Copper-Enhanced Gold Label. *Talanta* **2008,** *77*, 727–732.
43. Sim, J. H.; Kwak, Y. H.; Choi, C. H.; Paek, S.-H.; Park, S. S.; Seo, S. A Birefringent Waveguide Biosensor Platform for Label-Free Live Cell Detection of Listeria Monocytogenes. *Sens. Actuators, B: Chem.* **2012,** *173*, 752–759.

44. Park, M.-K.; Park, J. W.; Oh, J.-H. Optimization and Application of a Dithiobis-Succinimidyl Propionate-Modified Immunosensor Platform to Detect Listeria Monocytogenes in Chicken Skin. *Sens. Actuators, B: Chem.* **2012,** *171*, 323–331.
45. Wei, D.; Oyarzabal, O. A.; Huang, T.-S.; Balasubramanian, S.; Sista, S.; Simonian, A. L. Development of a Surface Plasmon Resonance Biosensor for the Identification of Campylobacter Jejuni. J. *Microbiol. Methods* **2007,** *69*, 78–85.
46. Waswa, J.; Irudayaraj, J.; Deb Roy, C. Direct Detection of *E. coli* O157: H7 in Selected Food Systems by a Surface Plasmon Resonance Biosensor. *LWT Food Sci. Technol.* **2007,** *40*, 187–192.
47. Seia, M. A.; Pereira, S. V.; Fontán, C. A.; De Vito, I. E.; Messina, G. A.; Raba, J. Laser-Induced Fluorescence Integrated in a Microfluidic Immunosensor for Quantification of Human Serum IgG Antibodies to Helicobacter Pylori. *Sens. Actuators, B: Chem.* **2012,** *168*, 297–302.
48. Pereira, S. V.; Messina, G. A.; Raba, J. Integrated Microfluidic Magnetic Immunosensor for Quantification of Human Serum IgG Antibodies to Helicobacter Pylori. *J. Chromatogr. B* Biofunctionalization of Metallic Nanoparticles and Microarrays for Biomolecular Detection. 878, 253–257.
49. Stege, P. W.; Raba, J.; Messina, G. A. Online Immunoaffinity Assay-CE Using Magnetic Nanobeads for the Determination of Anti-Helicobacter Pylori IgG in Human Serum. *Electrophoresis* **2010,** *31*, 3475–3481.
50. Molina, L.; Messina, G. A.; Stege, P. W.; Salinas, E.; Raba, J. Immuno-Column for On-Line Quantification of Human Serum IgG Antibodies to Helicobacter Pylori in Human Serum Samples. *Talanta* **2008,** *76*, 1077–1082.
51. Donaldson, L. J.; Rutter, P. D.; Ellis, B. M.; Greaves, F. E.; Mytton, O. T.; Pebody, R. G.; Yardley, I. E. Mortality from Pandemic A/H1N1 2009 Influenza in England: Public Health Surveillance Study. BMJ **2009,** *339*, b5213.
52. Kao, L. T.-H.; Shankar, L.; Kang, T. G.; Zhang, G.; Tay, G. K. I.; Rafei, S. R. M.; Lee, C. W. H. Multiplexed Detection and Differentiation of the DNA Strains for Influenza a (HIN1 2009) Using a Silicon-Based Microfluidic System. *Biosens. Bioelectron.* **2011,** *26*, 2006–2011.
53. Duffy, D. C.; McDonald, J. C.; Schueller, O. J.; Whitesides, G. M. Rapid Prototyping of Microfluidic Systems in poly (dimethylsiloxane). *Anal. Chem.* **1998,** *70*, 4974–4984.
54. Brennan, D.; Justice, J.; Corbett, B.; McCarthy, T.; Galvin, P. Emerging Optofluidic Technologies for Point-of-Care Genetic Analysis Systems: A Review. *Anal. Bioanal. Chem.* **2009,** *395*, 621–636.
55. Lee, K. G.; Lee, T. J.; Jeong, S. W.; Choi, H. W.; Heo, N. S.; Park, J. Y.; Park, T. J.; Lee, S. J. Development of a Plastic-Based Microfluidic Immunosensor Chip for Detection of H1N1 Influenza. *Sensors* **2012,** *12*, 10810–10819.
56. Fiebig, E. W.; Wright, D. J.; Rawal, B. D.; Garrett, P. E.; Schumacher, R. T.; Peddada, L.; Heldebrant, C.; Smith, R.; Conrad, A.; Kleinman, S. H. Dynamics of HIV Viremia and Antibody Seroconversion in Plasma Donors: Implications For Diagnosis and Staging of Primary HIV Infection. *AIDS* **2003,** *17*, 1871–1879.
57. Saliterman, S. *Fundamentals of BioMEMS and Medical Microdevices*; SPIE ress: Bellingham, 2006.
58. Ma, C.; Liang, M.; Wang, L.; Xiang, H.; Jiang, Y.; Li, Y.; Xie, G. MultisHRP-DNA-Coated CMWNTs as Signal Labels for an Ultrasensitive Hepatitis C Virus Core Antigen Electrochemical Immunosensor. *Biosens. Bioelectron.* **2013,** *47*, 467–474.

59. Parida, M.; Santhosh, S.; Dash, P.; Tripathi, N.; Saxena, P.; Ambuj, S.; Sahni, A.; Rao, P. L.; Morita, K. Development and Evaluation of Reverse Transcription-Loop-Mediated Isothermal Amplification Assay for Rapid and Real-Time Detection of Japanese Encephalitis Virus. *J. Clin. Microbiol.* **2006,** *44*, 4172–4178.
60. Frank, R.; Hargreaves, R. Clinical Biomarkers in Drug Discovery and Development. *Nat. Rev. Drug Discovery* **2003,** *2*, 566–580.
61. Yurkovetsky, Z. R.; Linkov, F. Y.; E Malehorn, D.; Lokshin, A. E. Multiple Biomarker Panels for Early Detection of Ovarian Cancer. *Future Oncol.* **2006**, 2, 733–741.
62. Hewitt, S. M.; Dear, J.; Star, R. A. Discovery of Protein Biomarkers for Renal Diseases. *J. Am. Soc. Nephrol.* **2004,** *15*, 1677–1689.
63. Chen, X.; Jia, X.; Han, J.; Ma, J.; Ma, Z. Electrochemical Immunosensor for Simultaneous Detection of Multiplex Cancer Biomarkers Based on Graphene Nanocomposites. *Biosens. Bioelectron.* **2013,** *50*, 356–361.
64. Hanash, S. M.; Baik, C. S.; Kallioniemi, O. Emerging Molecular Biomarkers—Blood-Based Strategies to Detect and Monitor Cancer. *Nat. Rev. Clin. Oncol.* **2011**, 8, 142–150.

CHAPTER 9

NANOTOXICITY ASSESSMENT

PRATIK KUMAR SHAH[1,*], JAIRO NELSON[2], and CHEN-ZHONG LI[2]

[1]*Bio-MEMS and Microsystems Laboratory, Department of Electrical and Computer Engineering, Florida International University, Miami, USA*

[2]*Bielectronics and Nanobioengineering Laboratory, Department of Biomedical Engineering, Florida International University, Miami, USA*

**E-mail: pshah003@fiu.edu*

CONTENTS

Each passing decade has provided a plethora of technological advancements in many fields of science. One of these ever-growing fields is in the creation and application of nanomaterials. Thousands of nanomaterials have been created over the past few decades that have improved the consistency of consumer goods, the speed of computer processors, and even the ways in which medicine can tackle ailments. The ever-increasing creation and application of nanomaterials will increase the likelihood of incidental exposure of the general public to these materials and for this reason it is imperative that the toxicity of these nanomaterials is characterized, elucidated, and quantified. This chapter provides a brief introduction to nanotoxicology, describing the common techniques used to determine the toxic nature of materials in laboratory experiments, and discusses the strengths and weaknesses some of these techniques pose even when conducting a well controlled toxicology study. In addition to the latter, newer and more creative complementary techniques to those already used for studying nanomaterial toxicity are also presented.

9.1 INTRODUCTION TO NANOTOXICITY

The application of nanotechnology has been an ever-increasing topic of discussion and interest over the past three decades due to the new functions these materials can serve. Over this period of time the research community has been responsible for creating and handling thousands of new nanomaterials with useful applications, but whose properties and effects on biological tissues are relatively new and unknown. Therefore, much effort is being placed into researching the unique physiochemical properties of nanomaterials and nanoparticles (NPs) in the range of (~1–100 nm).[1] It has been observed that particles of many materials exhibit different properties within the nanoscale range when compared to the same particles in bulkier forms (above the 100 nm threshold). The unique physical properties NPs exhibit at this nanoscale have served to inspire a plethora of functions in the research community and industrial sectors alike.

One would be surprised to discover the extent to which NPs are used in products across multiple industries. NPs' usage is increasing in household items like cosmetics, paints, electronics, and food products, as well as in medical, material, and even energy-producing technologies. Needless to say, the increased production and use of nanomaterials increases the probability that the population will be unintentionally exposed. For this reason, it is imperative to investigate at what level of exposure, if any, these new nanomaterials have the potential to become toxic to those who use them; making nanotoxicology a subject of proportional interest to the rise of NP applications.

Nanotoxicology needs to keep up with new materials being developed and expand our understanding of the long-term effects of NPs' exposure. Despite the many benefits nanomaterials have brought to the world, the exposure to these materials in acute quantities or over many years may have adverse effects on humans, animals, and the ecosystem.[2] The question is whether or not the risks associated with using a material in a specific industrial sector, disposing of it in a particular fashion, or using a material in medicine outweighs the benefits over the years. The main reason why continued studies in NPs' toxicity are so important is that the mechanism in which a nanomaterial becomes toxic is somewhat variable, affecting subsystems of biological processes and organelles in such a way that each type of tissue responds differently. The potential adverse interactions that take place between the organelles of a biological subsystem and the accumulated NPs in that tissue type may produce a biomarker indicative of NP toxicity. This biomarker may often be completely different from the biomarker released by another tissue type. This implies that there is no single analytical tool or assay that can provide a rapid all-encompassing assessment of NP nanotoxicity across multiple tissue types.

NPs are so small that they can gain entrance to the body in many different fashions, making them difficult to track and even more difficult to measure in vivo. While in vivo studies must be conducted, in vitro testing can provide the scientific community with a good approximation of the body's reaction to foreign NPs without risking the health of the patient. Many studies, both in vivo and in vitro, have discussed the possible routes of exposure to NPs and the associated risks. NPs can enter a biological system in a transdermal fashion, via wounds, through ingestion, and even through inhalation.[3] Inhaled NPs can pass through the epithelial tissue of the respiratory tract to gain access to the pulmonary–systemic circulatory system and eventually gain access to the lymphatic system as well. The relatively large surface area of these small NPs makes them more chemically reactive, allowing them to translocate to the central nervous system via nerve endings in the epithelia and olfactory bulb.[4,5] Certain NPs are even able to cross the blood–brain barrier (BBB)[6] and have been the subject of study for uses in therapeutic and diagnostic applications.[7,8] However, since NPs can also impair axonal signal transduction pathways, understanding the role of NP interactions with neurons is imperative prior to developing a treatment intended to bypass the

BBB. Figure 9-1 shows the different possibilities of nanomaterial exposure, their circulation in the body, and their effects on targeted organ dysfunctions.[9]

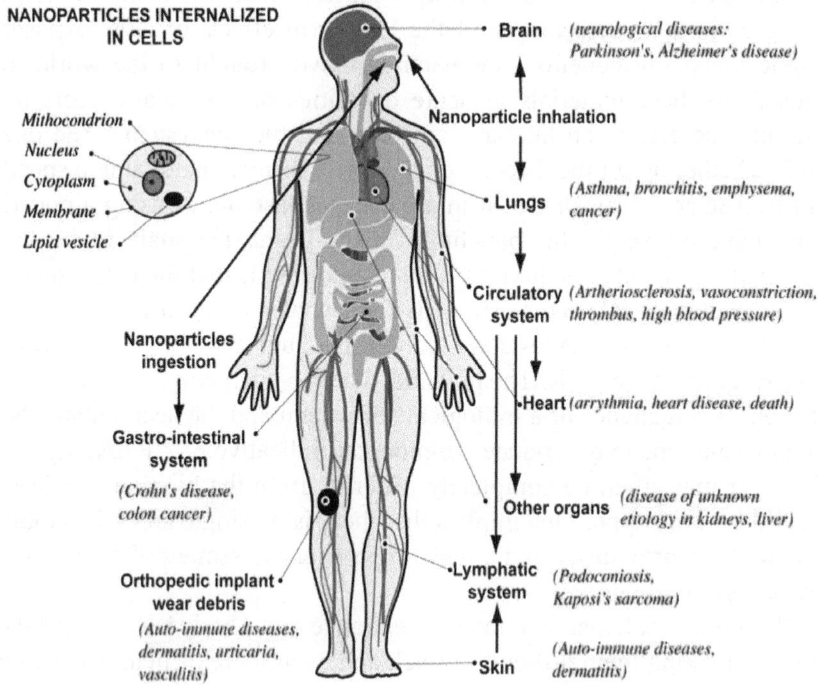

FIGURE 9-1 Schematics of human body with pathways of exposure to nanoparticles, affected organs, and associated diseases from epidemiological, in vivo, and in vitro studies. Reproduced from Ref. [9] with permission.

There are many effects NPs can have at the cellular level. Figure 9-2 highlights general possibilities of NPs' interferences with a biological cell. Toxic levels of an NP can manifest themselves by altering a cell's rate of proliferation,[10] morphology,[11] cell-to-cell communication,[12] and even by altering the path of differentiation.[13] While each of these can serve as biomarkers for nanotoxicity, it can be seen that some of these factors are easier to quantify than others. For example, the rate of cell proliferation in a culture can be easily quantified in real time via an impedance based cell-on-chip (COC) technique, but the change in a cell's morphology, while easily visible under a microscope, is typically not readily quantifiable because it cannot be measurably correlated to the NP concentration present. In many instances, the changes in certain molecular species present within a cell or tissue can serve as a biomarker for NP exposure. For example, NPs can damage cells by generating reactive oxygen species (ROS), physically exerting damage that leads to necrosis/apoptosis, and creating an imbalance in cytoplasmic Ca^{2+} concentration.[14–16] In the latter case, ROS can serve as a biomarker for the toxicity of the NP introduced since it is readily quantifiable.

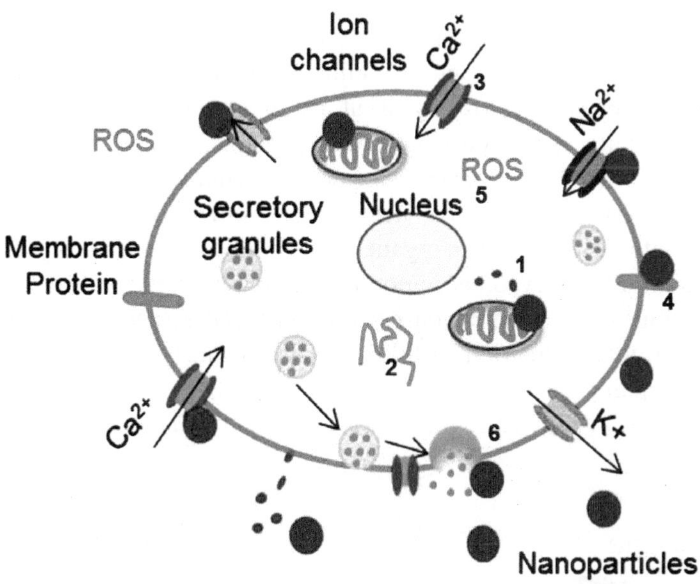

FIGURE 9-2 Schematic of nanoparticle (NP) interaction with a biological cell and possible interferences. Event 1 is represents the intra-cellular leaching of NPs and damaging mitochondrial DNA, event 2 represents the damage to cellular proteins, event 3 represents interference to ion channels, event 4 represents damage to membrane proteins and membrane structure, event 5 is an indicator of reactive oxygen species, and event 6 represents the interference with cellular mechanism, such as cell exocytosis cycle.

Due to the complexity of the interactions between NPs and tissues, some of the analytical challenges of nanotoxicology are (1) characterizing NPs within a biological system, (2) quantifying the rate at which cells and organelles uptake the NPs, (3) determining the localization of the NPs within cells and organelles after uptake, and (4) preventing NPs from interfering with the test method. As stated earlier, there is no singular tool to measure a tissue's response. The first step toward developing a method for quantifying a tissue's response to the presence of NPs is to understand where the particles accumulate once they are brought into a cell. Once the localization of NPs accumulation is known, the adverse effects on biological functions in the target tissue region can be speculated and relevant biomarkers can be elucidated for quantification. More importantly, the deviation from the tissue's normal function can then be quantified.

9.2 FACTORS OF NANOMATERIAL TOXICITY

Even though the introduction of this chapter discussed a few examples of nanotoxicity, it did not focus on providing the reader with the most basic explanation of what nanotoxicity is at its most fundamental level. We define nanotoxicity in the most basic way possible, "an NP or nanomaterial becomes toxic when its shape, size, surface charge, functional groups, chemical composition, reactivity, dose, or any other innate characteristic interferes[17] with the ability of an organelle's, cell's, or tissue's ability to function at optimal levels". Another important factor that plays a role in toxicity is the longevity of the interaction between NP–cell systems (Figure 9-3). How long can this interaction sustain without causing damage that the cell is unable to repair?

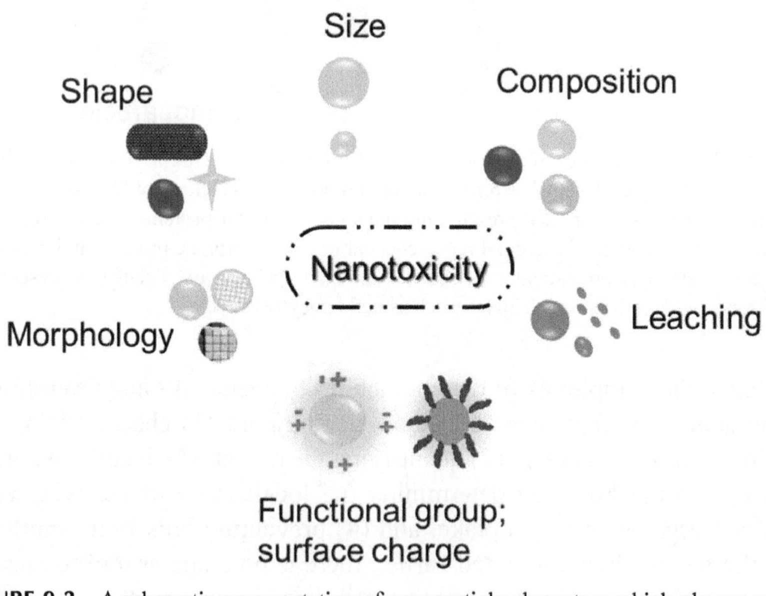

FIGURE 9-3 A schematic representation of nanoparticle characters which plays a role in defining their toxicity. These characteristics are their composition, size, shape, morphology, surface charge and active functional groups, and their leaching in solution.

First let us discuss the intrinsic properties, or chemical composition, of NPs and what research has dictated about their relative toxicities. Some NPs are inherently more toxic than others and examples of this are cadmium oxide (CdO) and copper oxide NPs. Copper oxide NPs, for example, can translocate from somatosensory neurons in the dermis to the dorsal root ganglion causing vacuoles, a disruption of the neural network, and neural detachment.[18] These NPs are far more toxic than more biocompatible materials such as gold (Au) or iron III oxide (Fe_2O_3). This may be due to the biocompatibility of the metal ion used. Gold is relatively nonreactive[19] and iron is already used in the heme groups of hemoglobin. This implies gold is actually relatively stable when introduced into a system, typically only interacting with other gold particles and leaving most of the cell's mechanisms unaffected. Iron, is more biocompatible and thereby less toxic, the body already uses iron in erythrocytes to bind oxygen and deliver it to the deoxygenated tissues. Since iron serves such an important function in the transfer of oxygen, the body has evolved mechanisms to manage excess iron, resulting in a material that can be cleared at a rate that prevents toxicity at the correct doses.[20]

The research being done on particle size and its potential for toxicity indicates that toxicity of an NP increases as the size of the particulate decreases.[21] It is believed that the reason for this is because the particles are too small to be managed well by the cell's uptake and excretion methods. Research shows that smaller NPs have a greater surface-to-volume ratio and thus a higher potential for reactivity with other molecules that enhance the NPs' intrinsic toxicity.[22] A nanomaterial can be too large or too long, as shown with long multiwalled carbon nanotubes (CNTs). It has been shown that a macrophage cannot fully engulf this foreign material if it is of a length longer than 20 μm,[23] of course, this material is no longer in the nanoscale and we do not focus on this mode of phagocytic stress, although a similar mechanism of stress may be in play at the nanoscale.

Besides the size, shape, and intrinsic properties of the nanomaterials being used, their surface properties may also alter their level of toxicity at any given dose regimen. The surface properties, of course, are directly related to defining the zeta potential (surface charge) and reactivity of the NP. However, a NP's seemingly lack of biocompatibility due to its surface properties is not always set in stone. It has been show that these properties can be fine-tuned to yield better interactions by adding a series of more biocompatible functional groups to an NP's surface or by adjusting its zeta potential. A study[24] showed that the surface charge of Au and Ag NPs, with respective sizes of ~26.5 nm and ~33.3 nm, and with a positive surface charge were more likely to be internalized by mast cells than their negatively charge counterparts. While only some examples are provided as to why characterizing NPs is so important in determining their toxic effects, all of these properties must be considered when assessing the safe use of these particles in medical or industrial settings. The assessment of NPs, rather, the assessment of how they exhibit toxicity, is the subject of Section 9.3.

9.3 ASSESSING NANOTOXICITY

As demonstrated earlier, an NP can become toxic to a biological tissue over a wide range of mechanisms and as such, there is a wide range of tests that can be employed to measure these effects. Many reports provide detailed explanations of conventional assessment techniques for quantifying nanotoxicity.[25–27] The methods of assessment that are discussed in this chapter are as follows:

1. Cell viability assays

   These essays measure the viability of the cell type in question using cellular staining techniques and enzymes in order to quantify the cell's membrane integrity, its metabolic rate, and proliferation rate.

2. Functional assays

   These assays detect toxicity at the genomic level such as the formation of ROS, DNA damage, and changes in gene expression.

3. Complementary assays

   These assays are recently developed novel techniques to overcome a few short comings of the traditional assays as well as to provide the measurement of biological parameters which were not possible to monitor by either viability or functional assays. These assay are also called biosensing devices based nanotoxicity assessment assays.[27]

- Lateral flow immunoassays (LFIA)

  These assays are chromatographic in nature and serve as an inexpensive way to measure target molecules without having to employ high-performance liquid chromatography (HPLC) with electrochemical detection (ECD) techniques. These assays can also detect molecules commonly measured using gas chromatography-mass spectrometry

(GCMS), HPLC tandem mass spectrometry, and enzyme-linked immunosorbent assay (ELISA) techniques.

- Atomic force microscopy (AFM)

  A technique sensitive to force that serves to gather the information on cell structure, topography, membrane nanostructures, and mechanics at a single-cell level.

- Carbon-fiber microelectrode (CFM) voltammetry and amperometry

  Uses a very small surface area carbon fiber embedded into a microtubule that can be used to quantify the concentrations of electrochemical species released by cells in response to stress via redox reactions.

- Microfluidic COC biosensors

  These chips can entrap a cell using electric potentials and employ electrochemistry to measure cell functions. This same chip can be used as an impedance-based sensor for measuring the proliferation characteristics of a cell culture under the influence of an introduced stress or stimulus.

9.3.1 CELL VIABILITY ASSAYS

Viability assays are used to measure the viability of a cell, or simply put, the cell's ability to sustain itself. The target of these assays is to quantify a cell's membrane integrity, metabolic rate, rate of proliferation, and whether its death is programmed (apoptotic) or causational (necrotic). These properties define a cell's overall viability or self-sustainability upon assessment. The first assay we describe is a dye based viability assay. These assays use dyes to stain organelles or even the entire cell under specific conditions to provide the observer with an idea of the cell's health. The majority of these assays work by including, converting, or excluding a dye into the cell that can be measured in a colorimetric or fluorescent fashion depending on whether the cell is alive or dead. An example of this class of assay is the diazo dye based trypan blue exclusion assay. This assay excludes all living cells from staining, essentially only leaving all dead tissue cells with a blue stain for identification. This assay characterizes the viability of cells within a population by allowing the observer to quantify how much of the tissue population died as a result of introducing an NP or stimulus. However, trypan blue assays are not without their false positives and false negatives and can sometimes be misleading.

For example, let us assume the introduction of NPs somehow increases the permeability of the cellular membrane. As a result, the dye would be taken up into the cell even though it is still viable, yielding a false positive. Conversely, let us assume a cell is in the earliest stages of apoptosis when the test is employed, at which point the cell's membrane is still intact and permeability remains normal. The cell could trigger apoptosis but the dye cannot penetrate the membrane until the programmed death runs its course, leading to a false negative because while the cell is technically alive during the implementation of the test, the cell is already fated die. This assay may indicate that the cell culture is alive (not stained blue) but not that metabolic activity has stopped. For this reason, multiple assays need to be run to get the gestalt picture of cellular viability. This is where the next assay we discuss becomes extremely useful.

This viability assay is called the Alamar Blue assay. This reagent is a water-soluble resazurin dye that is nontoxic to cells and can be measured by colorimetric change or fluorescence when it is in the presence of incubated cells that are metabolically active. This assay may serve as an extra step in verifying whether or not the trypan blue assay is yielding a true positive or negative result by determining if the cells truly are metabolically active, rather than just by whether or not their membranes are still intact. There are many other fluorescent and nonfluorescent based dye assays that measure cell viability, which we mention for reference only: the calcein AM and propedium iodide based assays, and the neutral red assay.

It can already be seen that there are many assays that can be used to quantify the viability of a cell culture population. Many of these assays overlap in function but each has its own strengths and weaknesses. The lactate dehydrogenase (LDH) assay, just like the trypan blue assay, has been used to identify cellular membrane integrity. LDH is an indicator for lytic cell death. LDH is released in a soluble form into the extracellular medium providing the observer with a biomarker for identifying damaged cellular membranes. Tetrazolium salt based assays such as MTT, MTS, and WST are commonly used to measure the rate of proliferation of a cell population in culture. This metabolic assay employs MTT, which is reduced by cells into a nonsoluble formazan dye, which can be quantified within a blue to purple spectral range. This dye is then made soluble by DMSO to attain an approximation of cell viability via metabolic rate. The more viable cells will exhibit a higher metabolic rate and will elicit a darker blue color. Alamar Blue offers a simple and sensitive alternative to MTT.

Another cell proliferation assay that will only be referenced is the [^{3}H] thymidine-based assay. Newly synthesized DNA will uptake ^{3}H, which can be correlated to a cell population's proliferation rate. Once again, this is an alternative test to be used to confirm the results of those mentioned earlier.

Why are there so many assays to measure what seems to be very similar viability parameters? Well, many researchers have studied the data obtained from cell viability assays and found inconsistencies between them. It is not possible for one assay to quantify a particular parameter of cell viability for all the various materials that can be introduced to the culture. Riviere and Zhang[28] found that many nanomaterials could interact with the dyes employed by the assay to quantify a cellular response. In some cases, a nanomaterial may even absorb the dye and skew the fluorescent and colorimetric data yielding inconsistent results between the various dyes and nanomaterials they employed in their studies. This leads us back to the point mentioned in the introduction, which states that "there is no singular golden standard for NPs toxicity assessment". It is always recommended that more than one dye-based assay be employed in a controlled study of nanotoxicity.

An example of the nanomaterial–dye interaction mentioned above is provided by Casey et al.[29] This group found that the CNTs they were using in their study were not only interfering with their Alamar Blue assay, but also with their subsequent attempts at using MMT and neutral red dyes. Situations like these raised concerns among the scientific community about how reliable conventional viability and cytotoxicity assays are when screening for the toxicity of current and future nanomaterials. This realization developed into an urgent need for alternative techniques to the ones available for assessing nanotoxicity and served to highlight that the limitations of dye-based techniques are greater than expected when NPs are the subject of the study.

Recently, studies have tried focusing on analyzing the toxic effects of NPs on a single cell rather than on a cell population. These single-cell studies become important when dealing with rare cells, neuronal cells, or cancerous cells as these cells does not always behave in a similar manner to their population, even in the same environmental conditions. In such studies, the nanotoxicity assessment assays previously mentioned are neither ideal, nor are they feasible, because they require a huge cell population. More importantly, they make assumptions of how an individual cell interacts with NPs by analyzing the population as a whole, which is not always ideal for certain studies. When single-cell cytotoxicity studies are required, a few optical cell viability assays can be combined with flow cytometry for higher throughput detection of cells. An example of this is Annexin V, which marks the outer surface of a cell's plasma membrane by competing with phosphatidylserine for active binding sites. Annexin V (or A5) is thus, a marker that binds to the plasma membrane's surface and is indicative of a cell that is in its early stages of apoptosis. The Annexin V marker can then be labeled with radioactive or fluorescent molecules to serve as a method for detecting apoptotic cells. Another assay with similar function to the Annexin V is the TUNEL assay. This assay is also used to visualize apoptotic cells but functions by labeling the terminal end of nucleic acids in fragmented DNA. The last assay to be discussed is the colony-forming assay, which indicates the ability of a single or relatively few cell(s) to form a colony over the period of a few days. This assay does not require additional labeling of markers or tagging, and directly measures the health cells even of small population.

9.3.2 CELL FUNCTIONAL ASSAYS

Functional assays serve to detect changes in gene expression, DNA damage, and the formation of ROS at the genomic level that result from nanotoxicity. Before we describe these assays we must first explain what ROS is and highlight the fact that ROS are oxidants, implying that when an oxygen species is produced it is highly reactive and must remove an electron from one of its surrounding molecules, which in turn generates another reactive species, an example of which is superoxide. Oxidative stress generates intracellular radicals that can damage nucleic acids (DNA and RNA), lipids, and proteins. There is a lot of hype in consumable products that advertise the antioxidant abilities for this very reason. Oxidative stress is a common observation when NPs are introduced to cell culture.[30, 31] The glutathione (GSH) assay is able to directly and indirectly measure intracellular ROS generation with a luminescent-based mode of detecting and quantifying the level of GSH within cells. Lipid peroxidation is also indicative of increased oxidative damage to cells which can be quantified via a lipid peroxidation assay. An example of this class of assay is the 2,7-Dichlorodihydrofluorescein (DCFH) assay, which operates by measuring the intracellular oxidation of DCFH in the presence of hydrogen peroxide and serves to measure the generation of free radicals in tissue cells.

Oxidative stress leads to cell and tissue inflammatory responses. While ROS generated oxidative stress biomarkers can be measured with the techniques mentioned earlier, it is also useful to measure cytotoxicity with respect to the biomarkers indicative of cellular inflammation. Cytokines, in particular, serve as a quantifiable biomarker for an inflammatory cellular response. These can be measured using ELISA and can be specific to the various cytokines released during cellular stress. The downfall of ELISA style immunoassays is their time-consuming nature and need for multiple steps for proper detection.

Damage to DNA can also be assessed at the single-cell level via a single-cell gel-electrophoresis method, also called the comet assay. This is a highly sensitive technique for DNA fragments and mitochondrial DNA greater than 50 kbp. In addition to DNA damage detecting assays, there are other oxidized guanine base techniques that can identify mutations in certain genes that result from oxidative stress. Immunohistochemistry and HPLC techniques are most useful in the latter type of studies. In addition, polymerase chain reaction (PCR) or reverse transcription-PCR arrays can be used to identify the panel of genes in question. In other studies, if the integrity of the chromosomes is in question, a karyotype analysis can provide the number of chromosomes affected by monitoring cells undergoing division for micronucleus anomalies in the presence of NPs. Newer and better techniques are currently gaining popularity to identify new parameters and also address few of the shortcomings of the latter assays. In Section 9.4 we address a few of the new outside-the-box techniques being employed by research laboratories around the world.

9.4 COMPLEMENTARY ASSAYS

As discussed in Section 9.3.2, there are many techniques that can be used to quantify cell viability under the influence of NPs. Each has its drawbacks and limitations that make them less ideal for certain types of studies involving smaller cell populations or in conditions that may yield erroneous results. For this reason, new technologies are being employed to mitigate these shortcomings. Many of these techniques rely on more sophisticated methods of detection that are device based. These relatively new techniques and how they can serve to evaluate yet to be discovered characteristics of NP–tissue interactions are addressed in a subsequent section of this chapter, as discussed below

9.4.1 LATERAL FLOW IMMUNOASSAYS

These assays are chromatographic in nature and serve as an inexpensive way to measure target molecules without having to employ HPLC with ECD techniques. These assays can also detect molecules commonly measured using GCMS, HPLC tandem mass spectrometry, and ELISA techniques. LFIA are typically composed of films (paper strips with small pads) that contain all the steps required to detect a molecule of interest in a colorimetric fashion.[32] A small drop of the media containing the molecule being tested is placed on one end of the film and the films' solvent will carry the contents to the opposing end where the results can be seen in color. These assays are more common than you would think by reading the description above. This assay type is commonly used in household tests for drug abuse, pregnancy, and glucose measurements.[32] These tests are now being fashioned to measure oxidative stress biomarkers induced by ROS species. Due to the time-consuming nature, expensive equipment, and specialized skill required to perform techniques like HPLC, LFIA are growing in popularity among the research community. 8-Hydroxyguanine and 8-Hydroxy-2′-deoxyguanosine (8-OHdG), its nucleoside, are of the most studied guanine bases using advanced HPLC, GCMS, and ELISA techniques. Our group sought to prove the effectiveness of the LFIA by creating an LFIA, integrated with a paper-thin CNT electrode (Figure 9-4), to accurately measure 8-OHdG concentration resultant from nanotoxicity at the genomic level.[33,34] The dual approach

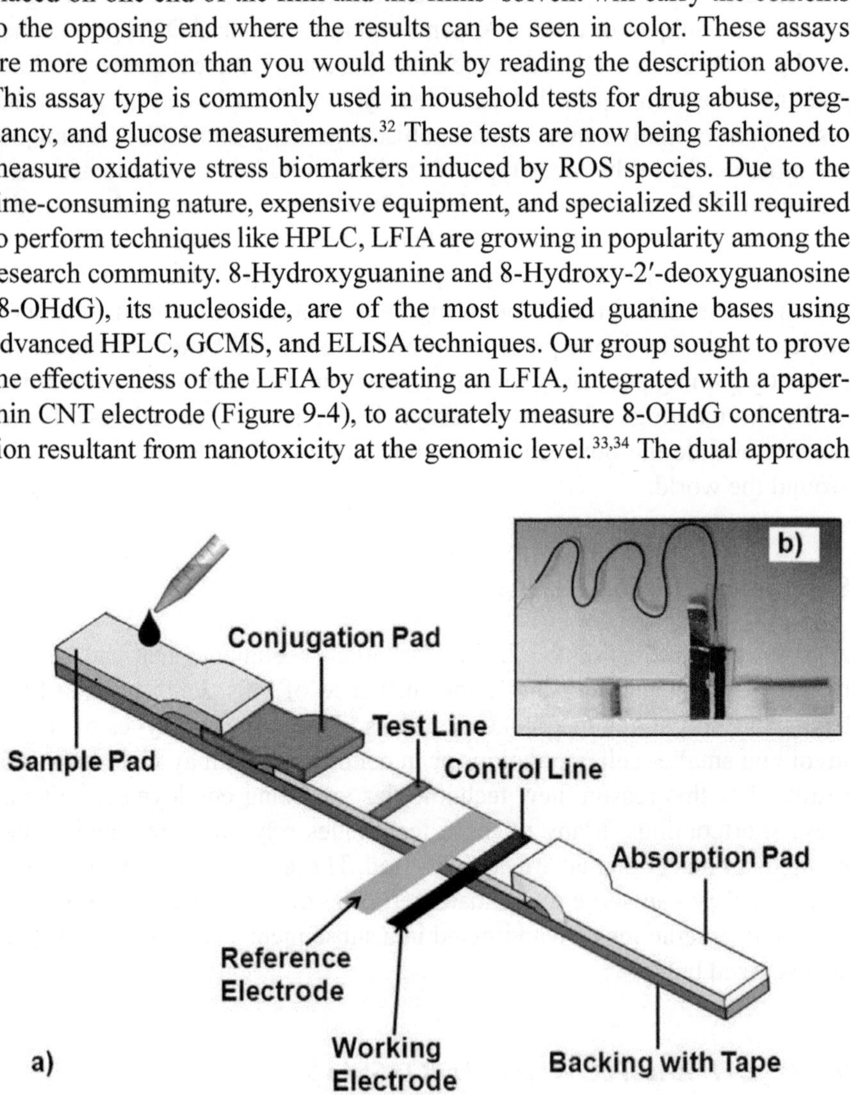

FIGURE 9-4 (A) A schematic and (B) photograph of a lateral flow immunoassay device integrated with a carbon nanotube electrode. Reproduced from Ref. [33] with permission of The Royal Society of Chemistry.

provides optical observation with highly sensitive electrochemical quantitative measurement of DNA damage biomarker, 8-OHdG. This approach is a simple and inexpensive analytical tool for detecting NP-induced toxicity. The other research groups are tailoring the LFIA platform to measure other biomarkers like cholesterol as well.

9.4.2 ATOMIC FORCE MICROSCOPY

AFM is an amazing tool for successfully elucidating the topography, structure, membrane nanostructures, and mechanics of mammalian cells at the single-cell level with nanoscale resolution under near-physiological conditions using a technique that is sensitive to atomically applied forces. The device can even measure the elasticity and adhesion of cells that comprise a tissue. The AFM technique, in combination with fluorescence microscopy, was used to scan and quantify the uptake and localization of silicon dioxide (SiO_2) NPs in a study by Blechinger et al.,[35] making this a useful technique for quantifying the mechanics of a cell influenced by NPs. Wu et al. also used this AFM technique to study the biophysical effects of vascular endothelial cells when exposed NPs to diesel fuel exhaust.[36] This allowed them to attain valuable data on the Young's modulus and topography of the cellular membrane under the diesel NP exposure.

9.4.3 CARBON-FIBER MICRO-ELECTRODE VOLTAMMETRY AND AMPEROMETRY

A CFM is a powerful tool to use in conjunction with voltammetric and amperometric techniques. These techniques employ electrochemistry to quantify the presence of a molecular species that can be oxidized or reduced. These techniques require a potentiostat and three electrodes: a working electrode (carbon or copper), a reference electrode, (silver–silver chloride), and a counter electrode (platinum or gold). A constant electrical potential or range of potentials can be applied to the working electrode to scan for a molecular species that will be oxidized or reduced at the said potential. These electrochemical techniques are further refined by making the working electrode's surface area smaller to yield highly sensitive results and low levels of noise under the right conditions. This can be achieved by creating a working electrode with a carbon fiber insulated by a capillary tube with an exposed fiber surface of ~5–10 μm in length, hence the name CFM.

The CFMs can detect the exocytosis of contents from a single cell and have the ability to detect diffusion limited current at extremely high scan rates to attain impeccable temporal resolution. Exploring the biophysical properties of cellular exocytosis with CFM amperometry has become an important tool in understanding cell-to-cell communication and quantifying NPs-induced stress biomarkers released by cells in real time. This technique is especially useful in studying neurotransmitter release patterns from neuronal tissues and single cells, both in vivo and in vitro. Marquise et al. performed a 48-h amperometry study to characterize the effect of gold NPs (12–46 nm) on the exocytosis of serotonin from a murine peritoneal mast cell in fibroblast culture.[37] It was observed that serotonin exocytosis increased per granule, there was an incremental expansion in the intracellular matrix, and a decrease in granule fusion and transport events. Love and Haynes,[38] evaluated the effects of Au (28 nm) and Ag (61 nm) citrate-reduced NPs on neuroendocrine cells. Their study used CFM amperometry to determine if there was a change in chromaffin exocytosis during the 24-h exposure to Au and Ag NPs.

As mentioned earlier, many consumer products contain NPs that many are not aware of. This is especially true for metal–oxide NPs (MO–NPs) like nonporous TiO_2 and nonporous SiO_2. These are used in cigarette filters, paints, and even some vitamins. A study conducted by Maurer-Jones et al. employed the CFM amperometry technique to explore the toxic effects of MO–NPs on immune cells and found that the presence of these NPs triggered functional changes in the secretions of chemical messengers from mast cell granules.[39] As seen in the examples above, the CFM electrochemical technique has many specific applications for real-time monitoring of vesicles that would otherwise not be possible using the assays described earlier.

9.4.4 MICROFLUIDIC COC APPROACH

Another promising nanotoxicity assessment tool that allows real-time and rapid multisampled analysis of cellular tissues in vitro is the microchip based biosensor approach. Chip based assessment systems allow the user flexibility in design. These chip-based systems look like chips that would be pulled out of one's computer since the chip's electrodes are typically etched or printed onto its surface just as a circuit board would be. The chip can thus be designed to contain many preprinted circuits that are independent from one another. From here a polystyrene wall can be built to surround an independent electrode set to serve as a well for a cell culture and its medium.

The cells will grow on the chip-electrode's surface and bridge the circuit between the working electrode and the grounding electrode. The cells can be measured by running a small current through them via the electrodes; however, as the cells grow, the resistance of the circuit changes and this impedance-based system can measure the growth or decay of a population in real time under NP's influence.

In a study conducted by our group, Hondroulis et al.,[40] a whole cell based electrical impedance sensing approach was developed and employed to assess the effects of silver and gold NPs, single-walled CNTs, and CdO in real time as described earlier. This is a noninvasive method of quantifying cellular functionality and viability under NP's exposure which can be used for many applications, especially in single-cell analysis. Our group recently developed such a chip, capable of selectively entrapping a single cell in a small microscale well using dielectrophoretic principles.[41] Figure 9-5 shows chip-based devices developed by our group for nanotoxicity assessment at the single-cell level.

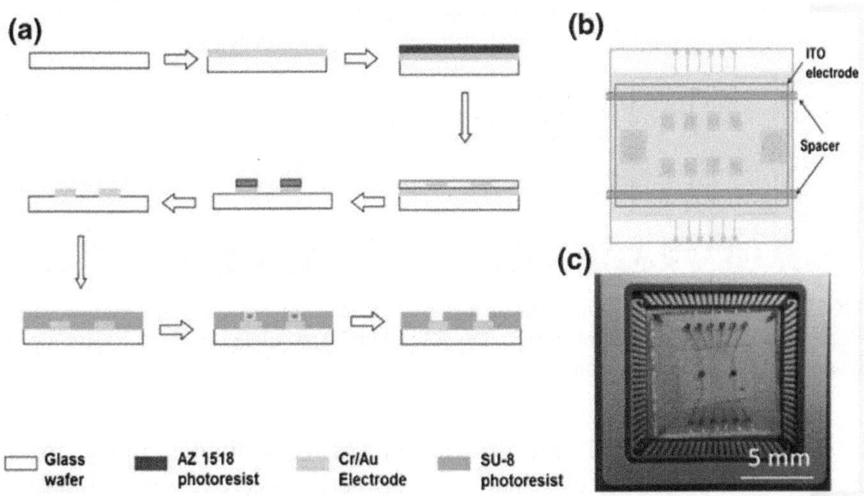

FIGURE 9-5 Chip based nanotoxicity devices. (A) Stepwise demonstration of lithography fabrication process, (B) chip assembly with a microfluidic channel and a top indium–tin oxide electrode for dielectrophoretical single cell trapping, and (C) the chip is connected to a power line carrier communication adapter. Reprinted from Ref. [41] with permission.

The same electrodes used to guide the cell into the microwells can be used to conduct electrochemical studies similar to those of the CFM techniques. The electrodes on the chip can also be triggered independently to capture single cells in wells while others conduct the study on biomarkers for NPs introduced to the system without the cumbersome setup of the CFM technique. This allows the study to be conducted far more easily under incubation and with little contact from the observer. This technique has been used to monitor the stress-induced release of catecholamines from rat pheochromocytoma, an adrenal tumorigenic cell line, when exposed to silver NPs.[42] The COC device in this experiment is used to identify the effects of 20 nm citrate coated silver NPs on exocytosis properties of a single PC12 cell, as shown in Figure 9-6.

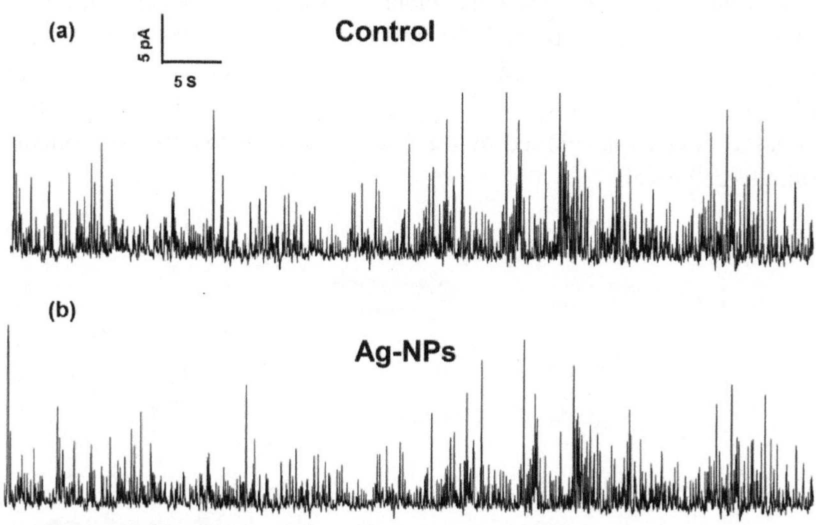

FIGURE 9-6 Amperometric measurements of catacalomines exocytosis from a single PC12 cell, where (A) is control and (B) is exposed with 20 nm silver nanoparticles. Reprinted from Ref. [42] with permission.

Electrochemical and optical measurement based fluidic chip platform have been explored for cell analysis over the last decade[43] which hold the potential to be used for nanotoxicity assessment. Microfluidic channels have become an important tool in the field of tissue engineering or fields that need to mimic the flow of a certain environment. Kim et al. conducted a study employing such a strategy to study the nanotoxicity of mesomorphous silica particles (<50 nm) on human endothelial cell that were cultured in a blood vessel mimicking environment thanks to the use of microfluidics.[44] This microfluidic vessel like environment allowed the cells to receive the nutrients they need in a bioreactor-styled fashion and allowed the research team to quantify the shear–stress characteristic of tissues that is typically difficult to reproduce in vitro. Microfluidics and COC techniques have increased the likelihood of successful studies that employ more accurate physiological environments. A larger role of microfluidic device based nanotoxicity assays are predicted[45] to handle the challenges of a huge number of NPs with various properties to be tested. An integrative approach of single cell, a small colony of cells and a tumor spheroid can provide valuable information of how NP interact with biological system from a single cell to the tissue level. Such a measurement device is shown in Figure 9-7.[45]

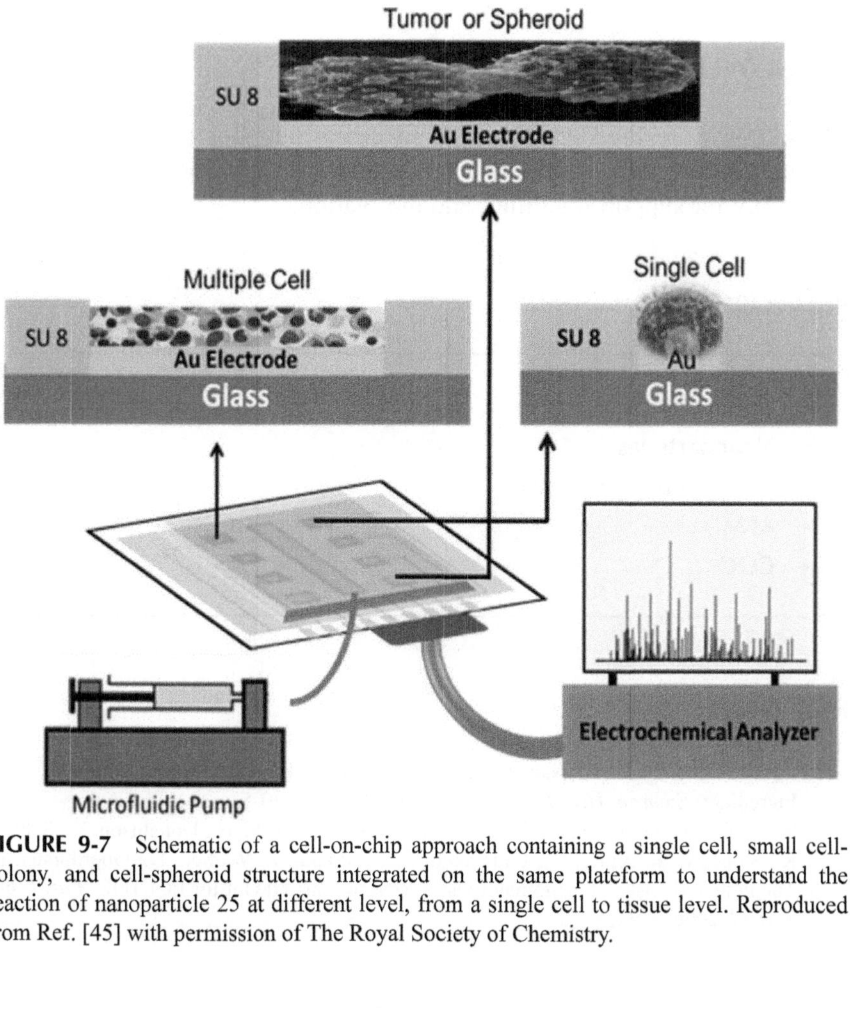

FIGURE 9-7 Schematic of a cell-on-chip approach containing a single cell, small cell-colony, and cell-spheroid structure integrated on the same plateform to understand the reaction of nanoparticle 25 at different level, from a single cell to tissue level. Reproduced from Ref. [45] with permission of The Royal Society of Chemistry.

CONCLUSION

With increasing applications of nanomaterials, the possibilities of their exposure to human and environment are abundance. In that case, accessing their toxicity should be the first step toward creating guidelines for safe use and disposal of nanomaterials. In this chapter, we have discussed various nanomaterial aspects responsible for toxic response toward biological systems. We also discussed various approaches to assess nanomaterial toxicity, such as viability assay and functional assays, along with the latest approaches to collect the information of nanomaterial toxicity not available by traditional assays.

ACKNOWLEDGMENT

We acknowledge the Department of Electrical and Computer Engineering and the Department of Biomedical Engineering of Florida International University for support, facilities, and motivation.

KEYWORDS

- **Nanotoxicity**
- **Nanoparticles**
- **Reactive oxygen species**
- **AFM**
- **COC**

REFERENCES

1. Kessler, R. Engineered Nanoparticles in Consumer Products: Understanding a New Ingredient. *Environ. Health Perspect.* **2011,** *119* (3), a120-5.
2. Borm, P. J.; Robbins, D.; Haubold, S.; Kuhlbusch, T.; Fissan, H.; Donaldson, K.; Schins, R.; Stone, V.; Kreyling, W.; Lademann, J.; Krutmann, J.; Warheit, D.; Oberdorster, E. The Potential Risks of Nanomaterials: A Review Carried Out for ECETOC. *Part Fibre Toxicol.* **2006,** *3*, 11.

3. Mossman, B. T.; Borm, P. J.; Castranova, V.; Costa, D. L.; Donaldson, K.; Kleeberger, S. R. Mechanisms of Action of Inhaled Fibers, Particles and Nanoparticles in Lung and Cardiovascular Diseases. *Part Fibre Toxicol.* **2007,** *4*, 4.
4. Oberdorster, G.; Sharp, Z.; Atudorei, V.; Elder, A.; Gelein, R.; Kreyling, W.; Cox, C. Translocation of Inhaled Ultrafine Particles to the Brain. *Inhal. Toxicol.* **2004,** *16* (6–7), 437–445.
5. Medina, C.; Santos-Martinez, M. J.; Radomski, A.; Corrigan, O. I.; Radomski, M. W. Nanoparticles: Pharmacological and Toxicological Significance. *Br. J. Pharmacol.* **2007,** *150* (5), 552–558.
6. Lockman, P. R.; Mumper, R. J.; Khan, M. A.; Allen, D. D. Nanoparticle Technology for Drug Delivery Across the Blood-Brain Barrier. *Drug Dev. Ind. Pharm.* **2002,** *28* (1), 1–13.
7. Silva, G. A. Neuroscience Nanotechnology: Progress, Opportunities and Challenges. *Nat. Rev. Neurosci.* **2006,** *7* (1), 65–74.
8. Jain, K. K., Applications of Nanobiotechnology in Clinical Diagnostics. *Clin. Chem.* **2007,** *53* (11), 2002–2009.
9. Buzea, C.; Pacheco, I.; Robbie, K. Nanomaterials and Nanoparticles: Sources and Toxicity. *Biointerphases* **2007,** *2* (4), MR17–MR71.
10. Shi, Z.; Huang, X.; Cai, Y.; Tang, R.; Yang, D. Size Effect of Hydroxyapatite Nanoparticles on Proliferation and Apoptosis of Osteoblast-Like Cells. *Acta Biomater.* **2009,** *5* (1), 338–345.
11. Buyukhatipoglu, K.; Clyne, A. M. Superparamagnetic Iron Oxide Nanoparticles Change Endothelial Cell Morphology and Mechanics via Reactive Oxygen Species Formation. *J. Biomed. Mater. Res. A* **2011,** *96* (1), 186–195.
12. Love, S. A.; Liu, Z.; Haynes, C. L. Examining Changes in Cellular Communication in Neuroendocrine Cells After Noble Metal Nanoparticle Exposure. *Analyst* **2012,** *137* (13), 3004–3010.
13. Tautzenberger, A.; Lorenz, S.; Kreja, L.; Zeller, A.; Musyanovych, A.; Schrezenmeier, H.; Landfester, K.; Mailander, V.; Ignatius, A. Effect of functionalised Fluorescence-Labelled Nanoparticles on Mesenchymal Stem Cell Differentiation. *Biomaterials* **2010,** *31* (8), 2064–2071.
14. Suh, W. H.; Suslick, K. S.; Stucky, G. D.; Suh, Y. H. Nanotechnology, Nanotoxicology, and Neuroscience. *Prog. Neurobiol.* **2009,** *87* (3), 133–170.
15. Fedorovich, S. V.; Alekseenko, A. V.; Waseem, T. V. Are Synapses Targets of Nanoparticles? *Biochem. Soc. Trans.* **2010,** *38* (2), 536–538.
16. George, S.; Pokhrel, S.; Xia, T.; Gilbert, B.; Ji, Z. X.; Schowalter, M.; Rosenauer, A.; Damoiseaux, R.; Bradley, K. A.; Madler, L.; Nel, A. E. Use of a Rapid Cytotoxicity Screening Approach To Engineer a Safer Zinc Oxide Nanoparticle through Iron Doping. *ACS Nano* **2010,** *4* (1), 15–29.
17. Chou, L. Y.; Ming, K.; Chan, W. C. Strategies for the Intracellular Delivery of Nanoparticles. *Chem. Soc. Rev.* **2011,** *40* (1), 233–245.
18. Prabhu, B. M.; Ali, S. F.; Murdock, R. C.; Hussain, S. M.; Srivatsan, M. Copper Nanoparticles Exert Size and Concentration Dependent Toxicity on Somatosensory Neurons of Rat. *Nanotoxicology* **2010,** *4* (2), 150–160.
19. Mieszawska, A. J.; Mulder, W. J. M.; Fayad, Z. A.; Cormode, D. P. Multifunctional Gold Nanoparticles for Diagnosis and Therapy of Disease. *Mol. Pharmaceutics* **2013,** *10* (3), 831–847.

20. Longmire, M.; Choyke, P. L.; Kobayashi, H. Clearance Properties of Nano-sized Particles and Molecules as Imaging Agents: Considerations and Caveats. *Nanomedicine (London, England)* **2008,** *3* (5), 703–717.
21. Nel, A.; Xia, T.; Madler, L.; Li, N. Toxic Potential of Materials at the Nanolevel. *Science* **2006,** *311* (5761), 622–627.
22. Donaldson, K.; Stone, V.; Tran, C. L.; Kreyling, W.; Borm, P. J. Nanotoxicology. *Occup. Environ. Med.* **2004,** *61* (9), 727–728.
23. Dorger, M.; Munzing, S.; Allmeling, A. M.; Messmer, K.; Krombach, F. Differential Responses of Rat Alveolar and Peritoneal Macrophages to Man-Made Vitreous Fibers In Vitro. *Environ. Res.* **2001,** *85* (3), 207–214.
24. Marquis, B. J.; Liu, Z.; Braun, K. L.; Haynes, C. L. Investigation of Noble Metal Nanoparticle Zeta-Potential Effects on Single-Cell Exocytosis Function In Vitro with Carbon-Fiber Microelectrode Amperometry. *Analyst* **2011,** *136* (17), 3478–3486.
25. Hillegass, J. M.; Shukla, A.; Lathrop, S. A.; MacPherson, M. B.; Fukagawa, N. K.; Mossman, B. T. Assessing Nanotoxicity in Cells In Vitro. *Wiley Interdiscip Rev. Nanomed. Nanobiotechnol.* **2010,** *2* (3), 219–231.
26. Marquis, B. J.; Love, S. A.; Braun, K. L.; Haynes, C. L. Analytical Methods to Assess Nanoparticle Toxicity. *Analyst* **2009,** *134* (3), 425–439.
27. Hondroulis, E.; Shah, P.; Zhu, X.; Li, C.-Z. Biosensing Devices for Toxicity Assessment of Nanomaterials. *Biointeractions of Nanomaterials*; CRC Press, **2014**, 117.
28. Monteiro-Riviere, N. A.; Inman, A. O.; Zhang, L. W. Limitations and Relative Utility of Screening Assays to Assess Engineered Nanoparticle Toxicity in a Human Cell Line. *Toxicol. Appl. Pharmacol.* **2009,** *234* (2), 222–235.
29. Casey, A.; Herzog, E.; Davoren, M.; Lyng, F. M.; Byrne, H. J.; Chambers, G. Spectroscopic Analysis Confirms the Interactions between Single Walled Carbon Nanotubes and Various Dyes Commonly Used to Assess Cytotoxicity. *Carbon* **2007,** *45* (7), 1425–1432.
30. Jones, C. F.; Grainger, D. W., In Vitro Assessments of Nanomaterial Toxicity. *Adv. Drug. Deliv. Rev.* **2009,** *61* (6), 438–456.
31. Chang, Y.; Yang, S. T.; Liu, J. H.; Dong, E.; Wang, Y.; Cao, A.; Liu, Y.; Wang, H. In Vitro Toxicity Evaluation of Graphene Oxide on A549 Cells. *Toxicol. Lett.* **2011,** *200* (3), 201–210.
32. Shah, P.; Zhu, X.; Li, C. Z. Development of Paper-Based Analytical Kit for Point-of-Care Testing. *Expert. Rev. Mol. Diagn.* **2013,** *13* (1), 83–91.
33. Zhu, X.; Shah, P.; Stoff, S.; Liu, H.; Li, C. Z. A Paper Electrode Integrated Lateral Flow Immunosensor for Quantitative Analysis of Oxidative Stress Induced DNA Damage. *Analyst* **2014,** 139, 2850–2857.
34. Zhu, X.; Hondroulis, E.; Liu, W.; Li, C. Z. Biosensing Approaches for Rapid Genotoxicity and Cytotoxicity Assays upon Nanomaterial Exposure. *Small* **2013,** *9* (9–10), 1821–1830.
35. Blechinger, J.; Bauer, A. T.; Torrano, A. A.; Gorzelanny, C.; Brauchle, C.; Schneider, S. W. Uptake Kinetics and Nanotoxicity of Silica Nanoparticles are Cell Type Dependent. *Small* **2013,** *9* (23), 3970–3980.
36. Wu, Y.; Yu, T.; Gilbertson, T. A.; Zhou, A.; Xu, H.; Nguyen, K. T. Biophysical Assessment of Single Cell Cytotoxicity: Diesel Exhaust Particle-Treated Human Aortic Endothelial Cells. *PLoS One* **2012,** *7* (5), e36885.

37. Marquis, B. J.; McFarland, A. D.; Braun, K. L.; Haynes, C. L. Dynamic Measurement of Altered Chemical Messenger Secretion After Cellular Uptake of Nanoparticles Using Carbon-Fiber Microelectrode Amperometry. *Anal. Chem.* **2008,** *80* (9), 3431–3437.
38. Love, S. A.; Haynes, C. L. Assessment of Functional Changes in Nanoparticle-Exposed Neuroendocrine Cells with Amperometry: Exploring the Generalizability of Nanoparticle-Vesicle Matrix Interactions. *Anal. Bioanal. Chem.* **2010,** *398* (2), 677–688.
39. Maurer-Jones, M. A.; Lin, Y. S.; Haynes, C. L. Functional Assessment of Metal Oxide Nanoparticle Toxicity in Immune Cells. *ACS Nano* **2010,** *4* (6), 3363–3373.
40. Hondroulis, E.; Liu, C.; Li, C. Z. Whole Cell Based Electrical Impedance Sensing Approach for a Rapid Nanotoxicity Assay. *Nanotechnology* **2010,** *21* (31), 315103.
42. Shah, P.; Zhu, X. N.; Chen, C. Y.; Hu, Y.; Li, C. Z. Lab-on-Chip Device for Single Cell Trapping and Analysis. *Biomed. Microdevices* **2014,** *16* (1), 35–41.
42. Shah, P.; Yue, Q.; Zhu, X.; Xu, F.; Wang, H.-S.; Li, C.-Z. PC12 Cell Integrated Biosensing Neuron Devices for Evaluating Neuronal Exocytosis Function Upon Silver Nanoparticles Exposure. *Sci. China Chem.* **2015**, *58* (10), 1600–1604.
43. Velve-Casquillas, G.; Le Berre, M.; Piel, M.; Tran, P. T. Microfluidic Tools for Cell Biological Research. *Nano Today* **2010,** *5* (1), 28–47.
44. Kim, T. H.; Kang, S. R.; Oh, B. K.; Choi, J. W. Cell Chip for Detection of Silica Nanoparticle-Induced Cytotoxicity. *Sensor Lett.* **2011,** *9* (2), 861–865.
45. Shah, P.; Kaushik, A.; Zhu, X.; Zhang, C.; Li, C. Z. Chip Based Single Cell Analysis for Nanotoxicity Assessment. *Analyst* **2014,** *139* (9), 2088–2098.

CHAPTER 10

NANOBIOSENSOR TECHNOLOGY FOR CARDIOVASCULAR DISEASES

RAJU KHAN[1*], and AJEET KAUSHIK[2]

[1]*Analytical Chemistry Division, CSIR-North East Institute of Science and Technology, Academy of Scientific and Innovative Research, Jorhat, Assam, India*

[2]*Center of Personalized Nanomedicine, Department of Immunology, Herbert Wertheim College of Medicine, Florida International University, Miami, USA*

**E-mail: khan.raju@gmail.com*

CONTENTS

Cardiovascular disease (CVD) is considered as a major threat to global health. Therefore, there is a growing demand for a range of portable, rapid, and low cost biosensing devices for the detection of CVD. Biosensors can play an important role in the early diagnosis of CVD without having to rely on hospital visits where expensive and time-consuming laboratory tests are recommended. A biosensor is a detecting device that combines a transducer with biologically sensitive and selective components. The interests in health-care system drive the development of various point-of-care (POC) systems to diagnose cardiac diseases. This chapter provides an overview of the available biosensor platforms for the detection of various CVD markers and considerations of future prospects for the technology are also addressed.

10.1 INTRODUCTION

The interests in health-care system drive the development of various POC systems to diagnose diseases such as diabetes, cancers, and cardiac diseases. The detection of glucose for diabetes is a representative commercialized POC system, which is very useful to determine blood sugar level at the comfort of our home. However, the diagnoses of other important diseases such as cancer or cardiac disease have not been realized in home care systems, because they require complicated processes to detect biomarkers. Therefore, many researchers have actively studied the mechanisms to develop a biosensing system so that such a system could be used for detection of cancer or cardiac disease by using various materials (e.g., nanomaterials). The use of nanomaterials and structures such as semiconductors and conducting polymer nanowires, and nanoparticles (NPs) (carbon nanotubes [NTs], silica NPs, dendrimers, noble metals NPs, gold nanoshells, superparamagnetic NPs quantum dots, polymeric NPs, etc.) for biosensor applications is expanding rapidly. However, some limitations remain with respect to the uniformity of production, biocompatibility of inorganic nanomaterials, and reliability of performance. The increasing need to monitor our health and environment in real time demands reliable miniature sensors that are able to detect a wide range of molecules. Sensitive, selective, and cost-effective analysis of biomolecules is important in clinical diagnosis and treatment. Biosensors are analytical devices constituted by a biorecognition element for a particular analyte and a transducer that converts the biorecognition event into a useful electrical signal.[1] The early diagnosis is crucial for patient survival and successful prognosis of the disease; therefore, sensitive and specific methods are required in this regard. Among the numerous diseases that affect humans, three are particularly relevant because of their worldwide incidence, prevalence, morbidity and mortality. They are diabetes, CVD, and cholesterol disorders.

Universally, CVD is recognized as the prime cause of death with estimates exceeding 20 million by 2015 due to heart disease and stroke. Facts regarding the disease, its classification, and diagnosis are still insufficient. However, understanding the issues involved in its initiation, symptoms, and early detection will reduce the high risk of sudden death associated with it. CVDs are the world's leading cause of death according to the World Health Organization statistics, and acute myocardial infarction (AMI) is the leading cause of morbidity and mortality among patients suffering from CVDs.[2] The determination of the levels of cardiac troponin I (cTnI), cardiac troponin T (cTnT), creatine kinase-MB (CK-MB), and myoglobin (Myo) has been used in the determination of AMI.[3] Among these cardiac markers, cTnI has been recognized as the gold standard for AMI diagnosis because of its superior cardiac specificity and selectivity.[4]

In recent years, the demand has grown in the field of medical diagnostics for simple and disposable devices that also demonstrate fast response times, are user friendly, cost efficient, and are suitable for mass production. Biosensor technologies offer the potential to fulfill these criteria through an interdisciplinary combination of approaches from nanotechnology, chemistry, and medical science. In summary, for a reader who is interested in understanding CVDs and new to the area of nanotechnology, this chapter will provide a mixed bag. Nanotechnology is having a rapidly expanding impact on all solutions can help the field, including biology and medicine. At the basic end of the biological spectrum, nanotechnology tools such as atomic force microscopy (AFM), optical tweezers, and NP-based reporters provide unique insight into biological processes. At the clinical end of the spectrum, nanotechnology provides novel opportunities for a wide range of applications, including drug and nucleic acid delivery, diagnostic imaging, regenerative medicine, and in vitro diagnostics. Among the numerous diseases affecting humans, three are relevant because of their worldwide incidence, prevalence, morbidity and mortality, namely diabetes, and CVD. In recent years, the demand has grown in the field of medical diagnostics for simple and disposable devices that also demonstrate fast response times, are user friendly, cost efficient, and are suitable for mass production. Biosensor technologies offer the potential to fulfill these criteria through an interdisciplinary combination of approaches from nanotechnology, chemistry, and medical science.

We also need to move our fundamental approach to health care from a reactive model to a wellness-oriented model. Here, the focus is on keeping people healthy for as long as possible with the least cost to the system. Sensors play an integral role in numerous modern industrial applications, including food processing and everyday activities such as transport, air quality, medical therapeutics, and many more. In this chapter, we explore a wide range of topics related to sensing, sensor systems, and applications for monitoring health, wellness, and the environment. This chapter will be particularly useful for clinical and technical researchers, engineers, students, and members of the general public who want to understand the current state of sensor applications in the highlighted domains.

10.2 BIORECOGNITION ELEMENTS AND TRANSDUCTION TECHNOLOGY

10.2.1 BIORECOGNITION ELEMENTS

A biosensor can be generally defined as a device that consists of a biological recognition system, often called a transducer. The interaction of an analyte with a bioreceptor is enabled in such a way so as to produce an effect that could be measured by the transducer, which in turn converts the information into a measurable effect, such as an electrical signal as shown in Figure 10-1. This is the conceptual principle of the biosensing process.

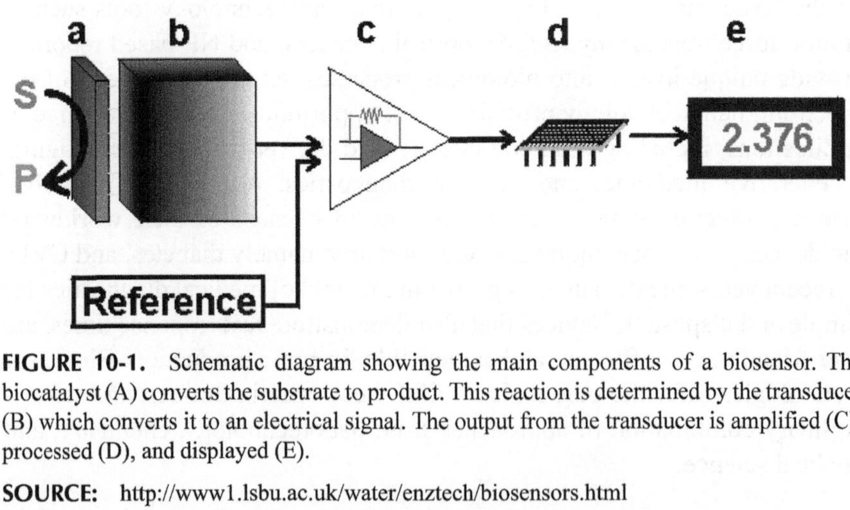

FIGURE 10-1. Schematic diagram showing the main components of a biosensor. The biocatalyst (A) converts the substrate to product. This reaction is determined by the transducer (B) which converts it to an electrical signal. The output from the transducer is amplified (C), processed (D), and displayed (E).

SOURCE: http://www1.lsbu.ac.uk/water/enztech/biosensors.html

Biosensors utilize biochemical mechanisms to identify an analyte of interest in chemical, environmental (air, soil, and water), and biological samples (blood, saliva, and urine). The sensor uses an immobilized biological material, which could be an enzyme, antibody, nucleic acid, or hormone, in a self-contained device. The biological material being used in the biosensor device is immobilized in a manner that maintains its bioactivity. Methods utilized include membrane (e.g., electroactive polymers) entrapment, physical bonding, and noncovalent or covalent binding. The immobilization process results in contact being made between the immobilized biological material and the transducer. When an analyte comes into contact with the immobilized biological material, the transducer produces a measurable signal, such as current, change in mass, or a change in color. Indirect methods can also be utilized, in which a biochemical reaction occurs between the analyte and sensor material, resulting in a product. During the reaction, measurable quantities such as heat, gas (e.g., oxygen), electrons, or hydrogen ions are produced, and can be measured. The development of biosensors was first reported in the early 1960s.[5] Biosensors have now seen an explosive growth and have a wide variety of applications primarily in two areas, viz. medical diagnosis monitoring and environmental sensing applications. The use of biosensors has increased steadily since Yellow Springs Instruments produced the first commercially successful glucose biosensor in 1975.[6] A biochip is a collection of miniaturized test sites (microarrays) arranged on a solid substrate that permits many tests to be performed at the same time in order to achieve higher throughput and speed. Typically, a biochip's surface area is no larger than a fingernail. Like a computer chip that can perform millions of mathematical operations in 1 s, a biochip can perform thousands of biological reactions, such as decoding genes, in a few seconds.

10.2.2 TRANSDUCERS FOR BIOSENSORS

The transduction process in a biosensor involves converting the biological activity that the sensor has measured via a bioreceptor into a quantifiable signal, such as current, an optical signal, or a change in measurable mass. Therefore, the biological component has to be immobilized on the transducer's surface. The most commonly utilized transducer mechanisms are electrochemical, optical, piezoelectric, and thermometric. There are three common electrochemical sensing approaches used in biosensors. Biosensors have attracted considerable attention and have found extensive applications in clinical laboratories, food industry, bioprocess control, and environmental monitoring. The biological recognition components result in the high specificity and sensitivity of the biosensors. They are usually immobilized in intimate contact with an appropriate transducer so that when the analyte is introduced into the system, the specific interaction between the analyte and the recognition component will produce a response signal in the transducer that can be used for detection purposes. Among these transducers, electrochemical transducers are promising sensing devices.[7,8] They have extremely high sensitivity, amenable to miniaturization, and easy to integrate into a circuit to perform continuous or automatic periodic sensing.[9,10] In addition, the equipment required for electrochemical analysis is simple and cheap compared to most other analytical techniques. Furthermore, they are more suitable for miniaturization and circuit integration, and the continuous electrical response allows for online control. With the development of microelectronics, integrated circuit industry, nanotechnology, and microfluidics, electrochemical transducers could be one of the most promising candidates for the future sensor developments.

Nanotechnology is gaining importance in all fields of application including the biological and biochemical sensor industries since a variety of dimension-dependent properties and nanoscopic phenomena are associated with small size. For example, the large surface-to-volume ratio is one of the most obvious properties of nanomaterials. There are other properties, such as optical, mechanical, ferroelectric, and ferromagnetic, associated with the nanoscaled dimensions.[11] How to offer the biosensor industry the benefits of nanotechnology is still under investigation in order to fulfill the demands of more accurate, faster and in situ continuous monitoring.

10.3 BIOSENSORS FOR CVDS APPLICATIONS

10.3.1 CVD AND CURRENT CARDIAC BIOMARKERS

CVD is the leading cause of death for both men and women e worldwide. It is important to learn about your heart to help prevent heart disease. However, if you have CVD, you can still live a healthier and more active life by learning about your disease and treatments and by becoming an active participant in your health care. CVD includes a condition called atherosclerosis that develops when a plaque builds up in the walls of the arteries. This plaque narrows the arteries and makes it difficult for blood to flow through and causes a heart attack or stroke. CVD can be caused by a range of factors and disorders that include genetic disorders, gender, age, high blood pressure and cholesterol, diabetes, obesity and overweight, smoking, and stress. The causes of CVD are more diverse that clinical testing becomes increasingly complex. There are a number of diseases associated with CVD that affects different parts of the body. One of the most important reasons of the increasing incidences of CVDs and cardiac arrest is hypercholesterolemia, that is, increased concentration of cholesterol in blood.[12] Aspartate transaminase (AST) was found to be elevated in patients with AMI in 1954 and was the first cardiac biomarker to be used in clinical practice. AST is found in the heart, liver, skeletal muscle, kidneys, and brain and is currently used clinically as a marker for liver status.[13] CVD is a major global cause of mortality in the developed countries. AMI is the major cause of death and disability worldwide with an ongoing increase in incidence. The risk of death is highest within the first few hours from AMI onset. The term AMI should be used when there is evidence of myocardial necrosis in a clinical setting consistent with acute myocardial ischemia. cTn is the most widely established and useful biomarker for myocardial injury. The cTn complex is made up of two subunits—C, I and T—which together control calcium-mediated interaction of actin and myosin, leading to the contraction and relaxation of striated muscle.[14] A label free electrochemical immunosensor has been reported based on an ionic organic molecule and chitosan stabilized-gold NPs (CTS-AuNPs) for the detection of cTnT.[15] A label free electrochemical immunosensor based on an ionic organic molecule ((E)-4-[(4- decyloxyphenyl)diazenyl]-1-methylpyridinium iodide) and CTS-AuNPs was developed for the detection of cTnT. The film of ionic organic molecules acts as a redox probe and from its electrochemical response the presence of cTnT antigens, which interact specifically with the anti-cTnT antibody immobilized on the surface of the immunosensor, can be detected. Under optimized conditions, using square-wave voltammetry (SWV) (a frequency of 100 Hz, an amplitude of 100 mV and an increment of 8 mV) and an incubation time of 10 min, the proposed immunosensor showed linearity in the range of 0.20–1.00 ng mL^{-1} cTnT, with a calculated limit of detection of 0.1 ng mL^{-1}. The proposed immunosensor shows some advantages when compared to other sensors reported in the literature, especially with regard to the detection limit and the time of incubation. A study of the inter-day precision (n = 8) showed a coefficient of variation of 3.33%. A schematic illustration of the steps involved in the construction of the proposed immunosensor is shown in Figure 10-2.

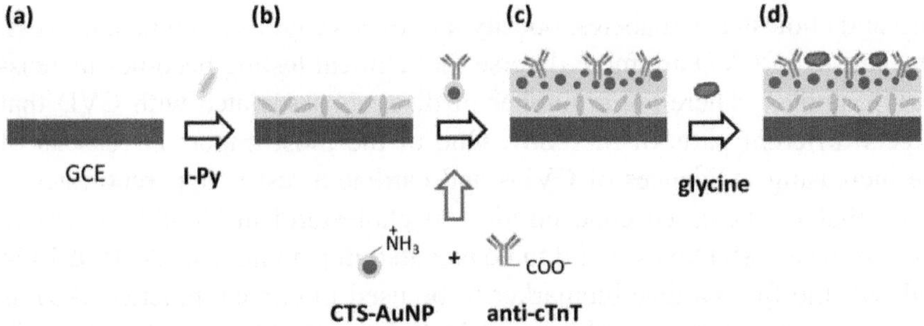

FIGURE 10-2 Schematic illustration of the successive steps involved in the construction of the immunosensor. (A) Bare glassy carbon electrode, (B) formation of the (E)-4-[(4-Decyloxyphenyl)diazenyl]-1-methylpyridinium iodide I-Py film, (C) adsorption of the chitosan stabilized-gold nanoparticles with the immobilized anti- cardiac troponin T antibody, and (D) blocking with glycine. Reprinted with permission from Ref. [15]. Copyright @ 2014 The Royal Society of Chemistry.

A fully automated fluid control system with disposable cartridge was developed for the surface acoustic wave-based POC testing (POCT), and the simultaneous detection of three cardiac markers in human blood, cTnI, CK-MB, and myoglobin, was made.[16] Several other cardiac-specific biomarkers that have emerged as reliable candidates and types of CVDs involved are described in Table 10-1.

After an AMI has occurred, cardiac biomarkers such as myoglobin, cTnI, cTnT, and CK-MB are released into the blood stream and therefore their detection in serum can aid an accurate AMI diagnosis.[17]

Several methods of cTnT detection for AMI diagnosis have been described in the literature, including electrochemiluminescence immunoassay, enzyme-linked immunosorbent assay (ELISA), radioimmunoassay, and immuno-chromatographic tests.[18] The application of the ionic liquid crystal (E)-1-decyl-4-[(4-decyloxyphenyl)diazenyl] pyridinium bromide (Br-Py) and AuNPs stabilized in a water-soluble 3-n-propyl-4-picolinium silsesquioxane chloride ($Si4Pic^{+}Cl^{-}$) matrix for the development of a new electrochemical immunosensor for the cardiac protein troponin was reported.[19] The formation of immune complex was monitored by SWV. This immune complex formation is responsible for the partial blockage of the electroactive surface compared with that obtained for the blank assay, and the different steps of the sensor construction are shown in Figure 10-3. The immunosensor was optimized and applied in the detection and quantification of cTnT.

TABLE 10-1 A summary of cardiac markers' detection on different transduction platforms and their detection range reported in the literature

Cardiac biomarker	Type of cardiovascular diseases involved	Linearity of analyte	Sensitivity	Limit of detection	References
Cardiac troponin I (cTnI)	Detection of acute myocardial infarction (AMI)	0.05 to 10 ng/mL	1.7 μg/mL	0.1 ng/mL	Ref. [34]
Monoclonal anti-cMb antibody (Mab1)	Human cardiac myoglobin (cMb)	–	–	0.18 ng mL^{-1}	Ref. [35]
Monoclonal antibodies (MAbs) of immunoglobulin G (IgG)	Detection of IgG and myoglobin (Myo)	–	–	3 ng/mL for IgG and 1.4 ng/mL for Myo	Ref. [21]
Anti-Myo antibody	AMI	Between 20 ng/ml and 230 ng/ml			Ref. [36]
cTnI, creatine kinase-MB (CK-MB), and b-type natriuretic peptide (BNP)	Detection of Myo	–	–	Myo (100 pg/mL), cTnI (250 fg/mL), CK-MB (150 fg/mL), and BNP (50 fg/mL)	Ref. [37]
cTnI	Detection of AMI	0.01–0.1 ng mL^{-1}	–		Ref. [38]
Monoclonal anti-cTnT antibody	Detection of cardiac troponin T (cTnT)	0.20–1.00 ng mL^{-1}	–	0.10 ng mL^{-1}	Ref. [15]
Guanidine hydrochloride (GdnHCl)	Detection of cardiac Myo	Between 0.001 mg mL^{-1} and 0.1 mg mL^{-1}	–	50 nM	Ref. [28]
Biotinylated-MAb against cTnI	AMI biomarker cTnI	1000 ng mL^{-1} down to 0.1 ng mL^{-1}	–	0.06 ng mL^{-1}	Ref. [39]
Human IgG	AMI Myo	4.93–6.38 nM	Highest affinity (K_d = 4.93 nM)	10 pM	Ref. [25]

TABLE 10-1 *(Continued)*

Cardiac biomarker	Type of cardiovascular diseases involved	Linearity of analyte	Sensitivity	Limit of detection	References
Anti-cTnT MAbs	Detection of cTnT	Between 0.1 ng mL^{-1} and 0.9 ng mL^{-1}	–	0.076 ng mL^{-1}	Ref. [19]
CTnI monoclonal antibodies	Detection of cTnI	0.092–46 ng/mL	–	0.092 ng/mL	Ref. [02]
Mouse anti-human Myo	AMI Myo	10–400 ng/mL	–	5 ng/mL	Ref. [40]

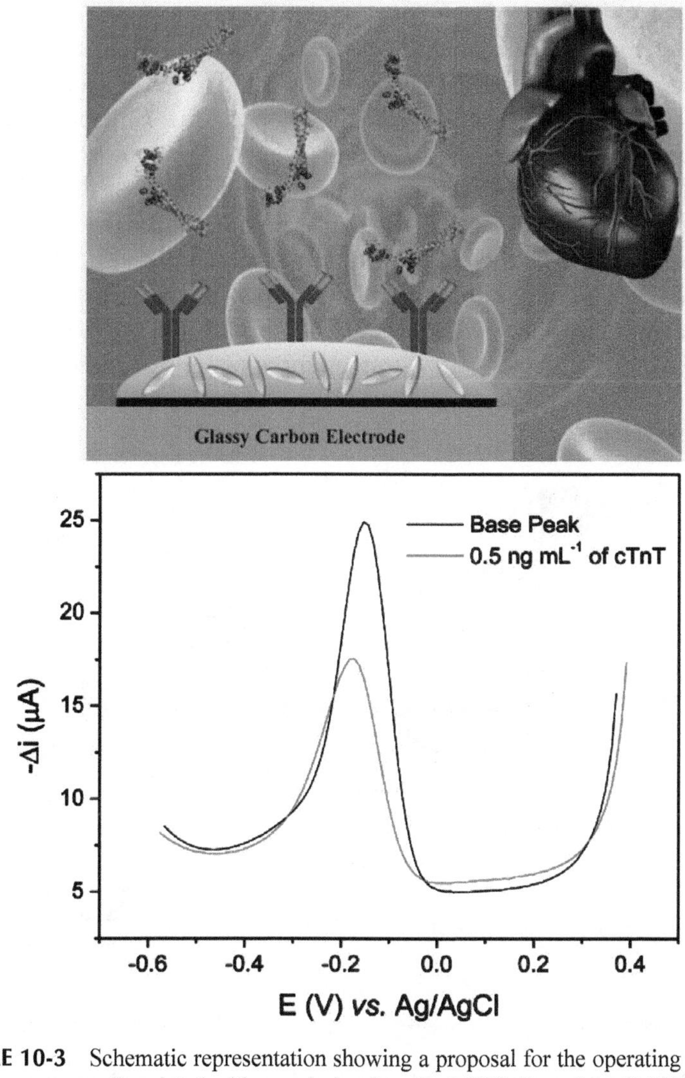

FIGURE 10-3 Schematic representation showing a proposal for the operating principle of the immunosensor. Reprinted with permission from Ref. [27]. Copyright @ 2007 The Royal Society of Chemistry.

Among the cardiac markers, Myo, although not cardiac specific, is one of the very first markers to increase after AMI.[20] Therefore, it is essential to reliably and sensitively detect Myo. At present, some approaches, including the electrochemical method,[21] fluorescent method,[22] surface plasmon resonance,[23] and colorimetric method[24] were developed for Myo detection. Most of these approaches are antibody based detection methodologies which utilize anti-Myo antibody to recognize the target Myo. A microfluidic SELEX system with an uncomplicated chip was introduced to screen the aptamer for Myo detection. In this microfluidic SELEX system, the positive and negative selection units were integrated in the same channel; thus, both positive selection and negative selection could be achieved simultaneously. This integration will simplify the operation process; thus, time can subsequently be saved. Negative selection is important for the specificity of aptamer screening.[25] The working principle of positive and negative selection units integrated microfluidic chip for aptamer screening is schematically demonstrated in Figure 10-4. This work may provide not only an alternative for Myo detection but also a new strategy for aptamer selection.

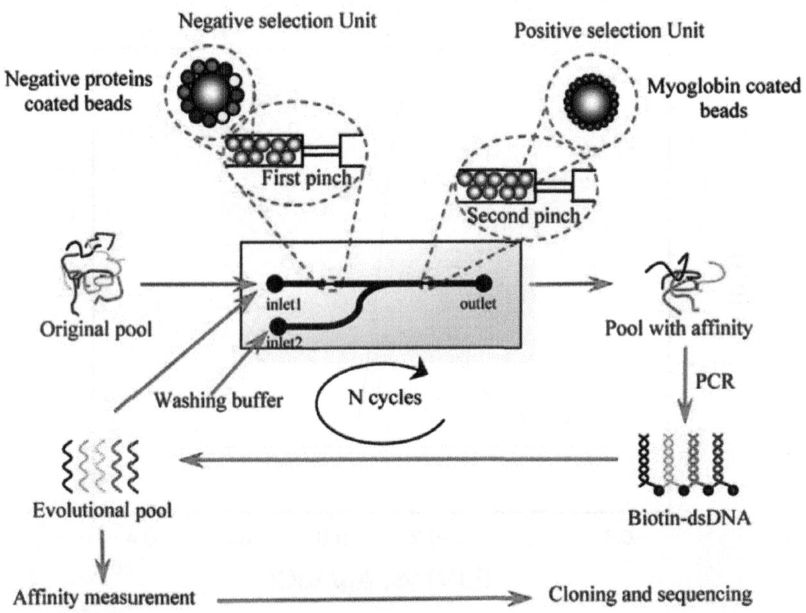

FIGURE 10-4 Myoglobin-aptamer selection based on a positive and negative selection units integrated microfluidic chip. Reprinted with permission from Ref. [25]. Copyright @ 2014 American Chemical Society.

10.3.2 POC CARDIAC MARKERS

Archakov et al. have described in a review article that nanotechnology in clinical proteomics is a new medical research direction, dealing with the creation and application of nanodevices for performing proteomic analyses in the clinic. Nanotechnological progress in the field of AFM makes it possible to perform clinical studies on the revelation, visualization, and identification of protein disease markers in particular of those with the sensitivity of 10^{-17} M that surpasses by several orders the sensitivity of commonly adopted clinical methods. At the same time, implementation of nanotechnological approaches into diagnostics allows for the creation of new diagnostic systems based on the optical, electro-optical, electromechanical, and electrochemical nanosensoric elements with high operating speed. We have shown that the combination of biospecific fishing with AFM allows the creation of diagnostic systems for the revelation and visualization of hepatitis B and C markers.[26] AFM images of hepatitis C markers—hepatitis C virus core antigen (HCVcoreAg), anti-HCVcoreAg, and also HCVcoreAg–anti-HCVcore complexes were taken. It was found that the heights of the complexes formed (3–7 nm) exceeded the heights of the isolated antigen (1.5–2 nm) and antibody (1–1.5 nm) molecules. We have created an AFM biochip with an immobilized anti-HCVcore. Upon incubation of the AFM biochip with the immobilized anti-HCVcore in HCVcoreAg solution, a number of anti-HCVcore–HCVcore Ag complexes of 3–7 nm emerged (Figure 10-5).[27]

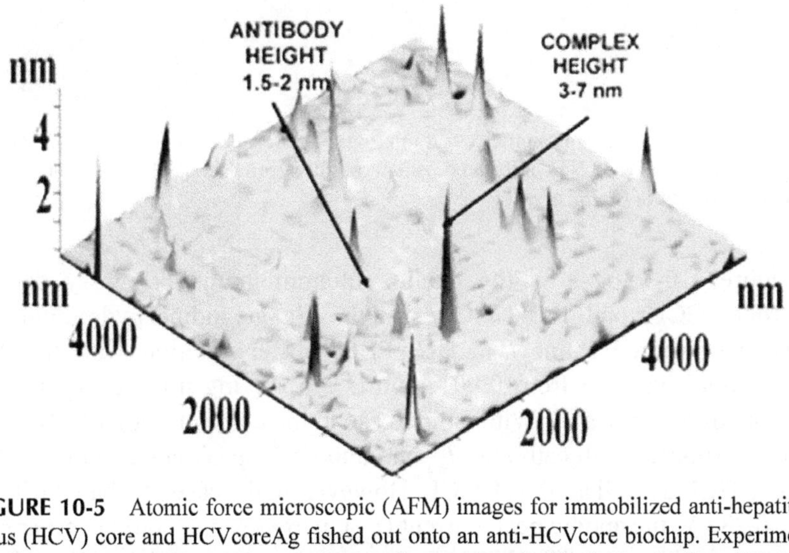

FIGURE 10-5 Atomic force microscopic (AFM) images for immobilized anti-hepatitis C virus (HCV) core and HCVcoreAg fished out onto an anti-HCVcore biochip. Experimental conditions: 1μM (2 mL) HCVcoreAg in phosphate-buffered saline/tablets buffer, pH 7.4, was placed onto the AFM biochip with immobilized anti-HCV core, then incubated for 2 min, and rinsed in distilled water; $T = 25$ °C. Image area is 5×5 μm. Reprinted with permission from Ref. [27]. Copyright @ 2007 The Royal Society of Chemistry.

An alternative antibody-free strategy for the rapid electrochemical detection of cardiac myoglobin was fabricated using hydrothermally synthesized TiO_2 NTs (Ti-NTs). The tubular morphology of the Ti-NT led to facile transfer of electrons to the electrode's surface, which eventually provided a linear current response (obtained from cyclic voltammetry) over a wide range of ~~Mb~~Myo concentration. The sensitivity of the Ti-NT-based sensor was remarkable and was equal to 18 μA mg mL^{-1} (detection limit = 50 nM). This, coupled with the rapid analysis time of a few tens of minutes (compared to a few days for ELISA), demonstrates its potential usefulness for the early detection of AMI.[28] Figure 10-6 shows the transmission electron microscopic images of the Ti-NTs. The NTs were 100–150 nm in length with an inner diameter of 5–6 nm and an outer diameter of 10 nm.

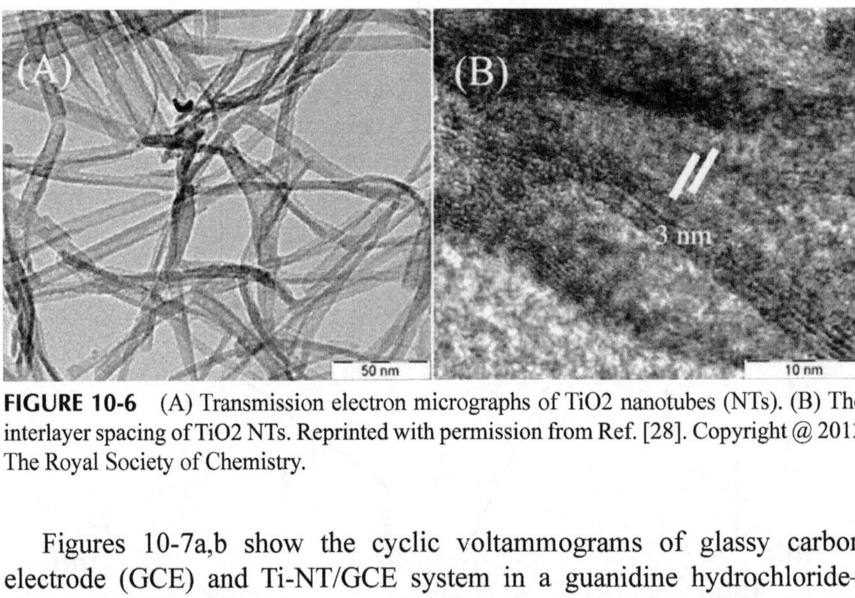

FIGURE 10-6 (A) Transmission electron micrographs of TiO2 nanotubes (NTs). (B) The interlayer spacing of TiO2 NTs. Reprinted with permission from Ref. [28]. Copyright @ 2013 The Royal Society of Chemistry.

Figures 10-7a,b show the cyclic voltammograms of glassy carbon electrode (GCE) and Ti-NT/GCE system in a guanidine hydrochloride–phosphate-buffered saline (GdnHCl–PBS) buffer solution with the Mb concentration varying from 0.001 $mgmL^{-1}$ to 0.1 mg mL^{-1} using a fixed scan rate of 500 mV s^{-1}.[28] With an increase in the concentration of Myo, the current response (both cathodic, $I_{p,c}$ and anodic $I_{p,a}$ peak currents) increases for bare GCE as well as Ti-NT/GCE. However, in the case of a bare GCE, at $E_{p,a}$ = 0.283 V, the anodic peak current ($I_{p,a}$) increases in a nonlinear manner. However, in the case of Ti-NT/GCE, the current increases almost linearly in the 50 nM to 6 μM concentration range. From the slope of linear fit of the anodic peak current ($I_{p,a}$) versus the Mb concentration data for Ti-NT/GCE in Figure 10-7c, the sensitivity was estimated to be approximately 18.44

μA mg mL^{-1}. The detection limit was found to be 60 nM (1 μg mL^{-1}) with Ti-NT/GCE.

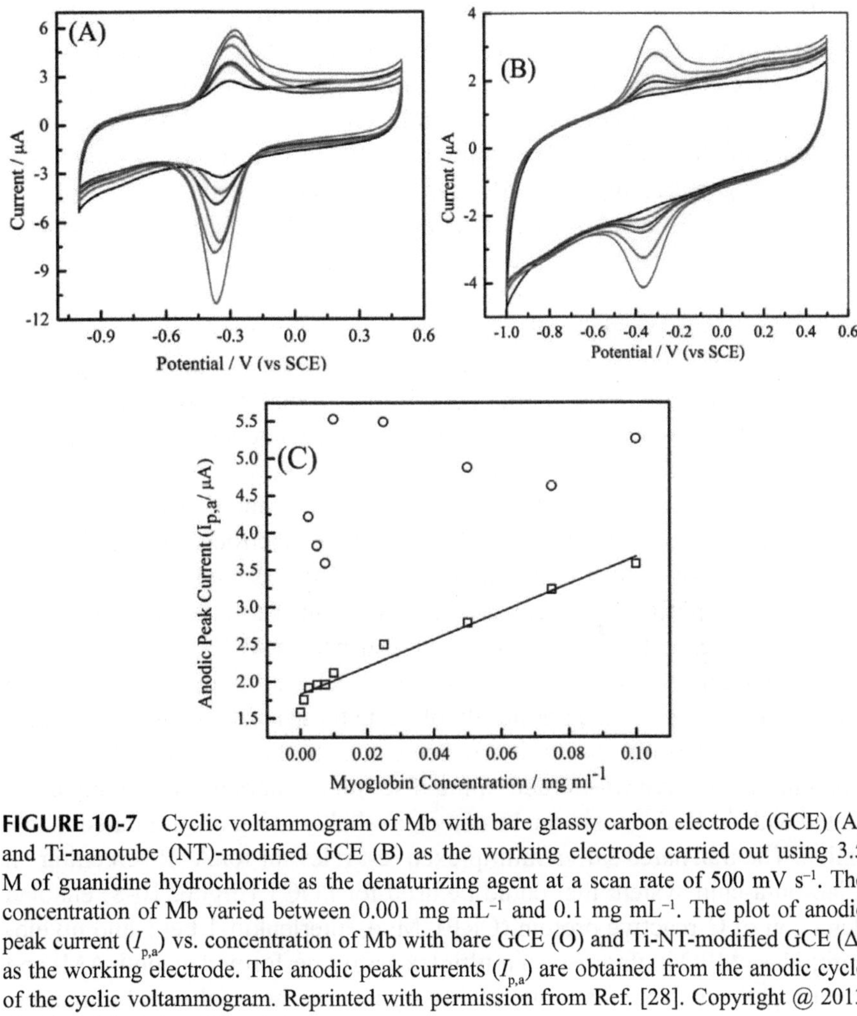

FIGURE 10-7 Cyclic voltammogram of Mb with bare glassy carbon electrode (GCE) (A) and Ti-nanotube (NT)-modified GCE (B) as the working electrode carried out using 3.5 M of guanidine hydrochloride as the denaturizing agent at a scan rate of 500 mV s^{-1}. The concentration of Mb varied between 0.001 mg mL^{-1} and 0.1 mg mL^{-1}. The plot of anodic peak current ($I_{p,a}$) vs. concentration of Mb with bare GCE (O) and Ti-NT-modified GCE (Δ) as the working electrode. The anodic peak currents ($I_{p,a}$) are obtained from the anodic cycle of the cyclic voltammogram. Reprinted with permission from Ref. [28]. Copyright @ 2013 The Royal Society of Chemistry.

10.3.3 BIOSENSORS IN CVD

cTn is the preferred biomarker for detecting acute myocardial injury (MI) and infarction. Apple et al. [29]have studied whether multiple biomarkers of numerous pathophysiological pathways would increase the diagnostic accuracy for detecting MI. International associations of cardiology, laboratory medicine, epidemiology, and emergency medicine have issued guidelines that have designated cTn as the preferred biomarker, both for aiding in MI diagnosis and for risk stratification in patients with suspected acute coronary syndrome (ACS), and have recommended that independent studies be conducted to validate all cTn assays after clearance by the US Food and Drug Administration before they are to be clinically accepted.[30] ACS refers to a constellation of clinical symptoms caused by acute myocardial ischemia. Owing to their higher risk for cardiac death or ischemic complications, patients with ACS must be identified among the estimated 8 million patients with nontraumatic chest symptoms presenting for emergency evaluation each year in the United States.[31] Myocardial necrosis is accompanied by the release of structural proteins and other intracellular macromolecules into the cardiac interstitium as a consequence of compromise of the integrity of cellular membranes. Such biomarkers of myocardial necrosis include cTnI and cTnT, CK, myoglobin, lactate dehydrogenase, and so on. The diagnosis of acute, evolving, or recent MI requires (in the absence of pathologic confirmation) findings of a typical rise and/or fall of a biomarker of necrosis, in conjunction with clinical evidence (symptoms or electrocardiogram [ECG]) that the cause of myocardial damage is ischemia. Because recognition of acute MI is important to prognosis and therapy, measurement of biomarkers of necrosis is essential in all patients with suspected ACS.

The first time reported the measurement of salivary biomarkers associated with AMI and explore the possibility of using these new salivary biomarkers in novel nano-biochip ensembles for screening chest pain patients for AMI.[32] The initial objective was to determine if serum biomarkers commonly associated with AMI diagnosis can be detected reliably using unstimulated whole saliva. The translated saliva multiplexed test relevant for AMI screening on a lab-on-a-chip system that could be used for POCT (Figure 10-8) enabled detection of C-reactive protein (CRP), Myo, interleukin-1 beta, and myeloperoxidase MPO in fluorescent multiplex assays performed on both AMI and control patients. Note that high fidelity signals for these salivary biomarkers are extracted with the nanobiochip approach whereby the high surface area of the beads, high local concentration of capture antibody, and pressure drive transport of samples and reagents provide an integrated approach that allows for these salivary biomarker measurements to be completed in POC settings. Novel cardiac biomarkers demonstrated significant differences in concentrations between patients with AMI and controls without AMI. The saliva based biomarker panel of CRP, Myo, and myeloperoxidase exhibited significant diagnostic capability ($AUC = 0.85$, $P<0.0001$) and in conjunction with ECG

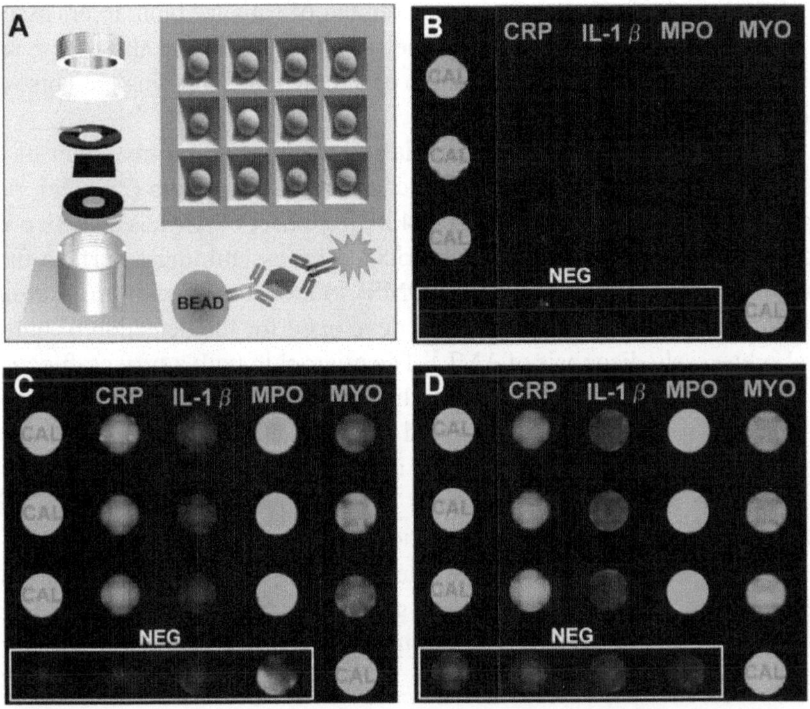

FIGURE 10-8 Multiplex lab-on-a-chip (LOC) for acute myocardial infarction (AMI) screening. (A). A scanning electron micrograph of the silicon microchip is shown with the LOC fluidic compartment on the left. An immuno-schematic depicts the sandwich type immunoassay detection modality and the analyte of interest (C-reactive protein, interleukin-1 beta, myoglobin [CRP, IL-1β, Myo], or MPO antigens are represented in blue). Examples of fluorescence micrographs of an LOC multiplex assay for CRP, IL-1β, Myo, and MPO are shown for non-AMI control (B), non-ST-segment elevation MI i.e., non-ST-elevation myocardial infarction (NSTEMI) (C), and ST-segment elevation MI (STEMI) (D) patients. Reprinted with permission from Ref. [32]. Copyright @ 2009 US National Library of Medicine National Institutes of Health.

ABBREVIATIONS: *NEG*, negative; *CAL*, calibrator.

yielded strong screening capacity for AMI ($AUC = 0.96$) comparable to that of the panel (brain natriuretic peptide, troponin-I, CK-MB, myoglobin; $AUC = 0.98$) and far exceeded the screening capacity of ECG alone (AUC approximately 0.6). It is reported that about 13.2 million individuals in the United States have coronary artery disease, 7.8 million have suffered an AMI, and 6.8 million have symptoms of angina pectoris. To detect or exclude myocardial necrosis in such patients, serial measurements of biochemical markers in serum are required, which often include Myo, CK-MB, total CK, and cTnT and cTnI. The process of measurement of salivary biomarkers associated with AMI helps in exploring the possibility of using these new salivary biomarkers in novel nanobiochip ensembles for screening patients with chest pain for AMI.

Early diagnosis of AMI is essential for the successful treatment of the disease. The levels of human cTnI, a cardiac muscle protein (29 kDa), rises rapidly in the blood stream within 3–4 h after onset of AMI. A sensitive and rapid electrochemical microchip fabricated by assembling a surface-functionalized polydimethylsiloxane (PDMS) microchannel with an interdigitated array (IDA) gold electrode was developed for the detection of human cTnI in the early diagnosis of AMI.[33] The microchip with a surface-functionalized PDMS channel before assembling with the IDA chip could enhance the electron transfer between the gold electrode surface and enzyme-generated electro-active species, resulting in a great increase in sensitivity. The gold surface topology of the former microchip successively treated with silane, protein G, and anti-cTnI solutions was examined by AFM in a scan area of 30 μm × 30 μm to validate the possibility of generating fouling via the binding of them on the working gold electrode surface patterned with widths of 10 μm and gaps of 5 μm (Figure 10-9).

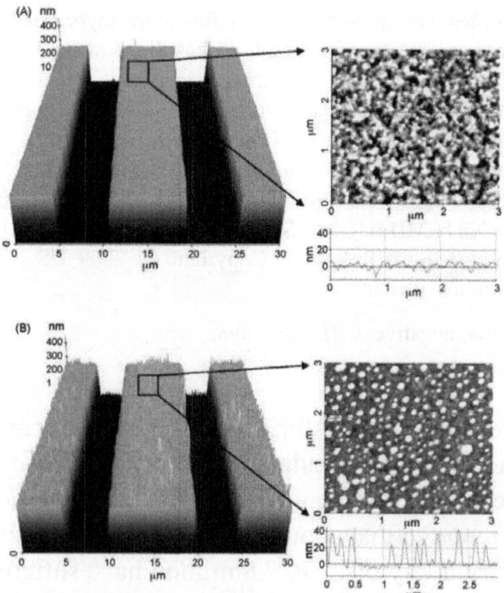

FIGURE 10-9 Atomic force microscopic images of gold surface of the microchip before (A) and after (B) successive treatment of silane, protein G, and anti-cardiac troponin I solutions. Reprinted with permission from Ref. [33]. Copyright @ 2007 Elsevier.

10.4 CHALLENGES AND FUTURE PROSPECTS

In this chapter, we hope that this brief overview has illustrated that biosensors have achieved considerable success both in the commercial and academic arenas and that the need for new, easy-to-use, home, and decentralized diagnostics is greater than ever. The enormous success of the cardiovascular marker and sensor serves as a model for future possibilities and should not overshadow the multifarious other applications that this versatile technology can address. The impact of this freedom of patient access and mobility of data, enabled by information technology, is likely to stimulate a demand for more analytical data and enable the patient or healthy subject to add data themselves, aided by a new range of over-the-counter biosensors. The expanding market generated by this boom in personal diagnostics will stimulate the development of new, inexpensive sensor platforms that can compete effectively to meet consumer needs. Definitively, the great progress achieved in the last decade in materials science and biochemistry is impelling the development of novel biosensors capable of detecting the CVDs with high sensitivity, high selectivity, and long-term stability. In accordance with our review presented in this chapter, biosensors require multiple physicochemical properties that a unique material used as mechanical and biochemical support of the analyte cannot provide by itself; hence, a more sophisticated design will be the future trend to guarantee the efficiency and accuracy desired for each application and range of detection required. Although, limits of detection in the nanomolar range have been achieved until now, it is desirable that the newest biosensors can identify CVDs in the nano and picomolar ranges. The introduction of nanobiosensors has allowed the development of user friendly and in-field application devices. Due to the reproducibility problems, biosensors based on nanostructured materials will be introduced in the market until their stability can be guaranteed. In this chapter, we have reviewed the advances of biosensors in detecting CVDs and the future trends for their research and technological development.

ACKNOWLEDGMENTS

The authors are thankful to the Director, Council of Scientific and Industrial Research-North East Institute of Science and Technology, Jorhat, Assam, India for permission and providing the facilities. RK is also grateful to the Department of Science and Technology, India, for financial support under the DST-RFBR project (INT/RUS/RFBR/P-153). AK acknowledges the

National Institutes of Health grants, namely RO1-DA027049, R21-MH 101025, RO1-MH085259, and RO1-DA 034547.

KEYWORDS

- **Cardiovascular disease**
- **Acute myocardial infarction**
- **Biochip**
- **Immunosensor**
- **Biosensor**

REFERENCES

1. Rivas, G. A.; Rubianes, M. D.; Rodríguez, M. C.; Ferreyra, N. F.; Luque, G. L.; Pedano, M. L.; Miscoria, S. A.; Parrado, C. Carbon Nanotubes for Electrochemical Biosensing. *Talanta* **2007,** *74*, 291.
2. Kong, T.; Su, R.; Zhang, B.; Zhang Q.; Cheng, G. CMOS-Compatible, Label-Free Silicon-Nanowire Biosensors to Detect Cardiac Troponin I for Acute Myocardial Infarction Diagnosis. *Biosens. Bioelectron.* **2012,** *34*, 267–272.
3. Bhayana, V.; Henderson, A. R. Ibopamine Substitution in a Dopamine-Dependent Patient. *Lancet* **1993,** *342*, 1554.
4. Bodor, G. S.; Porter, S.; Landt, Y.;Ladenson, J. H. Development of Monoclonal Antibodies for an Assay of Cardiac Troponin-I and Preliminary Results in Suspected Cases of Myocardial Infarction. *Clin. Chem.* **1992,** *38*, 2203–2214.
5. Clark, L. C.; Jr. Lions. C. Electrode systems for continuous monitoring in cardiovascular surgery, *Ann Acad. Sci.* **1962,** *102*, 29.
6. Setford, S. J.; White, S. F., Bolbot, J. A. Measurement of Protein Using an Electrochemical Bi-Enzyme Sensor. *Biosens. Bioelectron.* **2002,** *17* (1–2), 79–86.
6. Goepel, W. Chemical Sensing, Molecular Electronics, and Nano- Technology: Interface Technologies Down to the Molecular Scale. *Sens. Actuators, B* **1991,** *4*, 7–21.
7. Sethi, R. S. Transducer Aspects of Biosensors. *Biosens. Bioelectron.* **1994,** *9*, 243–264.
8. Ghindilis, A. L.; Atanasov, P.; Wilkins, M.; Wilkins, E. Immunosensors—Electrochemical Sensing and Other Engineering Approaches. *Biosens. Bioelectron.* **1998,** *13* (1), 113–131.
9. Warsinke, A.; Benkert, A.; Scheller, F. W. Electrochemical Immunoassays. *Fresenius J. Anal. Chem.* **2000,** *366*, 622–634.
10. Poole, C. P.; Owens, F. J. *Introduction to Nanotechnology*; John Wiley & Sons: New Jersey, 2003.
11. Franco M, Cooper R, Bilal U, Fuster V. Challenges and Opportunities for Cardiovascular Disease Prevention. *Am. J. Med.* **2011,** *124* (2):95–102.

12. Dolci, A.; Panteghini, M. The Exciting Story of Cardiac Biomarkers: From Retrospective Detection to Gold Diagnostic Standard for Acute Myocardial Infarction and More. *Clin. Chim. Acta.* **2006,** *369*, 179–187.
13. Daubert, M. A.; Jeremias, A. The Utility of Troponin Measurement to Detect Myocardial Infarction: Review of the Current Findings. *Vasc. Health Risk Manag.* **2010,** *6*, 691–699.
14. Brondani, D.; Piovesan, J. V.; Westphal, E.; Gallardo, H.; Dutra, R. A. F.; Spinelli, A.; Vieira, I. C. A Label-Free Electrochemical Immunosensor Based on an Ionic Organic Molecule and Chitosan-Stabilized Gold Nanoparticles for the Detection of Cardiac Troponin T, *Analyst,* **2014,** *139*, 5200.
15. Choi, Y. S.; Do, J. P.; Lee, H J.; Lee, S. S.; Lee, J.; Lee, Y.HoS.; Kim, K.; Lee, J. N.; Han, K.Y.; Park, J C. *15th International Conference on Miniaturized Systems for Chemistry and Life Sciences*; Seattle: Washington, DC, 2011, October 2–6.
16. Pedrero, M.; Campuzano, S.; Pingarrón, J. M. Electrochemical Biosensors for the Determination of Cardiovascular Markers: A Review. *Electroanalysis* **2014,** *26*, 1132–1153.
17. Dutra, R. F.; Mendes, R. K.; Silva, V. L.; Kubota, L. T. Surface Plasmon Resonance Immunosensor for Human Cardiac Troponin T Based on Self-Assembled Monolayer. *J. Pharm. Biomed. Anal.* **2007,** *43*, 1744–1750.
18. Zapp, E.; da Silva, P. S.; Westphal, E.; Gallardo, H.; Spinelli, A.; Vieira, I. C. Troponin T Immunosensor based on Liquid Crystal and Silsesquioxane-Supported Gold Nanoparticles. *Bioconjug. Chem.* **2014,** *25* (9): 1638–1643.
19. Weber, M.; Rau, M.; Madlener, K.; Elsaesser, A.; Bankovic, D.; Mitrovic, V.; Hamm, C. Diagnostic Utility of new Immunoassays for the Cardiac Markers cTnI, Myoglobin and CK-MB Mass. *Clin. Biochem.* **2005,** *38*, 1027–1030.
20. Lee, I.; Luo, X.; Cui, X. T.; Yun, M. Highly Sensitive Single Polyaniline Nanowire Biosensor for the Detection of Immunoglobulin G and Myoglobin. *Biosens. Bioelectron.* **2011,** *26*, 3297–3302.
21. Darain, F.; Yager, P.; Gan, K. L.; Tjin, S. C. On-Chip Detection of Myoglobin Based on Fluorescence. *Biosens. Bioelectron.* **2009,** *24*, 1744–1750.
22. Masson, J. F.; Battaglia, T. M.; Khairallah, P.; Beaudoin, S.; Books, K. S. Quantitative Measurement of Cardiac Markers in Undiluted Serum. *Anal. Chem.* **2007,** *79*, 612–619.
23. Zhang, X.; Kong, X.; Fan, W.; Du, X. Iminodiacetic Acid-Functionalized Gold Nanoparticles for Optical Sensing of Myoglobin via Cu2+ Coordination. *Langmuir* **2011,** *27*, 6504–6510.
24. Wang, Q.; Liu, Wei.; Xing, Y.; Yang, X.; Wang, K.; Jiang, R.; Wang, P.; Zhao, Q. Screening of DNA Aptamers against Myoglobin Using a Positive and Negative Selection Units Integrated Microfluidic Chip and Its Biosensing Application. *Anal. Chem.* **2014,** *86*, 6572–6579.
25. Archakov, A. I.; Ivanov, Yu. D. Analytical Biotechnology for Medicine Diagnostic. *Mol. Biosyst.,* 2007, *3*, 336–342.
26. Archakov, A. I.; Ivanov Y. D. Analytical Nanobiotechnology for Medicine Diagnostics. *Mol. BioSyst.* **2007,** *3*, 336–342.
27. Mandal, S. S.; Narayan, K. K.; Bhattacharyya, A. J. Employing Denaturation for Rapid Electrochemical Detection of Myoglobin Using TiO_2 Nanotubes. *J. Mater. Chem. B,* **2013,** *1*, 3051–3056.
28. Apple, F. S.; Smith, S. W.; Pearce L. A.; Murakami, M. M. Assessement of the Multiple-Biomarker Approach for Diagnosis of Myocardial Infarction in Patients Presenting with Symptoms of Acute Coronary Syndrome. *Clin. Chem.* **2008,** 55, 93–100.

29. Thygesen, K.; Alpert, J. S.; White, H. D. Universal Definition of Myocardial Infarction. *Eur. Heart. J.* **2007,** *28*, 2525–2538.
30. Storrow, A. B.; Gibler, W. B. Chest Pain Centers: Diagnosis of Acute Coronary Syndromes. *Ann. Emerg. Med.* **2000,** *35*, 449–461.
31. Floriano, P. N.; Christodoulides, N.; Miller, C. S.; Ebersole, J. L.; Spertus, J.; Rose, B. G.; Kinane, D. F.; Novak, M. J.;Steinhubl, S.; Acosta, S.; Mohanty, S.; Dharshan, P.; Yeh, C.-ko.; Redding, S.; Furmaga, W.; McDevitt, J. T. Use of Saliva-Based Nano-Biochip Tests for Acute Myocardial Infarction at the Point of Care: A Feasibility Study. *Clin. Chem.* **2009,** *55*, 1530–1538.
32. Ko, S.; Kim, B.; Jo, S. S.; Ohc, S. Y.; Park, J. K. Electrochemical Detection of Cardiac Troponin I Using a Microchip with the Surface-Functionalized poly(dimethylsiloxane) Channel. *Biosens. Bioelectron.* **2007,** *23*, 51–59.
33. Kim, W.; Kim, B.; Kim, A.; Huh, C.; Ah, C.; Kim, K.; Hong, J.; Park, S.; Song, S.; Song, J.; Sung, G. Response to Cardiac Markers in Human Serum Analyzed by Guided-Mode Resonance Biosensor. *Anal. Chem.* **2010,** *82*, 9686–9693.
34. Gnedenko, O.; Mezentsev, Y.; Molnar, A.; Lisitsa, A.; Ivanov, A.; Archakov, A. Highly Sensitive Detection of Human Cardiac Myoglobin Using a Reverse Sandwich Immunoassay with a Gold Nanoparticle-Enhanced Surface Plasmon Resonance Biosensor. *Analytica Chimica Acta* **2013,** *759*, 105–109.
35. Darain, F.; Yager, P.; Gan, K.; Tjin, S. On-chip Detection of Myoglobin Based on Fluorescence. *Biosens. Bioelectron.* **2009,** *24*, 1744–1750.
36. Lee, I.; Luo, X.; Huang, J.; Cui, X.; Yun, M. Detection of Cardiac Biomarkers Using Single Polyaniline Nanowire-Based Conductometric Biosensors. *Biosensors* **2012,** *2*, 205–220.
37. Mohammed, M. I.; Desmulliez, M. P. Lab-on-a-Chip Based Immunosensor Principles and Technologies for the Detection of Cardiac Biomarkers: A Review. *Lab on a Chip* **2011,** *11*, 569–595.
38, Li, F.; Yu, Y.; Cui, H.; Yang, D.; Bian, Z. Label-Free Electrochemiluminescence Immunosensor for Cardiac Troponin I Using Luminol Functionalized Gold Nanoparticles as a Sensing Platform. *Analyst* **2013,** *138*, 1844.
39. Suprun, E. V.; Shilovskaya, A. L.; Lisitsa, A. V.; Bulko, T. V.; Shumyantseva, V. V.; Archakov, A. I. Electrochemical Immunosensor Based on Metal Nanoparticles for Cardiac Myoglobin Detection in Human Blood Plasma. *Electroanalysis* **2011,** *23*, 1051–1057.

CHAPTER 11

NANOBIOSENSING PLATFORMS FOR *IN VIVO* APPLICATIONS

KRATI SHARMA

Rare Genomic Institute, Hanover, MD 21076,USA

E-mail: krati.sharma@raregenomics.org

CONTENTS

Significant driving forces for successful in vivo implantation have become a major challenge in lucrative market of genetic diagnostics, genome sequencing, DNA damage detection, and high-throughput screening for drug discovery. From coupling of DNA probes to evaluation of optical, electrochemical, mass-sensitive and bio-microelectromechanical systems (bio-MEMS) biosensors, different strategies have been employed from laboratory to field. Most of the in vivo applications of biosensors stand against the extravagant claims and predictions due to poor sensitivity observed on administration, limited resolution of the signal in real samples and poor storage stability. Nevertheless, despite these unsuccessful applications, many methods are introduced by exploiting novel biological and biomimic systems to produce fast, accurate, and inexpensive genosensors. Selected applications along with current state-of-the art principles of biosensing in vivo are discussed in this chapter with special reference to disease detection, therapeutics, and role of contrast agents.

11.1 INTRODUCTION

The research in nanobiosensing technology has generated significant interest in recent years in global emerging diseases and health problems. Tremendous advances in the development of novel biological sensing tools with high sensitivity and selectivity have opened up the potential diagnostic applications in the areas of biomarker discovery, cancer diagnosis, and early detection of other life-threatening diseases. Biosensors consists of a biomarker immobilized onto the surface of a signal transducer which detects biochemical and biophysical signals associated with a single biological cell or a molecule. Modern biosensors exploit micro- and nanofabrication technologies along with diverse sensing strategies such as optical, electrical, and mechanical transducers. These promising techniques from laboratory to field have potential applications in clinical diagnosis such as biochips, glucose sensors, biomarkers for disease detection, point-of-care diagnostics and so on. Small, fast, and pioneering biosensors help in detecting biomarkers specific to certain diseases, monitoring health conditions, to know onset of a disease, its progression and possible treatment. The important fact to bear in mind is that successful sensors are environment friendly in the context of their applications. The successful implantation of these sensors needs to be modeled as per thorough information on the target and expected effects at cellular and molecular level. This requirement restricts uses of any supporting reagent in the test medium. Therefore, most of the developments made in vitro are rejected when used in vivo. It is also common to observe decreased sensitivity of a biosensor relatively in a short period. Although the limits of current molecular diagnostics demand safety studies for in vivo uses, on comparing with other techniques in clinical testing such as microarray or proteomic analyses, biosensors are less expensive and less technically demanding. Improved understanding on the underlying mechanism of disease initiation, disease progression, and therapeutic response along with identification of biomarkers has emerged a new hope in promoting personalized medicine. Finding novel yet successful strategies can serve as diagnostic or therapeutic indicators for the new emerging concept, Personalized Medicine.

11.2 TECHNIQUES IN BIOSENSING-IN VIVO

Biosensors based on nucleic acid immobilization are rapidly being developed to meet goal of bringing fast, accurate and advance techniques for gene testing, disease detection, biomarkers generation, and so on. Functional nanoparticles (NPs) in conjugation with biological molecules (peptides, proteins, nucleic acids, etc.) are incorporated in biosensors for detecting multiple signals, imaging, and sensitivity. Due to the submicron size of the particles, nanosystems have revolutionized the field of biological analysis of multiple substances in vivo. The goal of this chapter is to connect with recent genosensor and lab-on-chip techniques applied in vivo to have an enormous effect on disease detection and therapeutics. Although there are advantages in biosensors designing and sensing efficacy, certain improvements are still needed in existing methods and technologies to allow diagnosis and prognosis of the disease treatment. With big time applications of nanobiosensing on introducing appliances from laboratory to field, continuous monitoring of life threatening diseases has become possible. Though lowering the foot falls of existing methods, and developing a proficient approach in therapeutics yet to achieve success in real time applications. After completion of the Human Genome Project, doors are opened to explore the wide diagnostics methods from detection of mutations to personalized medicine. Over the past decade, intense activities were aimed at developing DNA biosensors and gene chips. Such devices offer much promise for obtaining the sequence-specific information in a faster, simpler, and cheaper way compared to traditional hybridization assays. These DNA microarray and biosensor technologies are rapidly advancing applications ranging from genetic testing to gene expression and drug discovery. Further scaling down, particularly of the support instrumentation, should lead to hand-held DNA analyzers. Innovative efforts, coupling fundamental biological and chemical sciences with technological advances in the fields of micromachining and microfabrication should lead to even more powerful devices that will accelerate realization of large-scale genetic testing. A wide range of new gene chips and DNA biosensors (Figure 11.2a) are thus expected to reach the market in the coming years. While offering remarkable tools for genetic analysis, proper applications of these new devices would still need a solid intellectual input.

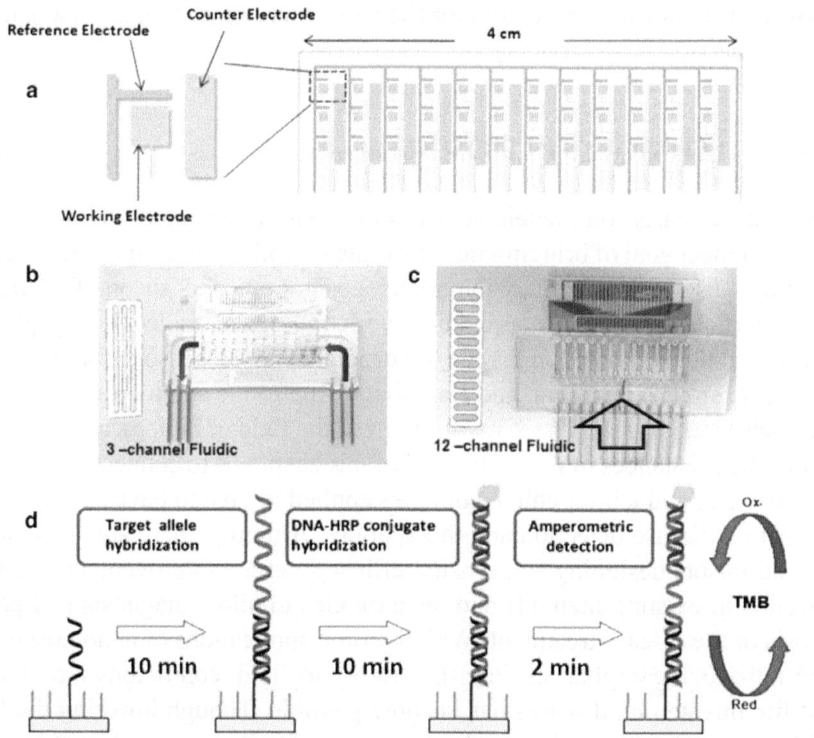

FIGURE 11.2a Diagram-based explanation for a genosensor.

Immobilization of a nucleic acid probe onto the transducers mainly depends on the surface physiochemical properties. The favorable change in probe's conformation brings rapid binding with target. This requires various schemes for enhancing immobilization of the probe on a transducer. These include the use of thiolated DNA for complex formation onto gold transducers (gold electrodes or gold-coated piezoelectric crystals), covalent linkage to the gold surface via functional alkanethiol-based monolayers, the use of biotylated DNA for complex formation, covalent coupling to functional groups on carbon electrodes, or a simple adsorption onto carbon surfaces.[1] Recent advances in the realm of clinical diagnosis have enhanced real-time genome analysis applications including genetic polymorphism, point mutations, and detection of pathogenic species. For example, peptide nucleic acids (PNAs) are DNA mimics in which nucleases are attached with neutral pseudopeptide backbone.[2] This molecular backbone possesses various electrical characteristics which makes it unique for electrochemical genosensing. Such use of electrochemical properties and surface-confined PNA recognition features are further used in sequence-specific DNA biosensors design for biomarkers detection and genetic mutations.

Similarly, optical biosensors have attracted much attention in disease detection and therapeutic response. DNA optical biosensors detect the emitted signal from a fluorescent label with the help of optic fiber. The mechanism involves the single-stranded DNA probe at the end of fiber optical part and monitoring the fluorescent changes resulting from the complex formation between fluorescent compound and double-stranded DNA hybrid. The first DNA optical biosensor used ethidium bromide as an indicator in the work by Krull and co-workers.[3] Optical biosensors integrated with evanescent wave devices can offer real-time label free optical detection of DNA hybridization.[4,5] These biosensors read changes in the interfacial refractive index using surface plasmon resonance (SPR) technique. Such changes in refractive index are seen due to surface binding reaction. In vivo, label-free optical detection can be made through changes in other optical properties. For example, a novel particle based calorimetric detector monitors the change in distance due to hybridization event, which results in change in optical properties of gold NPs (Au NPs).[6] Another novel approach using molecular beacons (MBs) resulted into high specificity and sensitivity.[7] In the experiment, the aptamer attached with methylthioninium chloride at the end immobilized on the gold electrode in target's absence will remain in one of the three characteristics stem structure. However, as soon as the target molecule—in this case the platelet-derived growth factor (PDGF)—approaches, it reshapes into the three-stem structure and results in low distance between the electrodes and the MB.

Electrochemical biosensors make another successful category for detection of sequence-specific biosensing of DNA.[7,8] The hybridization process is generally detected by increased current signal of a redox indicator (that recognizes the DNA duplex) or change in electrochemical parameters after hybridization. Salient features such as stable secondary structure, easily synthesized genome, and so on have led to the idea of selecting new nucleic acid ligands called aptamers. Aptamers (Apts) are artificially synthesized single strand of DNA or RNA ligands that can be generated against amino acids, drugs, and proteins.[9] Several reviews on aptamers have appeared in the literature for their use in clinical analysis and as therapeutic agents. With respect to the application in therapeutics, an aptamer has recently been approved by the US Food and Drug Administration (FDA) for treatment of ocular vascular disease.[10] Mass spectroscopy or a biosensor uses aptamers as a powerful tool. The affinity and specificity are the key features of aptamers that helps in successful recognition of molecule of interest. Aptamer's affinity ranges from micromolar to the nanomolar level and can be distinguished between closely related targeted sequences. Aptamers remains in unfolded state in solution, but fold upon binding with their molecular targets respectively into molecular architectures. This unique feature is the principal mechanism applied in development of electrochemical sensors[11] and hence used in the use of PDGF DNA aptamer[12] (Figure 11.2b). Many assays on thrombin-binding aptamer were developed for detection of thrombin which were not very successful till this new amplification strategy was introduced. In this new amplification strategy, an assay was coupled with an electrochemical assay which resulted into better sensitivity towards detection of thrombin[13,14]

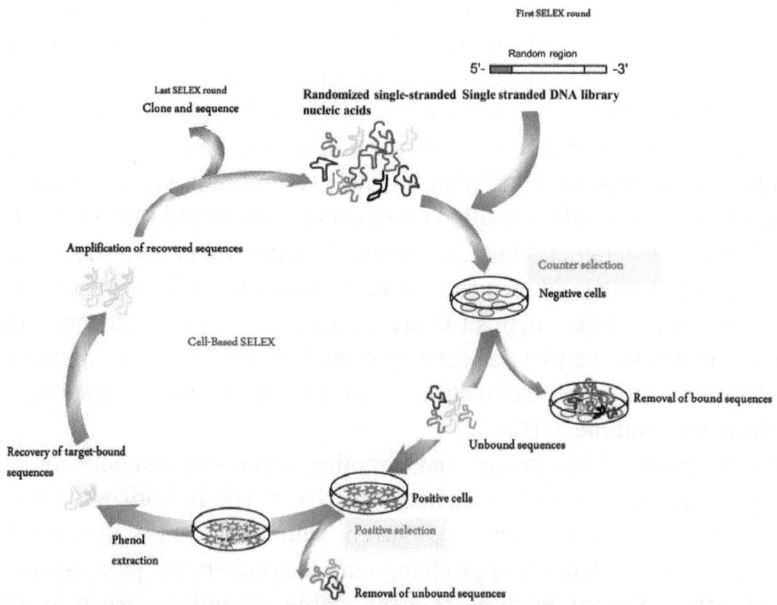

FIGURE 11.2b Diagram-based explanation for DNA-based aptamer.

Similar strategy was employed in detection of coeliac disease (CD). A research on CD depicted that human leukocyte antigen (HLA)-DQ2 heterodimer exists in 95% of the CD patients. This heterodimer is present either in cis- or trans conformation. In order to detect medium-to-high resolution HLA-DQ2/DQ8 heterodimer, an electrochemical assay was developed using 36 electrodes coupled with 10 sequence specific probes. Many probes were required for some alleles due to high numbers of HLAS alleles and complex HLA system. Majority of the hybridized probes were detected in less than 25 min on all of the 36 electrodes.[20] Another HLA genotyping method is based on the polymerase chain reaction (PCR). This includes sequence-specific primer-PCR,[15] sequence-specific oligonucleotide probe-PCR,[16] single strand conformation polymorphism,[17] restriction fragment length polymorphism,[18] and sequencing based typing[19] and are expensive, labor intensive, time consuming, and require extensive post-PCR manipulation.

Among label free detection methods, mass-sensitive devices constitute another useful method which depends on the use of quartz crystal microbalance (QCM) transducers. This system consists of an oscillating crystal with the DNA probe immobilized on its surface. Detecting capabilities of this system depends on mass and oscillating frequency. Oscillating frequency is inversely proportional to increase in mass due to hybridization frequency. This technique has been recently applied in finding taxane resistivity to human mammary epithelial tumor cells.[21] Considering all methods for obtaining hybridization pattern, researchers believe that fluorescence imaging and mass spectroscopy are the most commonly used methods. Although these above discussed techniques were successful in finding single target but a basic fault in genome processing does not involve a single element of protein or DNA. It is comprised of an interconnected biological networks. This generated the crucial need for development of multiplexed biosensors[22] and hence, led to the growing interest in multiplexed biosensors. These multiplexed biosensors were developed to detect biomarkers which could distinguish normal and pathogenic biological processes. Further, multiple biosensors with DNA microarrays were developed for analysis of complex DNA samples.[23,24] Various terminologies, like gene chips or biochips or DNA arrays, are often used for such microarray based sensors. A DNA microchip is built on molecular biology concepts, advance microfabrication technique, surface and analytical chemistry, software, robotics, and automation. Bioinformatics tools were also integrated to relate complexity of data into useful information. The Affymetrix gene chip (Affymetrix Inc., Santa Clara, CA, USA) can be used for screening cancer and cancer gene identification. Complementary

DNA (cDNA) microarrays are collected in libraries and are developed by applying reverse transcription method on messenger RNA (mRNA). These single-stranded cDNA from the cDNA libraries were further utilized in various applications such as disease detection, genome sequencing, screening of molecular interaction, biomarker discovery, monitoring signaling pathways, and screening of protein changes during disease progression. Similarly Hyseq Inc. (Sunnyvale, CA, USA) is collaborating with PerkinElmer Inc. to design a universal microarray DNA "superchip" system. In this system, a fluorescence indicates the side-by-side binding of targeted DNA with the tagged probe and the associated software helps in printing the sequence as per detection. The same chip is very useful in detecting mutations related to genetic diseases, cancer, or other infectious diseases.

Investigators soon discovered another useful approach, modeling of biomimics based on the biomolecules. These modeled biomimics induce easy immobilization on the sensor surface along with adding labels for the marker's detection.[25] To fully understand the mechanism, microfluidics and MEMS called bio-MEMS or lab-on-a-chip became the main center of attraction for measuring DNA, gene transcripts (mRNA), proteins, electrolytes and small molecules. Researchers found that sensors with moving mechanical structures and complex design limit their reliability in subcutaneous implants. This phenomenon was observed during glucose measurement. Therefore, a fully implantable MEMS dielectric affinity glucose biosensor with a perforated electrode was developed and embedded into a suspended diaphragm. It allowed clear glucose diffusion monitoring. The ability to integrate all the platforms on a single chip allows significant advantages with respect to speed, cost, sample preparation consumption, contamination, efficiency, and automation. Miniturization also enabled successful advancement in point-of-care testing. Nanogen Inc. (San Diego, CA, USA) developed an electronic sample preparation microfabricated chip-based device[27] which was demonstrated on dielectrophoretic separation of *Escherichia coli* from blood cells.[28] Integrated electric feature allowed to make changes in electric field based on the binding strengths between complete match or single nucleotide match. After the isolation, cells were lysed using different high-voltage pulses. On concluding advantages of such microfabrication and micromachining technologies on an electronic chip, many drawbacks were resolved successfully which were confronted during traditional hybridization methods. However, such biosensors couldn't fulfill real time in vivo monitoring. This was clearly observed in glucose biosensor. Lack of glucose diffusion passage renders this method unsuccessful in real time environment.[26]

11.3 APPLICATIONS OF NANOBIOSENSING-IN VIVO

11.3.1 BLOOD GLUCOSE MONITORING

Real time blood glucose monitoring, aside from a clinical analyzer, is desired for triggering proper alarm in detecting hyperglycemia and hypoglycemia. In vivo glucose biosensors have been found highly adaptive in diabetes management. The first such device application for in vivo glucose monitoring was demonstrated by Schichiri et al. in 1982.[29] Diverse and major health improving efforts in sensor development have built a potential market for glucose sensors. The first application is based on enzyme-catalyzed reaction monitored electrochemically using a small portable device. This unit comes along with testing strips where the patient puts a drop of blood for glucose detection. Based on the chip programming, the calibrated result displays on the screen and, if required, result can be stored into the device. In case of insulin dependent diabetes, few devices are developed for more intense reading to avoid large variations (at least one measurement for every 1–5 min is needed). Microdialysis or ultrafiltration technology is employed to solve the problem of diffusion of low-molecular weight substances from the sample (Figure 11.3.1.). These substances cross the sensor's outer membrane which results in reducing sensitivity. Recently Europe has

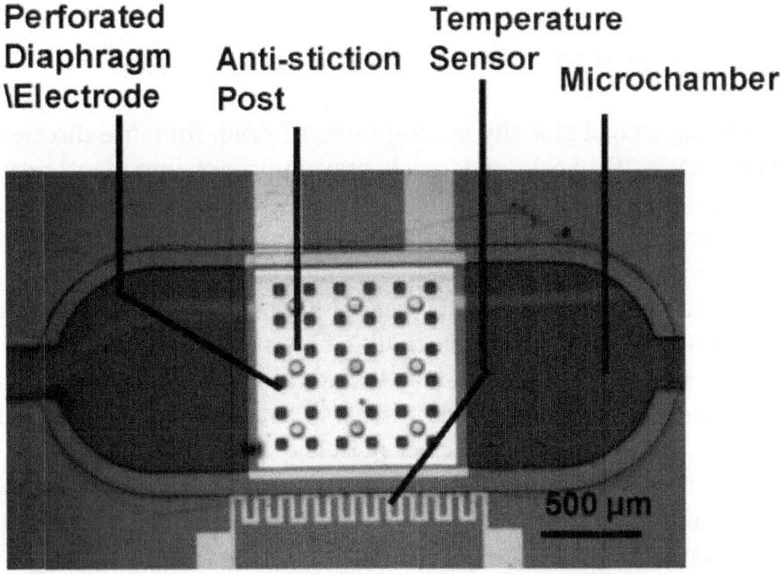

FIGURE 11.3.1 A microelectromechanical systems dielectric affinity glucose biosensor.

improved this method by performing microdialysis technique followed by enzyme-based system. Recent reviews on the optical based electrochemical methods on glucose monitoring have shown progress on measuring oxygen consumption based on enzyme catalysis reaction.[34,35] The main approaches being used in biological system monitoring are near-infrared (NIR) spectroscopy, body fluid's analysis, microcalorimetry, enzyme electrodes, optical sensors, and so on. However, how far the methods are accepted is solely depends on the successful implantation of clinically accurate device. And the successful administration depends on many factors such as biocompatibility, stability, in vivo calibration,[30,31] and miniature size with minimal discomfort. But most of such devices implanted in the subcutaneous tissue are found unstable and not sufficiently reliable due to no biocompatibility, no enzyme stability, eliciting acute inflammation response, and so on.[32,33.] Hence, alternatively, non-invasive glucose sensing has received tremendous growing attention. Researchers also incorporated the largest defending organ for glucose monitoring, skin, using sonophoresis and iontophoresis for glucose extraction. Although limitations to results are expected due to diet variations and regular body fitness exercises. The third application of glucose sensing approaches is artificial pancreas development for continuous glucose monitoring. There are recently no implanted sensors although different techniques are under the phase of development which can keep implants for longer time in the body.

11.3.2 BIOMARKER FOR DETECTION OF CANCER

Cancer is the second globally leading cause of death from life-threatening diseases. Early diagnosis and quick prevention are important parts of cancer treatment and many immunoassay methods have been developed for detection of cancer biomarkers. Breast cancer, cervical cancer and gastric cancer are the leading causes of morbidity and mortality worldwide. There are many causes for cancer disease such as genetic mutations or environmental factors, including exposure to harmful chemicals, radiations, bacteria, infections, and toxins (aflatoxin).[36] Biomarkers are common indicators of a particular disease. Cancer biomarkers are widely used in oncology in order to detect presence of various tumor growths. Finding cancer biomarkers in genes play an important role in clinical diagnoses and evaluation of treatment for patients. Many immunoassay methods have been developed for detection of cancer biomarkers. As the factors are not limited, it's treatment is not limited. It demands complex clinical testing and range of biomarkers for disease diagnosis. So far, biomarkers are generated for finding DNA mutations, RNA, proteins (enzymes and glycoproteins), hormones and related molecules, molecules in cell signaling, and modeled molecules. For example, survivin is highly important antiapoptotic protein in cell division and apoptosis. It's expression is often found high in human cancers, therefore, it is considered to be a good biomarker in carcinomas therapy. Prostrate-specific antigen is an identified biomarker to screen prostrate cancer to detect early stage of the disease and to watch reoccurrence of the disease after treatment.[37] The research has advanced toward POC testing as well.[38]

11.3.3 BIOMARKERS FOR DETECTION OF CARDIOVASCULAR DISEASE

One of the major risks of high cholesterol in blood or hypercholesterolemia is cardiovascular disease (CVD). It is highly preventable, yet, it is also a major cause for human death. This demands regular cholesterol monitoring to prevent one-selves from the life-threatening diseases. Biomarkers have become increasingly important in dealing with CVD and hence, cardiac troponin (cTnl) is easily detected and useful as a biomarker in therapeutics. A non-myocardial tissue-specific marker might also be helpful such as C-reactive protein (CRP). Ambiguity in CRP functions may result in CVD, myocardial infarction, kidney malfunction, obesity, high body mass index, and correlated development of cancer. Hence, CRP is a very crucial marker for such clinical diseases. CRP, which signals developing atherosclerosis or acute ischemia is one of the acute phase proteins which along with other proteins such as CRP, mannose-binding protein, complement factors, serum amyloid A, fibrinogen, retinal binding protein, ceruloplasmin, and antithrombin play an important role in signaling inflammation. Among these, CRP is the most sensitive marker of inflammation. During inflammation, phosphocholine present on the surface of dead or dying cells binds at the CRP active site and activates the complement pathway which in turn increases the physiological level of CRP from 2 mg/L to 300 mg/L in 6–8 h. During the evaluation of three cardiac biomarkers, cTnl, creatinine kinase isozyme MB (CK-MB), and myoglobin, it was found that the use of cTnl was high compared to the other two biomarkers in the diagnosis of chest pain and myocardial infarction.[39] However, use of all the three may be helpful in diagnosis of patients with acute coronary syndrome.[40] Most recently, a nonenzymatic sandwich-type electrochemical immunosensor is fabricated for detection of gastric cancer biomarker CA72-4 using dumbbell-like PtPd–Fe_3O_4 NP s as a novel kind of label.[41]

11.3.4 BIOMARKERS FOR DETECTION OF NEURODEGENERATIVE DISEASE

SPR imaging technique showed a better correlation with the existing enzyme-linked immunosorbent assay (ELISA) kit for detection of potential biomarker, insulin-like growth factor-binding protein 7 (IGFBP7) (Figure 11.3.4a). IGFBP7 simulates cell adhesion, effects apoptosis, regulation of cell growth, and angiogenesis[42] (Figure 11.3.4b). A study has shown that IGFBP7 level increases in patients with Alzheimer's disease[43]. This protein is also implicated in some cancers. Hence, it serves as an important tool for both disease detection and relevant therapy.[44] Recently a highly sensitive biosensor is developed for neuropathy target esterase (NTE) which plays an important role in delaying neurotoxicity. NTE is a target protein for neuropathic organophosphorus (OP) compounds that produce OP compound-induced delayed neurotoxicity. This tyrosine carbon-paste biosensor system assayed for NTE in whole human and hen blood. NTE assay plays a promising role as a biomarker for neuropathic OP compounds. It is also predicted for early diagnosis.[45] Fluorescence resonance energy transfer (FRET) based flow cytometry biosensor was recently engineered for advancing prediction on neurodegenerative diseases which confirmed its sensitivity and specificity to tau and synnuclein fibrils. The tau proteins are found abundant in neurons and stabilized microtubules. Pathological aggregation of tau protein in the human brain leads to the neurodegenerative diseases. Using the biosensor, researchers have proved that "tau" protein seeding from a disease

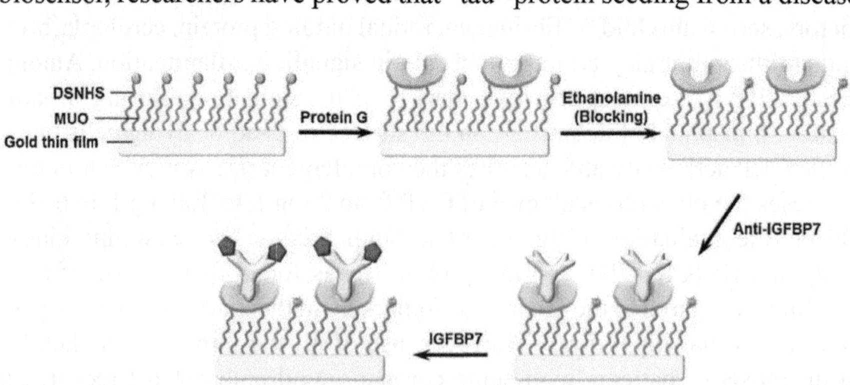

FIGURE 11.3.4a Assay for AntiIGFB7.

may be used as a robust marker for tauopathy and plays an important role in neurodegeneration. However, the study on robust assay is still to find ways to measure proteopathic seeding activity in biological specimens.[46]

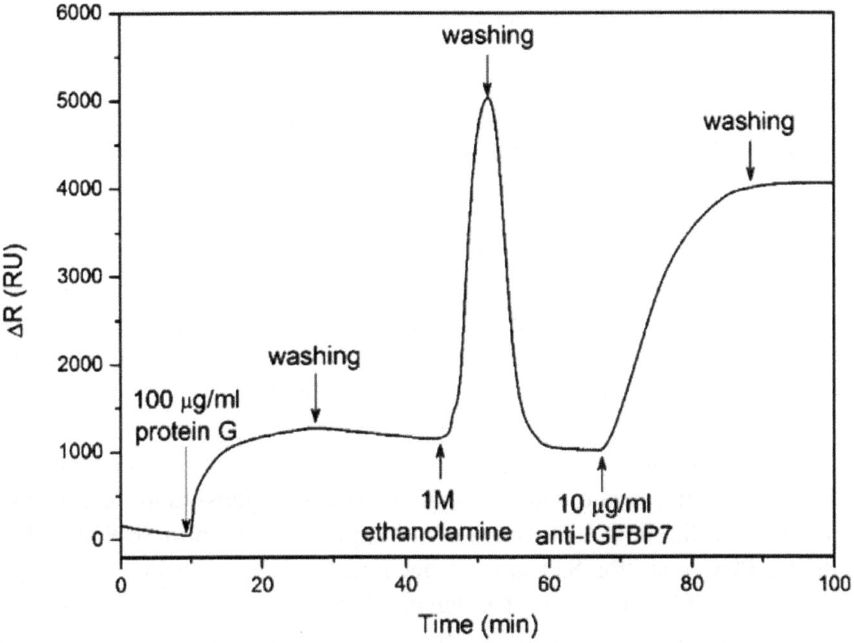

FIGURE 11.3.4B AntiIGFB7 reaction

11.3.5 BIOMARKERS FOR DETECTION OF OTHER DISEASE

Biomarkers kidney injury marker-1 and neutrophil gelatinase-associated lipocalin can diagnose patients quickly and accurately.[47–49] Recent discovery involves measuring blood urea nitrogen and serum creatinine, but these biomarkers are nonfunctional until 50% of the kidney function is impaired.[50] This shows that this marker is definitely not helpful in diagnosing before the onset of acute kidney injury (AKI). SPR imaging technology was also reported for the study of interactions between targeted molecules in human serum and peptides spanning the HnRNPA2/B1, the common antigenic target to define specific markers for monitoring diseases such as systemic lupus erythematosus (SLE), rheumatoid arthritis (RA)[51,52] as well as autoimmune hepatitis (AIH).[53] HnRNPA2/B1 is an important protein in mRNA processing and biogenesis. It is the most viable autoantigen[54] for having features such as spliced transcripts variants, ability to bind proteins and RNA, and is highly conserved. SPR imaging technology is used to measure dissociation rate constants of complexes formed after binding. Statistical analysis of the results have shown that peptide P7 (AA55-70) could be a specific biomarker for AIH patients, compared to SLE and RA, and most autoantibodies that react with P7 belong to the IgM isotope[55].

11.3.6 POINT-OF-CARE DIAGNOSTICS

Monitoring different diseases using a single assay can help prevent disease and prolong life. Early diagnosis can decrease the time and cost required after the onset of a disease. For example, standard laboratory tests for detecting sexually transmitted infections (STIs) typically take 2 days to 2 weeks where a 20-min POC test is ideal.[56] This POC is a small, portable, fast and highly accurate, and easy to use device with an integrated biosensor system.[57] Currently there are a few good performing, commercialized products for measuring STIs, measuring renal injury biomarkers associated with AKI and measurement of cardiac biomarkers for diagnosis and treatment of myocardial infarction.[58] There are some improvements in these POCs. The existing POC tests for STI are not accurate enough according to clinical requirements. Gonorrhea and *Chlamydia* followed by herpes simplex virus and seroconversion for human immunodeficiency virus (HIV) are the series of priority in development of POC for STIs.[59–61] Oral fluid, often called "the mirror of the body", is a perfect medium and led to the origin of salivary-based diagnostics. Many foreign agents in saliva are rare to detect which initially created a doubt on the success of salivary-based diagnosis. But now monitoring of HIV-1 and HIV-2 as well as various drugs including marijuana, cocaine, and alcohol has become possible. Comparative analysis of plasma-based detectors to saliva based depicts that extracellular proteins are predominant in human salivary proteome (HSP). However, further decoding of mRNA present in the saliva led to the demonstration of four genes (IL8, OAZ1, SAT, and IL1B) from the normal salivary transcriptome core. Thus the salivary transcriptome proposes the combined advantages of biomarker discovery in a noninvasive biofluid.[62] However, in case of tumor cells early diagnosis, it was not successful due to less concentration of these cells in blood. This barrier was overcome by a microfluidic chip-based micro-hall detector. It screens a single cell in the presence of many blood cells or free reactants which can be fabricated onto a mobile platform for POC use. With advancement in obtaining quick results, a low-cost dongle was developed with all mechanical, optical, and electronic functions of a laboratory-based ELISA. It performs triplexed immunoassay for HIV antibody, treponemal-specific antibody for syphilis, and nontreponemal antibody for active syphilis infection with high sensitivity and specificity.[63]

11.3.7 DEFECTS DETECTION IN CELL SIGNALING

The precise location and nature of the molecular interactions in living cells became a key consideration for delivering a "sensor" in the living cell without interfering with the cell's function. In order to understand the physical interactions between proteins involved in a typical bimolecular process, FRET has proven its impact on most of the applications. Further, employing labeling techniques of cellular components such as the nuclei, mitochondria, and cytoskeleton enables their localization within fixed and living preparations. Fluorescence based micro-encapsulated particles can deliver various biologically relevant cations and anions, oxygen, nitric oxide (NO), and glucose. FRET on conjugation with bioluminescence resonance energy transfer and quantum dots (QDs) helped in estimating protease enzymatic activity in its physiological roles. It also made possible early detection of primary tumors. Such systems offer highly sensitive and real time monitoring strategy for developing drugs and discovering role of enzymes. Moreover, combination of QD-RET along with protease-modulated sensors will also ease therapeutic approaches. Cell's ability to perceive and correctly respond to micro-environment is the basis of development, tissue repair, and immunity. Any minute error in cellular information processing can lead to diseases such as cancer, autoimmune diseases, and diabetes. Cell's communication take place at the varying proximity signaling molecules interact with target cell as a ligand to cell surface receptors to process the information. Therefore, understanding physiochemical parameters has become a major need in disease detection. For example, to detect CVDs, FRET based 3′,5′-cyclic guanosine monophosphate (cGMP) biosensor cGi500 was designed for monitoring cGMP signals. cGMP is an important second messenger that internalizes the message carried by intercellular messengers such as peptide hormone, NO and can function as an autocrine signal. Currently, this biosensor is the most useful tool for cGMP imaging in murine smooth muscle cells and cerebrellar granule neurons.[64] Based on the most recent research trends to ease the analysis of cGMP signaling in vivo, two methods namely epifluorescencebased ratio imaging and multiphoton-based microscopy are described. This has made monitoring of NO-induced cGMP transients and vasodilation possible in mice by applying both the methods.[65] G-protein-coupled receptors (GPCRs) are the most conserved and diverse group of membrane receptors. They play an incredible role in an array of functions in the human body and are the great source for drug designing for various diseases. GPCR shares a seven transmembrane spanning architecture for signaling. A novel technology, cellular dielectric spectroscopy (CDS) in CellKey system, helps in the detection of GPCRs' activation and distinguishes signals between different subtypes of the G protein (Gs, Gi/o, and Gq). CDS categorizes into two types: (1) an optical-based biosensor and (2) an electrical impedance based biosensor. The electrical impedance based biosensor has been specifically designed for GPCR detection and can differentiate signals between Gs, Gi/o, and Gq.[66] The CellKey™ system can detect multiple types of GPCR signaling in a single assay in less time with high throughput. The CellKey™ system differentiates between each of the three major G protein-dependent pathways (Gs, Gq, and Gi/o) with different kinetics, and is more specific for Gi/o protein-couple receptors.[67] The CellKey™ system possesses higher sensitivity than other traditional methods.[68] Electrochemical based biosensors detect many valuable signaling molecules. AmiNO-IV sensor is applied to NO record measurements in the blood of humans and biological tissues, brain, or lung in vivo. NO is a signaling molecule for many vascular diseases in vivo and in humans, and quantization of NO using an electrochemical electrode sensor helps in finding endothelial availability of NO in endothelial dysfunction and CVDs.[69] Important biomarkers in this regard are summarized in Table 11-1.

TABLE 11-1 [Reproduced from Mascini M.; Tombelli S. *Biomarkers* **2008,** 13 (7–8), 637–657]

Diseases	Biomarkers
Acute myocardial infarction	cardiac troponin t, cardiac troponin i
Atherosclerosis	Platelet-derived growth factor
Fibrosis	Progesterone
MALIGNANT DISEASES	
Anaemia cancer	Ferritin
Brease	ER,PR, HER2, CA15-3, CA125, CA27.29, CEA BRCA1, BRCA2, MUC-1, CEA,
Bladder	NY-BR-1, ING- 1
Cardiovascular disease DISEASE	Thrombin

TABLE 11-1 *(Continued)*

Diseases	Biomarkers
Cervix	P53, Bcl-2, Brn-3a, MCM, SCC-Ag, TPA, CYFRA 21-1, VEGF, M-CSF
Chronic myelogenous leukemia	BCR/ABL gene
Colon	HNPCC, FAP, CEA, CA19-9, CA24-2, p53
Esophagus	SCC
General physical stress	CRP
Hepatitis	Hepatitis B virus
Human papilloma virus	L1 viral region
Leukemia	Chromosomal aberrations
Liver	AFP, CEA
Lung	NY-ESO-1, CEA, CA19-9, SCC, CYFRA21-1, NSE
Melanoma	Tyrosinase, NY-ESO-1
Oral	IL-8
Ovarian	CA125, AFP, hCG, p53, CEA
Pancreas	CA19-9, CEA, MIC-1
Prostrate	PSA,PAP
Solid tumors	Circulating tumor cells in biological fluids, expression of targeted growth factor receptors
Stomach	CA72-4, CEA, CA19-9

11.3.8 DEFECTS DETECTION IN NUCLEIC ACID

DNA microarrays (or DNA chips) are in high demand for large-scale transcriptional profiling and single nuclear polymorphism discovery. In order to meet the requirements in viral diagnosis, DNA biosensors are already in use for hepatitis B virus[70,71] or papilloma virus[72,73] or finding disease-causing mutations in the genome sequences such as leukemia[74] or breast cancer.[75] A recent in vivo study performed on the nude mice injected with tumors. This study administered a protocol to enhance accumulation of liposomes-encapsulating doxorubicin (DOX) in the tissues relative to free DOX or passively targeted plain liposomes. Results from the study showed growth termination in tumor. This study also assisted in overcoming the multidrug resistance of cancer cells.[76] However, this study also showed some side effects which could lead to cardiotoxicity. Electro-catalytic activity of silver nanocrystals (Ag-NCs) in genetic analysis and clinical biomedical applications provides a new accepted technique for detection of microRNA with high sensitivity and selectivity using a label-free method. The strategy involved an MB probe and the efficient catalytic characteristics of Ag-NCs toward H_2O_2 reduction. The MBs are hairpin-shaped molecules integrated with a mass of suppressed fluorophores which restore their fluorescence on binding with targeted specific sequence of oligonucleotides.[77] Electrochemical biosensors can detect DNA damage at lower nickel concentrations for toxic analysis. However, in vivo study on *Caenorhabditis elegans* nematode germ line exposed to $NiCl_2$ concluded that programmed cell death is a result of genetic DNA damage. The damaged DNA was immobilized on pyrolytic graphite electrodes using the layer-by-layer technique. In the presence of a redox mediator, the guanines of DNA were oxidized and resulted in a peak current which supported the fact that extracted DNA from nematodes exposed to Ni was damaged.[78] One study explored the use of a complex biosensor system in finding the role of oxidative stress and associated changes due to oxidative stress in mitochondrial structure and function. This study was performed on patients with Kindler syndrome (KS). It is a syndrome that develops due to the mutation in FERMT1 (fermitin family member 1) gene. Redox biosensor experiments are conducted on a highly sensitive chimerical redox biosensor system (Grx1-roGFP2) and a second sensor containing the signal peptide directed to the mitochondrial (mito-Grx1-roGFP2) system showed a higher oxidized/reduced ratio in the KS keratinocytes[79].

11.3.9 CANCER CELL DETECTION

QD bioconjugates are excellent potential probes in noninvasive imaging methods and simultaneous detection of multiple tumor markers in in vivo diagnosis of cancer such as covalently linked QDs to three bioconjugated probes, arginine–glycine–aspartic acid (RGD) peptide (QD800-RGD), anti-epidermal growth factor receptor (EGFR) monoclonal antibody (MAb) QD840-anti-EGFR, and anticarcinoembryonic antigen related cell adhesion molecule-1 (CEACAM1) MAb (QD820-anti-CEACAM1) separately. Another sensor was developed using immunosensing electrodes fabricated on a glass substrate with different immobilized antigen. Using electrochemical enzyme-based competitive immunoassay, the biosensor was capable of measuring the concentrations of seven important tumor markers: AFP, ferritin,

CEA, hCG-beta, CA 15-3, CA 125, and CA 19-9. This brought multianalyte detection in single assay and significant advantages were observed over single-analyte tests.[80] Another combination of type 2CdTe/CdS QDs with cyclic RGD was used as a potential probe in early detection of pancreatic tumor cells for in vivo imaging.[81] QDs are also formulated with amphiphilic triblock copolymer for in vivo protection, targeting ligands for tumor antigen recognition, and multiple polyethylene glycol (PEG) molecules for improved biocompatibility and circulation. Antibody-conjugated QDs are used to target a prostrate-specific membrane antigen (PSMA). In the experiment, none of the anti-PSMA mAbs reacted with the noncancerous tissue. Anti-PSMA with extracellular domain binding mAbs were found with high intensified staining in tumor-associated tissues.[82] Simultaneous targeted cancer imaging, therapy, and real-time therapeutic monitoring can prevent over- or under treatment. This work describes the design of a multifunctional nanomicelle for recognition and precise NIR cancer therapy. The nanomicelle encapsulates a new pH-activatable fluorescent probe, a robust NIR photosensitizer R16FP, a newly screened cancer-specific aptamer for targeting viable cancer cells. The fluorescent probe can light up the lysosomes for real-time imaging. Upon NIR irradiation, R16FP-mediated generation of reactive oxygen species causes lysosomal destruction and subsequently triggers lysosomal cell death. Meanwhile, the fluorescent probe can show the cellular status and in vivo visualizes the treatment process. This protocol also provides molecular information for precise therapy and therapeutic monitoring.[83]

11.3.10 NEURODEGENERATIVE DISEASE DETECTION

Mitochondrial redox signal plays a central role in defining the status of neuronal physiology and disease. These redox signals are measured by ratiometric biosensors. A new technique called "multi-parametric imaging" is used on living mice to differentiate non-uniformity between the signals of neuronal physiology with the diseased one. The spontaneous contractions and reversible redox changes were more defined. Such changes were due to acute or chronic neuronal activity. Further permanent mitochondrial damage, change in calcium influx and change in mitochondrial membrane permeability after spinal cord injury were also come into notice.[84] Electrochemical-based biosensors have been extensively employed in finding valuable signaling molecules and neurotransmitters. One such sensor was used for glioma cells study[85] and increased emphasis on monitoring variations at the molecular level during sleep/awake cycle. A relatively new and interesting series of biosensors for measurement of adenosine is proposed by signaling the amount of peroxide from oxidation of glucose.[86] These biosensors have tremendous applications in neurochemical measurements. In the neuroscience and for in vivo monitoring processes, implanted sensors are better than imaging techniques due to their small size. It is implied that the sensor should be placed close to the source for clear observation. Also, the sensor's layers (outer layer, enzyme layer, and interference rejection layer) must be as thin as possible to have quick response time. However, the major challenge is receiving all of the functionalities which were integrated into these biosensors after implantation. These functions can be affected either by environment of the host or the sensor effects on the host. The adverse effect can cause blockages of analyte diffusion or acute inflammatory response which increases in an attempt to destroy the implant. Unstability can be observed within 3 days after sensor implantation, or may begin the healing process which will result in the formation of a fibrotic layer that isolates the sensor from the rest of the tissue. The another major effect is clot formation (thrombosis) in blood which can decrease the sensing capabilities of implanted sensors.[87]

11.3.11 BACTERIA AND VIRAL DETECTION

FRET in conjugation with NPs and a short primer assay manifested in the development of early secretory antigenic target-6 biomarker using a mixture of QDs and Au NPs conjugated with two sets of oligonucleotides, respectively in order to differentiate Bacillus Calmette–Guérin-vaccinated samples with MTBS -infected samples.[88] CdTe QDs are also employed for detection of hemagglutinin (HA) in the quantification of vaccine antigen and detection of influenza virus. In vivo application for detecting influenza virus can be improved by using QDs and magnetic NPs. The procedures such as automated robotic isolation and electrochemical analysis are incorporated into the successful study of binding between glycans and HA.[89]

11.3.12 THERAPEUTIC AND DRUG DELIVERY

Many drugs are effective in the early stages of treatment and on their receptive occurrence. However, certain patients develop drug resistance, thus rendering the therapy ineffective. For example, a popular MAb-based drug "Herceptin" for breast cancer developed drug-resistive activity. A study compared binding kinetic analysis in drug-sensitive and drug-resistive cells to examine the molecular scale origin of drug resistance. The label free SPR imaging technique detected two distinct populations of Her2. The first one has the same binding kinetics as the Her2-sensitive cells but the second population has dissociation process ten times higher which is likely responsible for the drug resistance.[90] Similarly, a research using the QCM biosensor technique studied the response of human mammary epithelial tumor cells to Taxanes. Taxanes are used for the treatment of many human cancers but developing resistance to one form of Taxane has raised a question on drug development. In order to assist this study, researchers compared the frequency and resistance effects on cell mass distribution and viscoelastic properties. The defined patterns for the particular drug resistivity can be a guideline for performing patient's biopsy to lead one step toward the development of personalized medicine.[21] In therapeutics, observed benefits of nanomaterials based nonviral gene delivery systems are more successful when compared to using viral vectors especially in terms of preparation, stability, modification and safe delivery. In order to ensure intracellular drug delivery into gastric cell line, a work was proposed with carbon dots (C-dots) based and PEI- absorbed nanocarrier to deliver siena against Survivin in human gastric cancer cell line MGC-803. C-dots were synthesized using microwave pyrolysis method.[91] Survivin is an antiapoptotic protein that is highly important in cell division, apoptosis, and its expression is often found to be high in human cancers. Therefore, it was used as a biomarker and also helpful in gastric carcinomas therapy. Aptamers were also proven helpful in cancer chemotherapy. Aptamers can bind to internalized cell surface receptors to deliver drugs and other desired molecules into the cells. Aptamers carrying cytotoxic drugs aim to kill cancer cells without harming noncancerous cells. NOX-A12 is a 45-nucleotide aptamer which inhibited binding of CXC12 (CXC chemokine ligand 12/stromal cell derived factor-1 (CXC12/SDF-1)) to its receptors.[93] CXC chemokine ligand 12/stromal cell derived factor-1 (CXC12/SDF-1) regulates migrations of leukemic cells to the bone marrow.[96] CXC12 binds with two chemokine receptors CXCR5 and CXCR7 which induces tumor growth and metastasis.[92] NOX-A12 interferes with the CXCL12 stability and diverts leukemic cells to re-enter the cell cycle and become available for chemotherapeutic attack. Aptamers, chemotherapeutics, and RNA drugs can be encapsulated into one nanovector that possesses multiple functionalities. Due to various characteristics, aptamer-based nanoparticles (particles with diameters ranging from 10 nm to 1000 nm) system will likely to be developed for further improvement in therapeutics and imaging systems.[94] For therapeutic examination of exogenous and endogenous cells which express GPCRs signaling, a novel technology called CDS is employed in CellKey™ system.

11.4 CONTRASTS AGENTS

11.4.1 GOLD

Au NPs exhibit bright NIR fluorescence emission of 700–900 nm and are less toxic.[95] These NPs can be adjusted into different shapes, sizes, and compositions to enhance contrast in optical and multispectral imaging. Since AuNPs are unstable and nontargeted, hydrophilic materials are coated on these particles to enhance their binding with biomolecules such as peptides, antibodies, nucleic acids, and small organic compounds, in vitro and in vivo. In the treatment of RA, the FDA has approved use of gold. Au NPs are less toxic and strongly tunable in the range of visible-to-NIR spectral region. With NIR dyes, Au NPs in conjugation with a fluorophore can simplify detection of deep tissue imaging.[96] A traditional method assayed two complementary sets of DNA sequences to target DNA immobilized on the surfaces of the Au NPs. Aptamer-modified Au NPs generated a high specific sensing system when assayed for platelet-derived growth factor receptors. Further change in color on linking of Apt-Au NPs with PDGF molecules added another feature.[97] Another electrochemiluminescence assay was made using a bio-bar code to hold aptamer/DNA primer/Au-Fe_3O_4 (TA/DP/Au-Fe_3O_4) nanoconjugate. This assay was developed to capture probe with 18mer. On applying isothermal conditions on this strategy, it was possible to detect certain cells such as Ramos cells even in the low concentration (as low as 16 cells) with an excellent selectivity.[98] Properties and applications of colloidal AuNPs play an important role in their changeable shapes. Two-photon scattering technique along with multifunctional AuNP conjugates resulted in highly selective and sensitive detection of breast cancer. In addition, gold nanorods (NRs) were used in multiple aptamer immobilizations to detect cancer cells. This increased the binding affinity of weak binding aptamers by ~26-fold compared to a single aptamer.[99]

11.4.2 SILICA NANO PARTICLES

Aptamer immobilization on silica NPs provides a biocompatible and versatile substrate for a biosystem assay.[100] Many essential features such as, easy separation on centrifugation,[101] number of fluorophores encapsulation inside single silica NPs which facilitate detection at lower concentrations,[102] and additional photostability using nonporous silica enhances their clinical utility. Development of mesoporous NPs improved drug delivery and oxygen sensing.[103,104] Various researches have emphasized on the development of such aptamer-immobilized NPs which can enhance signal monitoring. In order to meet this, a sandwich assay using aptamer functionalized silica NPs and cationic-conjugated polymer (CCP) was developed for optical detection of thrombin and lysosome. A complete set of fluorescein-labeled thrombin-binding aptamer was functionalized with silica NPs which after binding with target showed increase in brightness. This study depicts that energy transfer from CCP to fluorescein not only enhances signal monitoring but also allows the analyte extraction from a mixture of proteins. Similarly, flow cytometry with dye-doped silica NPs based method resulted in fast detection of cancer cells with high sensitivity.[105] Many aptamer-conjugated silica NPs produced low background signal, high signal enhancement, and efficient functionalization. Aptamer-conjugated silica NPs were also successful in multiplexed cancer cell detection.[106] Single excitation of three dyes co-encapsulated into the silica NPs emitted different signals which are also used in drug delivery methods.[107]

11.4.3 MAGNETIC NANO PARTICLES

Magnetic NPs are more useful and functional after modifying their surface by developing few atomic layers using inorganic NPs as their magnetic cores.[108] These inorganic NPs range from metals, alloys to metal-oxides, including the most focused metal–oxides—magnetite (Fe_3O_4) and maghemite (Fe_2O_3)—in clinical research.[109] The biocompatible metal NPs (MNPs) have certain magnetic properties and specificity. MNPs are mainly used in the field of biomedical separation, magnetic resonance imaging (MRI),[110] drug delivery,[111] and magnetic fluid hyperthermia therapy. Some assays are also developed using iron oxide NPs with aptamers which helped in detection of human R-thrombin protein. Another study developed an assay using MNPs with N-terminal A10 aptamer and a 57-base pair nuclease-stabilized 2′-fluoropyrimidine RNA molecule modified with C18-amine at the 3′-end for finding of PSMA-expressing prostate cancer cells. Apt-MNPs also act as a carrier for the chemotherapeutic agent, DOX. Apt-MNPS and Apt-FNPs after conjugation helped in early diagnosis of acute leukemia cells. For example, three sets of MNPs respectively conjugated with aptamer to develop an assay for three different types of cancer cells (CCRF-CEM acute leukemia cells, cell lines for Burkitt's lymphoma (Ramos), and non-Hodgkin's B-cell lymphoma (Toledo)). The whole assay was efficient for both detection and imaging.[112] Recent research used two MNPs with AS1411 aptamer (MF-AS1411) for multimodal targeting and imaging of nucleolin which is a cellular membrane protein highly expressed in cancer.[113]

11.4.4 SINGLE-WALLED CARBON NANOTUBES

Single-walled carbon nanotubes (SWCNTs) exhibit high aspect ratio, high surface area, electrical, thermal conductivity, and mechanical strength. These properties meet all the requirements need for an efficient biosensor. SWCNTs act as a potential transducer for biosensors elements. After several studies investigators found that the low detection limit can be overcome by pretreating carbon tubes with carbodiimidazole-activated Tween 20, and covalently immobilized thrombin aptamers. In fact, Apt-SWNT-FET assays were found better than IgE-mAB-modified SWNT-FET in IgE detection.[114] They were good quenchers for dye-labeled hairpin probe. Their capability of restoring the fluorescence signal depicts their high photophysical behavior which was studied both in the absence and presence of the target. The experimentation revealed that in the absence of the target it quenches fluorescence but in the presence of the target, aptamer binding with the target disturbs binding of the aptamer and SWCNT. This results in the dissociation of DNA from the SWCNT and hence, restores fluorescence.[115] Another assay with label-free and separation-free optical method for thrombin was developed using fluorescence signal of the Apt–SWCT which improved detection time.[116]

11.4.5 HYDROGELS

Hydrogels are cross-linked water loving, highly demanded, polymer structures in various biomedical and pharmaceutical applications.[117,118] A proper cross-linking of cDNA with acrydite-modified oligonucleotides can produce hydrogels.[119] An aptamer based reversible DNA-induced hydrogel system was employed in molecular recognition and separation technique. Two kinds of acrydite-modified DNA monomers were copolymerized with acrylamide monomers to obtain strands complementary to G1 and G2 for hydrogel formation. The complete assay brought DNA displacement to promote reversible hydrogel transition and an adenosine aptamer.[120] Aptamer-conjugated hydrogels are more efficient in the system for releasing the target.[121] They are also considered as a unique signal transduction for proper visualization.[122] In the demonstration of proper visualization methods from sol-to-gel and gel-to-sol, two acrydite-modified oligonucleotides, strand A and strand B were mixed in stoichiometric concentrations. Addition of a LinkerAdap results in the formation of a gelatin in this solution. Upon removal of the LinkerAdap, the gelatin gets dissolved and reverts back to its fluid state. Some studies even showed change in color during transition from one state to the other. Such change in color was observed during conversion of amylose to maltose in the presence and absence of iodine. Investigators noticed another detection feature, color-changing phenomenon.[123]

11.4.6 LIPOSOMES AND MICELLES

The liposomes-entrapping drugs are the most effective method in drug delivery. Drug delivery mainly depends on the rate by which a drug is delivered. Hence, the rate by which the drug is delivered has to be optimized.[124,125] Increase in microvascular permeability and defective lymphatic drainage results in enhancing permeability and retention effect. During occurrence of such effect, investigators found that the PEG-coated liposomes (Stealth) get accumulated in tumors.[126] Many molecules such as antibodies, folate, peptides, and nucleic acids are delivered through liposomes.[127,128] Further studies are required to determine whether folate targeting can confer a clear advantage in efficacy and/or toxicity to liposomal drugs. A system using folate-targeted Stealth Liposomal DOX (FTL-DOX) and cisplatin (FTL-cisplatin) proved advantageous for in-vitro experiments but in-vivo studies needs more specification for liposomes construction which highly depends on the tumor-model.[129] Hence, further studies are required for clear presentation of folate targeting benefits. Aptamer-binding liposomes improved retention time of a drug in plasma and enhanced inhibitory activity toward vascular endothelial growth factor (VEGF). This induced endothelial cell proliferation in vitro, compared with the free aptamer but in-vivo studies are still fragmentary. Dialkylglycerol (DAG) phosphoramidite containing a tetraethylene glycol spacer was synthesized and introduced at the 5′-end of the VEGF aptamer to make DAG-NX213, which was further incorporated into the liposome bilayer. For extending application of liposomes at the cellular level, 250 aptamers and fluorescein isothiocyanate-dextran molecules loaded inside liposomes released the loaded model drug within 30 min after incubation.[127] Recently, sgc8 aptamer binding with its target CEM cells or AS1411 aptamer are used in drug delivery methods for breast cancer. Polymeric micelles are another promising nonpolar water repellants carrier systems. More recently, micelles hybrid construction using hydrophilic oligonucleotides and hydrophobic polymers have drawn close attention.[130,131] For visual detection of target binding with DNA aptamer by reading change in turbidity, a sandwich structure was designed which contained anchored DNA, its partial cDNA, and **adenosine triphosphate** (ATP) aptamer. The aggregated micelles after adding ATP breaks the double bond of hybrid and releases the target molecule.[132] DNA–micelles with a modified recognition molecule presents another novel delivery system.[133] A well defined oligonucleotide–micelle structure using a DNA–diacyllipid conjugate enhanced coupling efficiency between synthetic DNA and the hydrophobic moiety.[134] To facilitate phosphoamidite chemistry which allows synthesis of a pyrene, a diacyllipid, and an aptamer phosphoramidite, a novel design was developed. This design was fabricated using highly hydrophilic single-stranded DNA aptamer, the pyrene molecule as a fluorescence reporter[135] and two C18 hydrocarbon tails which were highly hydrophobic in nature. The merits of this novel system greatly improved binding affinity, low K_{off} once on the cell membrane, rapid targeting ability, high sensitivity, and effective drug delivery. These features resulted from a specific interaction-induced nonspecific insertion process which, in turn, led to the fusion between Apt–micelles and the cell membrane. Moreover, the dynamic specificity of Apt–micelles in flow channel systems was demonstrated by flushing the Apt–micelles through a channel that included immobilized tumor cells on its surface.[136]

11.4.7 GADOLINIUM

Gadolinium (Gd), a lanthanide metal with seven unpaired electrons, posses high magnetic moment and water coordination. These features have been proven very effective in increasing proton relaxation time. Gd ions are strongly paramagnetic but highly toxic. Whereas gadolinium chelates are nontoxic and highly stable in the body.[137,138] Along with Gd–diethylenetriamine pentaacetic acid (DTPA), the first intravenous MRI contrast agent, many other chelates with varying biodistribution and specific application have been used clinically. Increased attention to overcome short retention time in blood and tissue, GD–DTPA was employed with the peptide CGLIIQKNEC (CLT1), a fibronectin–fibrin binding cyclic peptide. The chelate, Gd–DTPA-CLT1, was used in imaging of tumor tissues.[139,140]

Further, a CLT1 targeted contrast agent with Gd-tetraazacyclododecane-1,4,7,10-tetraacetic acid (GD-DOTA)43 monoamide chelates and two CLT1 molecules conjugated to a generation 3 (G3) polylysine dendrimer with a cubic silsesquioxane core (Gd–DOTA–G3–CLT1). This synthesized assay enhanced contrast in multimodalities (MRI and optical) imaging of prostrate tumor. GD–DOTA with cinnamoyl-Phe-d-Leu-Phe-d-Leu-Phe-Lys-NH2 (cFlFlFK) was used in MRI imaging of neutrophils at the sites of inflammation.[141] All the efforts made for contrast enhancing were achieved by DNA aptamer–DTPA–Gd(III) chelate. Another assay was demonstrated using 15mer thrombin aptamer for high resolution imaging.[142] Gadolinium in conjugation is a highly biocompatible and biodegradable nanoconjugate which serves as an efficient contrast agent for imaging and drug delivery. Highly effective novel cell-type-specific aptamers as caps were coupled with gadolinium-doped luminescent and mesoporous strontium hydroxyapatite nanorods (designated as Gd:SrHap nanorods) system. The Gd:SrHap-DOX-aptamer on releasing into cancer cells resulted in the pore opening and drug releasing. This can be used as an effective method for drug delivery in special cancer cell lines.[143]

11.4.8 11-CARBON

The central nervous system of mammals carries nicotine acetylcholine receptors (nACHRs) which facilitates the polarity of the cell membrane and activates the signal transduction pathways.[144] They are also believed to be involved in a variety of cognitive functions and memory,[145] neuroprotection in Parkinson's disease,[146] Alzheimer's disease,[147] and the progression of certain cancers.[148] Due to slow kinetics and low binding affinity, researchers concluded that there are no suitable tracers which can be used as a PET probe to study the features of nACHR. A traditional method utilized two ^{11}C-labeled compounds, 5-amino-7-(3-(411Cmethoxy)phenylpropyl)-2-(2-furyl)pyrazolo4,3-e-1,2,4-triazolo1,5-cpyrimidine[149] and 7-methyl-11C-(E)-8-(3,4,5-trimethoxystyryl)-1,3,7-trime thylxanthine[150] which was synthesized for imaging of A2A receptors but had limitations in their preparation. To overcome this barrier, investigators used^{123} I-labeled single-photon emission computed tomography (SPECT) tracer, 7-(2-(4-(2-fluoro-4-123Iiodophenyl)piperazin-1-yl)ethyl)-2-(furan-2-yl)-7H-pyrazolo4,3-e1,2,4triazolo1,5-cpyrimidin-5-amine (123IMNI-420) for evaluation of A2A receptors in two species of non-human primates.[151] Later it was used as a popular SPECT probe and in high demand due to easy availability at a central pharmacy, do not have to be generated on-site, and can be used for noninvasive imaging with SPECT.[151] Therefore, it was proposed that the SPECT probes would be more useful for the imaging of A2A receptors in the clinic.[151] On the basis of this proposal, a novel, ^{23}I-labeled SPECT tracer, 7-(2-(4-(2-fluoro-4-123Iiodophenyl) piperazin-1yl)ethyl)-2-(furan-2-yl)-7H-pyrazolo4,3-e1,2,4triazolo1,5-cpyrimidin-5-amine (123IMNI-420) was synthesized and evaluated for imaging of A2A receptors in humans with Parkinson's disease.[151] However, these radio chemicals cannot be used in most clinics because most clinics do not have the cyclotron facilities that are required for preparation of the ^{11}C-labeled compounds.

11.5 CONCLUSIONS

The rapid progress of nanomedicine has motivated development of genosensors by additions and refinements in concepts of integrated sensors, microfabrication, multiple analyte detection, and lab-on-a-chip. These technologies have addressed achievements in disease monitoring, biomarkers' detection, multiple target detection in single assay, diagnostics, and imaging. The research studies described in this chapter depict the broad potential of DNA sequence conjugated NPs in genosensors to enable early detection of tumors, continuous glucose monitoring, successful in vivo implantation, POC for HIV, neurodegenerative disease, and so on. However, the major challenge lies in achieving all functions of genosensors after in vivo implantation. To develop a robust assay using DNA sequence conjugated NPs, strategies need complete information on target and the environment which may hinder biosensor activities due to the factors such as cytotoxicity, membrane development around biosensors, noise development, and lack in specificities for target recognition. Therefore, surface physiochemical properties of NPs must be engineered when coupling DNA sequence for reducing nonspecific binding and maintaining essential conformation in the presence of the target. In the future, much more attention could be given to new approaches for developing novel assays for strengthening performance of genosensors in vivo.

ACKNOWLEDGMENT

The authors appreciate Rare Genomic Institute, Hanover, MD 21076, USA for support and motivation.

KEYWORDS

- **Nanobiosensing**
- **Biomarkers**
- **Nanoparticles**
- **Aptamers**
- **Nucleic acid**
- **SWCNTs, Liposomes**

REFERENCES

1. Wang, J. From DNA Biosensors to Gene Chips. *Nucleic Acids Res.* **2000,** *28* (16), 3011–3016.
2. Palchetti, I.; Mascini, M. Nucleic Acid Biosensors for Environmental Pollution Monitoring. *Analyst* **2008,** *133*, 846–854.
3. Piunno, P.; Krull, U.; Hudson, R.; Damha, M.; Cohen, H. Fiber-Optic DNA Sensor for Fluorometric Nucleic Acid Determination. *Anal. Chem.* **1995,** *67*, 2635–2643.
4. Watts, H.; Yeung, D.; Parkes, H. Real-Time Detection and Quantification of DNA Hybridization by an Optical Biosensor. *Anal. Chem.* **1995,** *67*, 4283–4289.
5. Thiel, A.; Frutos, A.; Jordan, C.; Corn, R.; Smith, L. *Anal. Chem.* **1997,** *69*, 4984–4956.
6. Storhoff, J.; Elghanian, R.; Mucic, C.; Mirkin, C.; Letsinger, R. One-Pot Colorimetric Differentiation of Polynucleotides with Single Base Imperfections Using Gold Nanoparticle Probes. *J. Anal. Chem. Soc.* **1998,** *120*, 1959–1964.
7. Wang, J. Electrochemical Biosensing of DNA Hybridization. *Chem. Eur. J.* **1999,** *5*, 1681–1685.
8. Mikkelsen, S. R. Electrochemical Biosensors for DNA Sequence Detection. *Electroanalysis* **1996,** *8*, 15–19.
9. Tombelli, S.,; Minunni, M.,; Mascini, M; Analytical applications of aptamers. *Biosens. Bioelectron.* **2005,** *20*, 2424–2434.
10. Cole, J. R.; Dick, L. W.; Morgan, E. J.; McGown, L. B. Affinity Capture and Detection of Immunoglobulin E in Human Serum Using an Aptamer-Modified Surface in Matrix Assisted Laser Desorption/Ionization Mass Spectrometry. *Anal. Chem.* **2007,** *79*, 273–279.
11. Willner, I.; Zayats, M. Electronic Aptamer-Based Sensors. *Angew. Chem. Int. Ed.* **2007,** *46*, 6408–6418.
12. Lai, R.; Plaxco, K. W.; Heeger, A. J. Effect of Molecular Crowding on the Response of an Electrochemical DNA Sensor. *Anal. Chem.* **2007,** *79*, 229–233.
13. Radi, A. E.; Acero Sanchez J. L.; Baldrich, E.; O'Sullivan, C. K. Reagentless, Reusable, Ultrasensitive Electrochemical Molecular Beacon Aptasensor. *J. Am. Chem. Soc.* **2006,** *128*, 117–124.

14. Zhang, H.; Wang, Z.; Li, X. F.; Le, C. Ultrasensitive Detection of Proteins by Amplification of Affinity aptamers. *Angew. Chem. Int. Ed.* **2006,** *45*, 1576–1580.
15. Olerup, O.; Zetterquist, H. HLA-DR Typing by PCR Amplification with Sequence-specific Primers (PCR-SSP) in 2 hours: An Alternative to Serological DR Typing in Clinical Practice including Donor-recipient Matching in Cadaveric Transplantation. *Tissue Antigens* **1992,** *39* (5), 225–235.
16. Saiki, R. K.; Bugawan, T. L.; Horn, G. T.; Mullis, K. B.; Erlich, H. A. Analysis of Enzymatically Amplified Beta-Globin and HLA-DQ Alpha DNA with Allele-Specific Oligonucleotide Probes. *Nature* **1986,** *324* (6093): 163–166.
17. Bannai, M.; Tokunaga, K.; Lin, L.; Ogawa, A.; Fujisawa, K.; Juji, T. HLA-B40, B18, B27, and B37 Allele Discrimination Using Group-Specific Amplification and SSCP Method. *Hum. Immunol.* 1996 ,46 (2), 107–113.
18. Michalski, J. P.; McCombs, C. C.; Arai, T.; Elston, R. C.; Cao, T.; McCarthy, C. F.; Stevens, F. M. HLA-DR, DQ Genotypes of Celiac Disease Patients and Healthy Subjects from the West of Ireland. *Tissue Antigens* **1996,** *47* (2), 127–133.
19. Sayer, D.; Whidborne, R.; Brestovac, B.; Trimboli, F.; Witt, C.; Christiansen F. HLA-DRB1 DNA Sequencing Based Typing: An Approach Suitable for High Throughput Typing Including Unrelated Bone Marrow Registry Donors. *Tissue Antigens* **2001,** *57* (1), 46–54.
20. Joda, H.; Beni, V.; Alakulppi, N.; Partanen, J.; Lind, K.; Strömbom, L.; Latta, D.; Höth, J.; Katakis, I.; O'Sullivan, C. K. *Anal Bioanal Chem.* **2014,** *406* (12), 2757–2769.
21. Dr. Braunhut, S. J.; McIntosh, D.; Vorotnikova, E.; Zhou, T.; Marx, K. A. Detection of Apoptosis and Drug Resistance of Human Breast Cancer Cells to Taxane Treatments using Quartz Crystal Microbalance Biosensor Technology. *ASSAY Drug Dev. Technol.* **2005,** *3* (1), 77–88. doi:10.1089/adt.2005.3.77.
22. Kopp, M. U.; de Mello, A.; Manz, A. Chemical Amplification: Continuous-Flow PCR on a Chip. *Science* **1998,** *280*, 1046–1050.
23. Joseph Wang. Survey and Summary from DNA Biosensors to Gene Chips. Service, R. *Science*, **1998,** *282*, 396–399.
24. Ramsey, G. DNA Chips: State-of-the Art. *Nat. Biotechnol.* **1998,** *16*, 40–45.
25. Liu, Y.; Li, X.; Zhang, Z.; Zuo, G.; Cheng, Z.; Yu, H. Nanogram Per Milliliter-Level Immunologic Detection of Alpha-Fetoprotein with Integrated Rotating-Resonance Microcantilevers for Early-Stage Diagnosis of Heptocellular Carcinoma. *Biomed. Microdevices.* **2009,** *11* (1), 183–191.
26. Cody Stringer, R.; Hoehn, D.; Grant, S. A. Quantum Dot-Based Biosensor for Detection of Human Cardiac Troponin I Using a Liquid-Core Waveguide. *IEEE Sens. J.*, **2008,** *8*, 295–300.
27 S'osnoweki, R. G.; Tu, E.; Butler, W.; O'Connell, J.; Heller, M. J. Rapid Determination of Single Base Mismatch Mutations in DNA Hybrids by Direct Electric Field Control. *Proc. Natl Acad. Sci. USA.* **1997,** *94*, 1119–1123.
28. Cheng, J.; Sheldon, E.; Wu, L.; Gerrue, L.; Carrino, J.; Heller, M.; O'Connell, J. Preparation and Hybridization Analysis of DNA/RNA from E. coli on Microfabricated Bioelectronic Chips. *Nat. Biotechnol.* **1998,** *16*, 541–545.
29. Shichiri, M.; Yamasaki, Y.; Hakui, N.; Abe, H. Wearable Artificial Endocrine Pancreas with Needle-Type Glucose Sensor. *Lancet* **1982,** *2*, 1129.
30. Henry, G. C. Getting Under the Skin: Implantable Glucose Sensors. *Anal. Chem.* **1998,** *70*, 594A.

31. Reach, G.; Wilson, G. S. Can Continuous Glucose Monitoring Be Used for the Treatment of Diabetes. *Anal. Chem.* **1992,** *64*, 381A.
32. Elbicki, J. M.; Weber, S. G. Glucose Microbiosensor Based on Glucose Oxidase Immobilized by AC-EPD: Characteristics and Performance in Human Serum and in Blood of Critically ill Rabbits. *Biosensors* **1989,** *4*, 251–257.
33. Gerritsen, M.; Jansen, J. A.; Kros, A.; Vriezema, D. M.; Sommerdijk, N. A. J. M.; Nolte, R. J. M.; Lutterman, J. A.; Van Hövell, S. W. F. M.; Van der Gaag, A. Influence of Inflammatory Cells and Serum on the Performance of Implantable Glucose Sensors. *J. Biomed. Mater. Res.* **2001,** *54*, 69.
34. Pasic, A.; Koehler, H.; Schaupp, L.; Pieber, T. R.; Klimant, I. Biosensors and Invasive Monitoring in Clinical Applications. *Anal. Bioanal. Chem.* **2006,** *386*, 1293–1302.
35. Pasic, A.; Koehler, H.; Schaupp, L.; Klimant, I. Handbook of Biophotonics. *Sens. Actua.* **2007,** *122*, 60–39.
36. Stephan, C.; Klaas, M.; Muller, C.; Schnorr, D.; Loening, S.; Jung, K. Interchangeability of Measurements of Total and Free Prostate-specific Antigen in Serum with 5 Frequently used Assay Combinations: an update. *Clin. Chem.* **2006,** *52*, 59–64.
37. Li, Z.; Wang, Y.; Wang, J.; Tang, Z.; Pounds, J. G.; Lin, Y. Rapid and Sensitive Detection of Protein Biomarker Using a Portable Fluorescence Biosensor Based on Quantum Dots and a Lateral Flow Test Strip. *Anal. Chem.* **2010,** *15;82* (16), 7008–7014.
38. Collinson, P.; Goodacre, S.; Gaze, D.; Gray, A.; Arrowsmith, C.; Barth, J.; Benger, J.; Bradburn, M.; Capewell, S.; Chater, T.; et al. Very Early Diagnosis of Chest Pain by Point-of-Care Testing: Comparison of the Diagnostic Efficiency of a Panel of Cardiac Biomarkers Compared with Troponin Measurement Alone in the Ratpac Trial. *RATPAC Trial. Heart* **2012,** *98*, 312–318.
39. MacDonald, S. P. J.; Nagree, Y. Rapid Risk Stratification in Suspected Acute Coronary Syndrome Using Serial Multiple Cardiac Biomarkers: A Pilot Study. *Emerg. Med. Australas.* **2008,** *20*, 403–409.
40. Wu, D.; Guo, Z.; Liu, Y.; Guo, A.; Lou, W.; Fan, D.; Wei, Q. Sandwich-Type Electrochemical Immunosensor Using Dumbbell-Like Nanoparticles for the Determination of Gastric Cancer Biomarker CA72-4. *Talanta* **2015,** *134*, 305–309.
41. Gommans, W. M.; Tatalias, N. E.; Sie, C. P.; Dupuis, D.; Vendetti, N.; Smith, L. et al. Screening of Human SNP Database Identifies Recoding Sites of A-to-I RNA Editing. *RNA* **2008,** *14* (10), 2074–85. doi:10.1261/rna.816908.
42. Agis-Balboa, R. C.; Arcos-Diaz, D.; Wittnam, J.; Govindarajan, N.; Blom, K.; Burkhardt, S. et al. A Hippocampal Insulin-Growth Factor 2 Pathway Regulates the Extinction of Fear Memories. *EMBO J.* August **2011,** *30* (19): 4071–4083.
43. Jang, D. H.; Choi, Y.; Choi, Y. S.; Kim, S. M.; Kwak, H.; Shin, S. H.; Hong, S. Sensitive and Selective Analysis of a Wide Concentration Range of IGFBP7 Using a Surface Plasmon Resonance Biosensor. *Colloids Surf B Biointerfaces*. **2014,** *123*, 887–891. doi: 10.1016/j.colsurfb.2014.10.037. Epub 2014 Oct 28.
44. Makhaeva, G. F.; Sigolaeva, L. V.; Zhuravleva, L. V.; Eremenko, A. V.; Kurochkin, I. N.; Malygin, V. V.; Richardson, R. J. Biosensor Detection of Neuropathy Target Esterase in Whole Blood as a Biomarker of Exposure to Neuropathic Organophosphorus Compounds. *J. Toxicol. Environ. Health A.* **2003,** *66* (7), 599–610.
45. Holmes BB.; Furman JL.; Mahan TE.; Yamasaki TR.; Mirbaha H.; Eades WC, Belaygorod L.; Cairns NJ.; Holtzman DM.; Diamond MI. Proteopathic Tau Seeding Predicts Tauopathy In Vivo. *Proc Natl Acad Sci U S A.* **2014,** *111* (41), E4376-E4385. doi: 10.1073/pnas.1411649111. Epub 2014 Sep 26.

46. Zhang, X. C.; Gibson, B.; Mori, R.; Snow-Lisy, D.; Yamaguchi, Y.; Campbell, S. C.; Simmons, M. N.; Daly, T. M. Analytical and Biological Validation of a Multiplex Immunoassay for Acute Kidney Injury Biomarkers. *Clin. Chim. Acta* **2013,** *415*, 88–93.
47. Sinha, V.; Vence, L. M.; Salahudeen, A. K.; Urinary Tubular Protein-Based Biomarkers in the Rodent Model of Cisplatin Nephrotoxicity: A Comparative Analysis of Serum Creatinine, Renal Histology, and Urinary KIM-1, NGAL, and NAG in the Initiation, Maintenance, and Recovery Phases of Acute Kidney Injury. *J. Investig. Med.* **2013,** *61*, 564–568.
48. Hoffmann, D.; Fuchs, T. C.; Henzler, T.; Matheis, K. A.; Herget, T.; Dekant, W.; Hewitt, P.; Mally A. Multiple Specificities of Autoantibodies Against hnRNP A/B Proteins in Systemic Rheumatic Diseases and hnRNP L as an Associated Novel Autoantigen. *Toxicology* **2010,** *277*, 49–58.
49. Klaassen, C. *Casarett & Doull's Toxicology. The Basic Science of Poisons,* 7th ed. Mcgraw-Hill: New York, 2013.
50. Siapka, S.; Patrinou-Georgoula, M.; Vlachoyiannopoulos, P. G.; Guialis, A. Multiple Specificities of Autoantibodies Against hnRNP A/B Proteins in Systemic Rheumatic Diseases and hnRNP L as an Associated Novel Autoantigen. *Autoimmunity* **2007,** *40* (3), 223–233.
51. Steiner, G.; Skriner, K.; Hassfeld, W.; Smolen, J. S. Clinical and Immunological Aspects of Autoantibodies to RA33/hnRNP-A/B Proteins—A Link Between RA, SLE and MCTD. *Mol. Biol. Rep.* **1996,** *23* (3–4), 167–171.
52. Huguet, S.; Labas, V.; Duclos-Vallee, J. C.; Bruneel, A.; Vinh, J.; et al. Heterogeneous Nuclear Ribonucleoprotein A2/B1 Identified as an Autoantigen in Autoimmune Hepatitis by Proteome Analysis. *Proteomics* **2004,** *4*, 1341–1345.
53. Backes, C.; Ludwig, N.; Leidinger, P.; Harz, C.; Hoffmann, J. et al. Immunogenicity of autoantigens. *BMC Genomics* **2011,** *12*, 340.
54. Beleoken, E.; Leh, H.; Arnoux, A.; Ducot, B.; Nogues, C.; De Martin, E.; Johanet, C.; Samuel, D.; Mustafa, M. Z.; Duclos-Vallée, J-C.; Buckle, M.; Ballot, E. PRi-based Strategy to Identify Specific Biomarkers in Systemic Lupus Erythematosus, Rheumatoid Arthritis and Autoimmune Hepatitis. *PLoS* **2013,** 8(12), e84600. .
55. Rompalo, A. M.; Hsieh, Y. H.; Hogan, T.; Barnes, M.; Jett-Goheen, M.; Huppert, J. S.; Gaydos, C. A. Point-of-Care Tests for Sexually Transmissible Infections: What Do 'End Users' Want? *Sex. Health* **2013,** *10*, 541–545.
56. Yager, P.; Domingo, G. J.; Gerdes, J. Point-of-Care Diagnostics for Global Health. *Ann. Rev. Biomed. Eng.* **2008,** *10*, 107–144.
57. Blick, K. E. The Benefits of a Rapid, Point-of-Care "TnI-Only" Zero and 2-Hour Protocol for the Evaluation of Chest Pain Patients in the Emergency Department. *Clin. Lab. Med.* **2014,** *34*, 75–85.
58. Hsieh, Y. H.; Hogan, M. T.; Barnes, M.; Jett-Goheen, M.; Huppert, J.; Rompalo, A. M.; Gaydos, C. A. Perceptions of an Ideal Point-of-Care Test for Sexually Transmitted Infections—A Qualitative Study of Focus Group Discussions with Medical Providers. *PLos One* **2010,** *5*, e14144.
59. Rompalo, A. M.; Hsieh, Y. H.; Hogan, T.; Barnes, M.; Jett-Goheen, M.; Huppert, J. S.; Gaydos, C. A. Point-of-Care Tests for Sexually Transmissible Infections: What Do 'End Users' Want? *Sex. Health* **2013,** *10*, 541–545.
60. Hsieh, Y. H.; Gaydos, C.; Hogan, T.; Uy, O.; Jackman, J.; Jett-Goheen, M.; Rompalo, A. Evaluation of Optical Detection Platforms for Multiplexed Detection of Proteins and

the Need for Point-of-Care Biosensors for Clinical Use. *Sex. Transm. Infect.* **2011,** *87*, A82–A83.

61. Segal, A.; David, T. Wong Salivary Diagnostics: Enhancing Disease Detection and Making Medicine Better. *Eur. J. Dent. Educ.* **2008,** *12* (Suppl 1): 22–29. doi: 10.1111/j.1600-0579.2007.00477.x.
62. Laksanasopin, T.; Guo, T. W.; Nayak, S.; Sridhara, A. A.; Xie, S.; Olowookere, O. O.; Cadinu, P.; Meng, F.; Chee, N. H.; Kim, J.; Chin, C. D.; Munyazesa, E.; Mugwaneza, P.; Rai, A. J.; Mugisha, V.; Castro, A. R.; Steinmiller, D.; Linder, V.; Justman, J. E; Nsanzimana, S.; Sia, S. K. A Smartphone Dongle for Diagnosis of Infectious Diseases at the Point of Care. *Sci Transl Med.* **2015,** *7* (273), 273re1. doi: 10.1126/scitranslmed.aaa0056.
63. Thunemann, M.; Fomin, N.; Krawutschke, C.; Russwurm, M.; Feil, R. Visualization of cGMP with cGi Biosensors. *Methods Mol Biol.* **2013,** *1020*, 89–120. doi: 10.1007/978-1-62703-459-3_6.
64. Thunemann, M.; Schmidt, K.; de Wit, C.; Han, X.; Jain, R. K.; Fukumura, D.; Feil, R. Correlative Intravital Imaging of cGMP Signals and Vasodilation in Mice. *Front Physiol.* **2014,** *5*, 394. Epub 2014 Oct 14.
65. Miyano, K.; Sudo, Y.; Yokoyama, A.; Hisaoka-Nakashima, K.; Morioka, N.; Takebayashi, M.; Nakata, Y.; Higami. Y.; Uezono, Y. History of the G Protein-Coupled Receptor (GPCR) Assays from Traditional to a State-of-the-Art Biosensor Assay. *J. Pharmacol. Sci.* **2014,** *126* (4), 302–309. doi: 10.1254/jphs.14R13CP. Epub 2014 Nov 22.
66. Miyano, K.; Sudo, Y.; Yokoyama, A.; Hisaoka-Nakashima, K.; Morioka, N.; Takebayashi, M.; Nakata. Y.; Higami, Y.; Uezono, Y. *J Pharmacol Sci.* **2014,** *126* (4), 302–309. doi: 10.1254/jphs.14R13CP. Epub 2014 Nov 2.
67. Peters, M. F.; Knappenberger, K. S.; Wilkins, D.; Sygowski, L. A.; Lazor, L. A.; Liu, J.; et al. Evaluation of Cellular Dielectric Spectroscopy, a Whole-Cell, Label-Free Technology for Drug Discovery on Gi-coupled GPCRs. *J. Biomol. Screen.* **2007,** *12*, 312–319.
68. Bokoch, G. M.; Katada, T.; Northup, J. K.; Ui, M.; Gilman, A. G. The Inhibitory Guanine Nucleotide-binding Protein (Ni) Purified from Bovine Brain is a High Affinity GTPase. *J. Biol. Chem.* **1985,** *260*(4), 2057–2063.
69. Saldanha, C.; de Almeida, J. P. L.; Silva-Herdade, A. S. Application of a Nitric Oxide Sensor in Biomedicine. *Biosensors* **2014,** *4*, 1–17. doi:10.3390/bios4010001.
70. Ding, C.; Zhao, C.; Zhang, M.; Zhang, S. Hybridization Biosensor Using 2,9-dimethyl-1,10-phenantroline Cobalt as Electrochemical Indicator for Detection of Hepatitis B Virus DNA. *Bioelectrochemistry* **2008,** *72*, 28–33.
71. Yao, C.; Zhu, T.; Tang, J.; Wu, R.; Chen, Q.; Chen, M.; Zhang, B.; Huang, J.; Fum, W. Biosensors for Hepatitis B Virus Detection. *Biosens. Bioelectron.* **2008,** *23*, 879–885.
72. Xu, L.; Yu, H.; Akhras, M. S.; Han, S. J.; Osterfeld, S.; White, R. L.; Pourmand, N.; Wang, S. X. Giant Magnetoresistive Biochip for DNA Detection and HPV Genotyping. *Biosens. Bioelectron.* **2008,** *24*, 99–103.
73. Dell'Atti, D.; Zavaglia, M.; Tombelli, S.; Bertacca, G.; Cavazzana, A. O.; Bevilacqua, G.; Minunni, M.; Mascini, M. Rapid Detection of Human Papilloma Virus Using a Novel Leaky Surface Acoustic Wave Peptide Nucleic Acid Biosensor. *Clinica. Chimica. Acta* **2007,** *383*, 140–146.
74. Chen, J.; Zhang, J.; Huang, L.; Lin, X.; Chen, G. Hybridization Biosensor Using 2-nitroacridone as Electrochemical Indicator for Detection of Short DNA Species of Chronic Myelogenous Leukemia. *Biosens. Bioelectron.* **2008,** *24*, 349–355.

75. Castaneda, M. T.; Merkoci, A.; Numera, M.; Alegret, S. Electrochemical Genosensors for Biomedical Applications Based on Gold Nanoparticles. *Biosens. Bioelectron.* **2007,** *22*, 1961–1967.
76. Liao, Z. X.; Chuang, E. Y.; Lin, C. C.; Ho, Y. C.; Lin, K. J.; Cheng, P. Y.; Chen, K. J.; Wei, H. J.; Sung, H. W. Trace and Label-Free microRNA Detection Using Oligonucleotide Encapsulated Silver Nanoclusters as Probes. *J Control Release.* **2015,** *pii*, S0168-3659(15)00082-6. doi: 10.1016/j.jconrel.2015.01.032.
77. Dong, H.; Jin, S.; Ju, H.; Hao, K.; Xu, L. P.; Lu, H.; Zhang, X. Trace and Label-Free microRNA Detection Using Oligonucleotide Encapsulated Silver Nanoclusters as Probes. *Anal Chem.* **2012,** *84* (20), 8670 –8674. doi: 10.1021/ac301860v. Epub 2012 Sep 28.
78. Huffnagle, I. M.; Joyner, A.; Rumble, B.; Hysa, S.; Rudel, D.; Hvastkovs, E. G. Dual Electrochemical and Physiological Apoptosis Assay Detection of In Vivo Generated Nickel Chloride Induced DNA Damage in Caenorhabditis Elegans. *Anal Chem.* **2014,** *86* (16), 8418–8424. doi: 10.1021/ac502007g. Epub 2014 Aug 5.
79. Zapatero-Solana, E.; García-Giménez, J.; Guerrero-Aspizua, S.; García, M.; Toll, A.; Baselga, E.; Durán-Moreno, M.; Markovic, J.; García-Verdugo, J.; Conti, C. J.; Has, C.; Larcher, F.; Pallardó, F. V.; Del Rio, M. Oxidative Stress and Mitochondrial Dysfunction in Kindler Syndrome. *Orphanet J. Rare. Dis.* **2014,** *9* (1), 211.
80. Wilson, M. S.; Nie, W. Multiplex Measurement of Seven Tumor Markers Using an Electrochemical Protein Chip. *Anal Chem.* **2006,** *78* (18), 6476–6483.
81. Yong, KT. Biophotonics and Biotechnology in Pancreatic Cancer: Cyclic RGD-peptide-Conjugated Type II Quantum Dots for In Vivo Imaging. *Pancreatology* **2010,** *10*, 553–564. doi: 10.1159/000283577.
82. Chang, S. S.; Reuter, V. E.; Heston, W. D. W.; Bander, N. H.; Grauer, L. S.; Gaudin, P. B. Five Different Anti-Prostate-Specific Membrane Antigen (PSMA) Antibodies Confirm PSMA Expression in Tumor-Associated Neovasculature. *Cancer Res.* **1999,** *59* (13), 3192–3198.
83. Tian, J.; Ding, L.; Ju, H.; Yang, Y.; Li, X.; Shen, Z.; Zhu, Z.; Yu, J. S.; Yang, C. J. A Multifunctional Nanomicelle for Real-Time Targeted Imaging and Precise Near-Infrared Cancer Therapy. *Angew Chem. Int. Ed. Engl.* **2014,** *53* (36), 9544–9549. doi: 10.1002/anie.201405490. Epub 2014 Jul 15.
84. Breckwoldt, M. O.; Pfister, F. M.; Bradley, P. M.; Marinković, P.; Williams, P. R.; Brill, M. S.; Plomer, B.; Schmalz, A.; St Clair, D. K.; Naumann, R.; Griesbeck, O.; Schwarzländer, M.; Godinho, L.; Bareyre, F. M.; Dick, T. P.; Kerschensteiner, M.; Misgeld, T. Multiparametric Optical Analysis of Mitochondrial Redox Signals During Neuronal Physiology and Pathology In Vivo. *Nat Med.* **2014,** *20* (5), 555–560. doi: 10.1038/nm.3520. Pub 2014 Apr 20.
85. Shibuli, K. An Electrochemical Microprobe for Detecting Nitric Oxide Release in Brain Tissue. *Neurosci. Res.* **1990,** *9*, 69–76.
86. Vallance, P.; Patton, S.; Bhagat, K.; Macallister, R.; Radomski, M.; Moncada, S.; Malinski, T. Direct Measurement of Nitric Oxide in Human Beings. *Lancet* **1995,** *346* (8968), 153–154.
87. Saldanha, C.; de Almeida, J. P. L.; Silva-Herdade, A. S. Redox Thiol Status Plays a Central Role in the Mobilization and Metabolism of Nitric Oxide in Human Red Blood Cells. *Cell Biol. Int.* **2009,** *33* (3), 268–275. doi: 10.1016/j.cellbi.2008.11.012. Epub 2008 Dec 11.

88. Shojaei, T. R.; Mohd Salleh, M. A.; Tabatabaei, M.; Ekrami, A.; Motallebi, R.; Rahmani-Cherati T.; Hajalilou, A.; Jorfi, R. Development of Sandwich-Form Biosensor to Detect Mycobacterium Tuberculosis Complex in Clinical Sputum Specimens. *Braz. J. Infect. Dis.* **2014,** *18* (6), 600–608. doi: 10.1016/j.bjid.2014.05.015. Epub 2014 Aug 30.
89. Krejcova, L.; Nejdl, L.; Hynek, D.; Krizkova, S.; Kopel, P.; Adam, V.; Kizek, R. Beads-Based Electrochemical Assay for the Detection of Influenza Hemagglutinin Labeled with CdTe Quantum Dots. *Molecules* **2013,** *18* (12), 15573–15586. doi: 10.3390/molecules181215573.
90. Wang, W.; Yin, L.; Gonzalez-Malerva, L.; Wang, S.; Yu, X.; Eaton, S.; Zhang, S.; Chen, H. Y.; LaBaer, J.; Tao, N. In Situ Drug-Receptor Binding Kinetics in Single Cells: A Quantitative Label-Free Study of Anti-Tumor Drug Resistance. *Sci Rep.* **2014,** *4*, 6609. doi: 10.1038/srep06609.
91. Dr. Braunhut, S. J.; McIntosh, D.; Vorotnikova, E.; Zhou, T.; Marx, K. A. Detection of apoptosis and drug resistance of human breast cancer cells to taxane treatments using quartz crystal microbalance biosensor technology. *ASSAY Drug Dev. Technol.* **2005,** *3* (1), 77–88. doi:10.1089/adt.2005.3.77.
92. Wang, Q.; Zhang, C.; Shen, G.; Liu, H.; Fu, H.; Cui, D. Fluorescent Carbon Dots as an Efficient Sirna Nanocarrier for its Interference Therapy in Gastric Cancer Cells. *J. Nanobiotechnol.* **2014,** *12*, 58 doi 10.1186/s12951-014-0058-0.
93. Duda, D. G.; Kozin, S. V.; Kirkpatrick, N. D.; Xu, L.; Fukumura, D.; Jain, R. K. CXCL12 (SDF1alpha)-CXCR4/CXCR7 Pathway Inhibition: An Emerging Sensitizer for Anticancer Therapies? *Clin. Cancer Res.* **2011,** *17* (8), 2074–2080.
94. Vater, A.; Klussmann, S. Toward Third-Generation Aptamers: Spiegelmers and Their Therapeutic Prospects. *Curr. Opin. Drug Discovery Dev.* **2003,** *6* (2), 253–261.
95. Chithrani, B. D.; Ghazani, A. A.; Chan, W. C. Determining the Size and Shape Dependence of Gold Nanoparticle Uptake into Mammalian Cells. *Nano Lett.* **2006,** *6* (4), 662–668.
96. Lee, H.; Lee, K.; Kim, I. K.; Park, T. G. Synthesis, Characterization, and In Vivo Diagnostic Applications of Hyaluronic Acid Immobilized Gold Nanoprobes. *Biomaterials* **2008,** *29* (35), 4709–4718.
97. Green, L. S.; Jellinek, D.; Jenison, R.; Ostman, A.; Heldin, C. H.; Janjic, N. Inhibitory DNA Ligands to Platelet-Derived Growth Factor B-chain. *Biochemistry-US.* **1996,** *35*, 14413–14424.
98. Chen, M.; Bi, S.; Jia, X.; He, P. Aptamer-Conjugated Bio-Bar-Code Au-Fe3O4 Nanoparticles As Amplification Station for Electrochemiluminescence Detection of Tumor Cells. *Anal Chim Acta.* **2014,** *837*, 44–51. doi: 10.1016/j.aca.2014.05.035. Epub 2014 Jun 2.
99. Huang, Y. F.; Chang, H. T.; Tan, W. H. Cancer Cell Targeting Using Multiple Aptamers Conjugated on Nanorods. *Anal Chem.* **2008,** *80*, 567–572.
100. Hilliard, L. R.; Zhao, X. J.; Tan, W. H. Immobilization of Oligonucleotides onto Silica Nanoparticles for DNA Hybridization Studies. *Anal. Chim. Acta* **2002,** *470*, 51–56.
101. Qhobosheane, M.; Santra, S.; Zhang, P.; Tan, W. H. Biochemically Functionalized Silica Nanoparticles. *Analyst* **2001,** *126*, 1274–1278.
102. Wittenberg, N. J.; Haynes, C. L. Using Nanoparticles to Push the Limits of Detection. *Nanomed. Nanobiotechnol.* **2009,** *1*, 237–254.
103. Vallet-Regí, M.; Balas, F.; Arcos, D. Mesoporous Materials for Drug Delivery. *Angew Chem. Int. Ed.* **2007,** *46*, 7548–7558.

104. Cheng SH.; Lee CH.; Yang CS.; Tseng FG.; Mou CY; Lo LW. Mesoporous Silica Nanoparticles Functionalized with an Oxygen-sensing Probe for Cell Photodynamic Therapy: Potential Cancer Theranostics. *J Mater Chem.* **2009,** 19:1252-1257
105. Estévez, M. C.; O'Donoghue, M. B.; Chen, X. L.; Tan, W. H. Highly Fluorescent Dye-doped Silica Nanoparticles Increase Flow Cytometry Sensitivity for Cancer Cell Monitoring.*Nano. Res.* **2009,** *2*, 448–461.
106. Chen, X. L.; Estévez, M. C.; Zhu, Z.; Huang, Y. F.; Chen, Y.; Wang, L.; Tan, W. H. Using Aptamer-Conjugated Fluorescence Resonance Energy Transfer Nanoparticles for Multiplexed Cancer Cell Monitoring. *Anal Chem.* **2009,** *81*, 7009–7014.
107. Zhu, C. L.; Song, X. Y.; Zhou, W. H.; Yang, H. H.; Wen, Y. H.; Wang, X. R. Targeted Drug Delivery: Concepts and Design. *J. Mater. Chem.* **2009,** *19*, 7765–7770.
108. Berry, C. C.; Curtis, A. S. G. Functionalisation of Magnetic Nanoparticles for Applications in Biomedicine. *J Phys D: Appl Phys.* **2003,** *36*, R198–R206.
109. Laurent, S.; Forge, D.; Port, M.; Roch, A.; Robic, C.; Elst, L. V.; Muller, R. N. Magnetic Iron Oxide Nanoparticles: Synthesis, Stabilization, Vectorization, Physicochemical Characterizations, and Biological Applications. *Chem. Rev.* **2008,** *108*, 2064–2110.
110. Fang, C.; Zhang, M. Q. Multifunctional Magnetic Nanoparticles for Medical Imaging Applications. *J. Mater. Chem.* **2009,** *19*, 6258–6266.
111. Veiseh, Gunn J. W.; Zhang, M. Q. Design and Fabrication of Magnetic Nanoparticles for Targeted Drug Delivery and Imaging. *Adv. Drug. Delivery Rev.* **2010,** *62*, 284–304.
112. Smith, J. E.; Medley, C. D.; Tang, Z. W.; Shangguan, D. H.; Lofton, C.; Tan, W. H. Aptamer-Conjugated Nanoparticles for the Collection and Detection of Multiple Cancer Cells. *Anal. Chem.* **2007,** *79*, 3075–3082.
113. Hwang, D. W.; Ko, H. Y.; Lee, J. H.; Kang, H.; Ryu, S. H.; Song, I. C.; Lee, D. S.; Kim, S. A Nucleolin-Targeted Multimodal Nanoparticle Imaging Probe for Tracking Cancer Cells Using an Aptamer. *J. Nucl. Med.* **2010,** *51*, 98–105.
114. Maehashi, K.; Katsura, T.; Kerman, K.; Takamura, Y.; Matsumoto, K.; Tamiya, E. Label-Free Protein Biosensor Based on Aptamer-Modified Carbon Nanotube Field-Effect Transistors. *Anal. Chem.* **2007,** *79*, 782–787.
115. Zhu, Z.; Tang, Z. W.; Phillips, J. A.; Yang, R. H.; Wang, H.; Tan, W. H. Regulation of Singlet Oxygen Generation Using Single-Walled Carbon Nanotubes. *J. Am. Chem. Soc.* **2008,** *130*, 10856–10857.
116. Chen, H.; Yu, C.; Jiang, C.; Zhang, S.; Liu, B. H.; Kong, J. L. A Novel Near-Infrared Protein Assay Based on the Dissolution and Aggregation of Aptamer-Wrapped Single-Walled Carbon Nanotubes. *Chem. Commun.* **2009,** 5006–5008.
117. Ravi Kumar, M. N. V.; Kumar, N.; Kashyap, N. Hydrogels for Pharmaceutical and Biomedical Applications. *Crit. Rev. Ther. Drug.* **2005,** *22*, 107–150.
118. Miyata, T.; Asami, N.; Uragami, T. A Reversibly Antigen-Responsive Hydrogel. *Nature.* **1999,** *399*, 766–769.
119. Nagahara, S.; Matsuda, T. Hydrogel Formation via Hybridization of Oligonucleotides Derivatized in Water-soluble Vinyl Polymers. *Polym. Gels. Netw.* **1996,** *4*, 111–127.
120. He, X.; Wei, B.; Mi, Y. Aptamer Based Reversible DNA Induced Hydrogel System for Molecular Recognition and Separation. *Chem. Commun.* **2010,** *46*, 6308–6310.
121. Soontornworajit, B.; Zhou, J.; Shaw, M. T.; Fanb, T. H.; Wang, Y. Hydrogel Functionalization with DNA Aptamers for Sustained PDGF-BB Release. *Chem. Commun.* **2010,** *46*, 1857–1859.
122. Yang, H. H.; Liu, H. P.; Kang, H. Z.; Tan, W. H. Aptamer-Conjugated Nanomaterials and Their Applications. *J. Am. Chem. Soc.* **2008,** *130*, 6320–6321.

123. Zhu, Z.; Wu, C. C.; Liu, H. P.; Zou, Y.; Zhang, X. L.; Kang, H. Z.; Yang, C. Y. J.; Tan, W. H. *Angew Chem. Int. Ed.* **2010,** *49*, 1052–1056.
124. Laginha, K. M.; Verwoert, S.; Charrois, G. J.; Allen, T. M. *Clin. Cancer Res.* **2005,** *11*, 6944–6949.
125. Johnston, M. J.; Semple, S. C.; Klimuk, S. K.; Edwards, K.; Eisenhardt, M. L.; Leng, E. C.; Karlsson, G.; Yanko, D.; Cullis, P. R. *Biochim. Biophys. Acta* **2006,** *1758*, 55–64.
126. Gabizon, A.; Catane, R.; Uziely, B.; Kaufman, B.; Safra, T.; Cohen, R.; Martin, F.; Huang, A.; Barenholz, Y. Poly(ethylene Glycol): Chemistry and Biological Applications. *Cancer Res.* **1994,** *54*, 987–992.
127. James, N. D.; Coker, R. J.; Tomlinson, D.; Harris, J. R.; Gompels, M.; Pinching, A. J.; Stewart, J. S. Development of Therapeutic Agents Handbook. *Clin. Oncol. (R. Coll. Radiol.)* **1994,** *6*, 294–296.
128. Weissmann, G.; Cohen, C.; Hoffstein, S. Introduction of Missing Enzymes into the Cytoplasm of Cultured Mammalian Cells by Means of Fusion-Prone Liposomes. *Trans. Assoc. Am. Physicians* **1976,** *89*, 171–183.
129. Ozawa, M.; Asano, A. The Preparation of Cell Fusion-Inducing Proteoliposomes from Purified Glycoproteins of HVJ (Sendai Virus) and Chemically Defined Lipids. *J. Biol. Chem.* **1981,** *256*, 5954–5956.
130. Leserman, L. D.; Weinstein, J. N.; Blumenthal, R.; Terry, W. D. Receptor-Mediated Endocytosis of Antibody-Opsonized Liposomes by Tumor Cells. *Proc. Natl. Acad. Sci. U. S. A.* **1980,** *77*, 4089–4093.
131. Straubinger, R. M.; Hong, K.; Friend, D. S.; Papahadjopoulos, D. Endocytosis of Liposomes and Intracellular Fate of Encapsulated Molecules: Encounter with a Low pH Compartment After Internalization in Coated Vesicles. *Cell* **1983,** *32*, 1069–1079.
132. Leamon, C. P.; Low, P. S. Delivery of Macromolecules into Living Cells: A Method that Exploits Folate Receptor Endocytosis. *Proc. Natl. Acad. Sci. U.S.A.* **1991,** *88*, 5572–5576.
133. Heath, T. D.; Fraley, R. T.; Papahdjopoulos, D. Antibody Targeting of Liposomes: Cell Specificity Obtained by Conjugation of F(ab')2 to Vesicle Surface. *Science* **1980,** *210*, 539–541.
134. Martin, F. J.; Hubbell, W.; Papahadjopoulos, D. Immunospecific Targeting of Liposomes to Cells: A Novel and Efficient Method for Covalent Attachment of Fab' Fragments via Disulfide Bonds. *Biochemistry* **1981,** *20*, 4229–4238.
135. Heath, T. D.; Montgomery, J. A.; Piper, J. R.; Papahadjopoulos, D. Antibody-Targeted Liposomes: Increase in Specific Toxicity of Methotrexate-Gamma-Aspartate. *Proc. Natl Acad. Sci. U.S.A.* **1983,** *80*, 1377–1381.
136. Papahadjopoulos, D.; Gabizon, A. Targeting of Liposomes to Tumor Cells In Vivo. *Ann. N. Y. Acad. Sci.* **1987,** *507*, 64–74.
137. Brasch, R. C. New Directions in the Development of MR Imaging Contrast Media. *Radiology* **1992,** *183* (1), 1–11.
138. Runge, V. M.; Gelblum, D. Y. Future Directions in Magnetic Resonance Contrast Media. Top *Magn. Reson. Imaging.* **1991,** *3* (2), 85–97.
139. Pilch, J.; Brown, D. M.; Komatsu, M.; Jarvinen, T. A.; Yang, M.; Peters, D.; Hoffman, R. M.; Ruoslahti, E. Peptides Selected for Binding to Clotted Plasma Accumulate in Tumor Stroma and Wounds. *Proc. Natl. Acad. Sci. U S A.* **2006,** *103* (8), 2800–2804.
140. Ye, F.; Wu, X.; Jeong, E. K.; Jia, Z.; Yang, T.; Parker, D.; Lu, Z. R. A Peptide Targeted Contrast Agent Specific to Fibrin-Fibronectin Complexes for Cancer Molecular Imaging with MRI. *Bioconjug Chem.* **2008,** *19* (12), 2300–2303.

141. Leung, K. *Molecular Imaging and Contrast Agent Database (MICAD)*; Bethesda, MD: National Center for Biotechnology Information (US); 2004–2013. 2013 May 01.
142. Bernard, E. D.; Beking, M. A.; Rajamanickam, K.; Tsai, E. C.; Derosa, M. C. Arget Binding Improves Relaxivity in Aptamer-Gadolinium Conjugates. *J. Biol. Inorg. Chem.* **2012,** *17* (8), 1159–1175. doi: 10.1007/s00775-012-0930-z. Epub 2012 Aug 19.
143. Li, Z.; Liu, Z.; Yin, M.; Yang, X.; Yuan, Q.; Ren, J.; Qu, X. Aptamer-Capped Multifunctional Mesoporous Strontium Hydroxyapatite Nanovehicle for Cancer-Cell-Responsive Drug Delivery and Imaging. *Biomacromolecules* **2012,** *13* (12), 4257–4263. doi: 10.1021/bm301563q. Epub 2012 Nov 16.
144. Albuquerque, E. X.; Pereira, E. F.; Alkondon, M.; Rogers, S. W. Mammalian Nicotinic Acetylcholine Receptors: From Structure to Function. *Physiol. Rev.* **2009,** *89* (1), 73–120.
145. dos Santos Coura, R.; Granon, S. Prefrontal Neuromodulation by Nicotinic Receptors for Cognitive Processes. *Psychopharmacology (Berl)* **2012,** *221* (1), 1–18.
146. Kawabata, J.; Suzuki, S.; Shimohama, S. α7 Nicotinic Acetylcholine Receptor Mediated Neuroprotection in Parkinson's Disease. *Curr. Drug. Targets.* **2012,** *13* (5), 623–630.
147. Hernandez, C. M.; Dineley, K. T. α7 Nicotinic Acetylcholine Receptors in Alzheimer's Disease: Neuroprotective, Neurotrophic or Both? *Curr. Drug. Targets.* **2012,** *13* (5), 613–622.
148. Schuller, H. M. Regulatory Role of the alpha7nAChR in Cancer. *Curr. Drug. Targets.* **2012,** *13* (5):680–687.
149. Ramlackhansingh, A. F.; Bose, S. K.; Ahmed, I.; Turkheimer, F. E.; Pavese, N. Adenosine 2A Receptor Availability in Dyskinetic and Nondyskinetic Patients with Parkinson Disease. *Brooks D.J. Neurology.* **2011,** *76* (21), 1811–1816.
150. Mishina, M.; Ishiwata, K.; Naganawa, M.; Kimura, Y.; Kitamura, S.; Suzuki, M.; Hashimoto, M.; Ishibashi, K.; Oda, K.; Sakata, M.; Hamamoto, M.; Kobayashi, S.; Katayama, Y.; Ishii, K. Adenosine A(2A) Receptors Measured with CTMSX PET in the Striata of Parkinson's Disease Patients. *PLoS One.* **2011,** *6* (2), e17338.
151. Tavares, A. A.; Batis, J.; Barret, O.; Alagille, D.; Vala, C.; Kudej, G.; Koren, A.; Cosgrove, K. P.; Nice, K.; Kordower, J. H.; Seibyl, J.; Tamagnan, G. D. Kinctic Modeling, Test-Retest, and Dosimetry of 123I-MNI-420 in Humans. *Nucl. Med. Biol.* **2013,** *40* (3), 403–409.

CHAPTER 12

WEARABLE NANOENABLED BIOSENSORS

APARAJITA SINGH*, SYED KHALID PASHA, and SHEKHAR BHANSALI

Electrical and Computer Engineering Department, Florida International University, Miami, FL USA

*E-mail: asing044@fiu.edu

CONTENTS

Demand for continuous and long term health monitoring has taken the personalized health care from the point-of-care (POC) sensors to wearable biosensors. With growing technological advancements the role of “nano” is becoming dominant for more sensitive, robust, and less intrusive wearable devices and sensors. In this chapter, we present the current state-of-the art nanoenabled wearable biosensors, in particular the popular strain sensors and electrochemical sensors. Wearable electrochemical sensors are relatively new area of research in the field of wearable sensors which poses unique challenges when compared to physical strain sensors. The challenges, advantages, and the underlying technology of the devices are highlighted. The biggest bottleneck for using the wearable technology is the “energy harvesting”. We summarize the state-of-the-art technologies which are promising solutions to this limitation and discuss the underlying concepts in brief.

12.1 FROM POC SENSING TO WEARABLE SENSING

Technology has given comfort as well as a complex lifestyle which has led to complex health conditions (Figure 12-1). The number of individual with several health conditions is increasing and stressing the health-care system. In many cases there is not even a proper health-care system. To ensure quality care for all the individuals including those in remote areas without having to stress the system or the individual it is technology where we seek refuge. These are technologies which make the POC sensing possible. POC biosensing allows detection of genetic and molecular biomarkers for identifying chronic diseases for timely diagnosis and treatment. With the POC devices it is possible to rapidly detect various biomarkers at lower detection levels within a small sample amount at laboratories, hospital, or even at home. Low cost personalized health care has revolutionized and continues to grow with POC’s fast analysis and high-sensitivity characteristics. Even with the personalized health-care facilities, it is hard to monitor chronic diseases which appear at irregular intervals, show no symptoms initially but can result in intense reactions sometimes. This makes continuous and long-term monitoring essential for early detection, diagnosis, and treatment. Thereby for a complete health care, various levels of health information, that is, from molecular level to body level is required. Wearable sensing technologies facilitate continuous and long-term body level health information without hindering the user’s daily activities. In this chapter, state-of-the-art wearable nanoenabled biosensors are presented with their underlying principle. An overview of the various advantages and challenges of these sensors is included.

12.2 SENSOR PLACEMENT

Out of the many considerations for a wearable sensor, predicting the proper place for a wearable sensor has proved to be a limitation. In an attempt to get to most accurate results as per the activity, researchers have developed various kinds of sensors which can be placed differently. For example, numerous accelerometer-based sensors have demonstrated that can be worn on hip, wrist, ankle, arm, thigh, waist, or even ear[1] to predict different activities. Table 12-1 summarizes some of the activities that can be detected by the existing wearable sensors in the market as per the location of the sensor.[2]

TABLE 12-1 Location of wearable sensor and type of detectable activity[2]

Location of sensor	Activity
Head	Force/Impact detection, electroencephalogram
Ear	Heart rate, metabolic rate, oxygen consumption, step count, calories
Shoulder	Movement, posture, muscle activity
Chest	Heart rate, breathing, movement, sleep tracking, body temperature, posture
Waist	Posture, sleep tracking, step count, balance
Arms	Skin temperature, movement, heart rate, blood pressure
Wrist	Heart rate, sleep tracking, step count, calories, oxygen level
Finger	Heart rate, blood oxygen
Legs	Balance, gait analysis, lactate threshold, pace, calories, heart rate, muscle activity, step count
Feet	Pressure, oxygen level, heart rate, step count, skin temperature, sleep tracking, calories, movement, gait, pace

12.3 NANOENABLED WEARABLE STRAIN SENSORS

The study of movement provides invaluable knowledge to accurately diagnose numerous medical conditions such as obesity, stroke, chronic pulmonary disease, osteoarthritis, sclerosis, and Parkinson's disease.[3] Figure 12-1 shows the vicious cycle of body movement and health conditions, thus

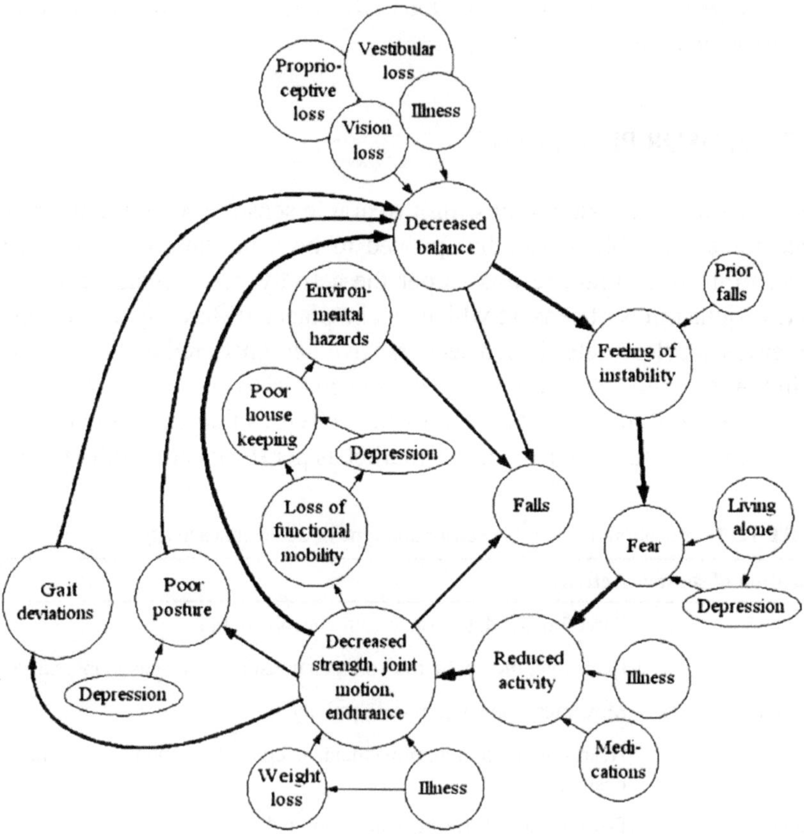

FIGURE 12-1 Vicious cycle of body motion and health conditions.Sensor Placement.

highlighting the importance of continuous monitoring of movement. The dominant sensors in continuously assessing the human movement or posture are accelerometers, gyroscopes, magnetometers, and pressure sensor.[2,4] These miniaturized, low-cost, and low power consuming wearable motion sensors have been enabled by microelectromechanical systems technology for biomechanical measurements. However, the sensor and electronic components cannot stretch, bend, twist, or deform in response to soft, conformable nature of the tissue and rotational motion of the joints. This limitation has led to research and development of ultralight, stretchable, and foldable sensors to be mounted on or near the skin. The main interest has been development of wearable motion, pressure, and strain sensor to sense joint and muscle movement for monitoring posture, movement, and breathing.[5] With the advances in materials science and nanotechnology, various wearable sensors have been reported from a range of nanostructured materials: nanowire,[6] grapheme,[5,7] nanotube,[8] and other nanoengineered structures.[9] This section gives an overview of the working principle of wearable strain sensors.

12.3.1 RESISTIVE STRAIN SENSOR

Subtle bodily motions associated with muscular motion, breathing, pulse, and speech require high sensitivity. Resistive type strain sensors have shown the highest sensitivity as compared to the capacitive type. In resistive type strain sensors, gauge factor (*G*) which is a measure of sensitivity describes how the relative resistance changes with strain (ε), that is, $\frac{\Delta R}{R} = \varepsilon G$. The gauge factor is low (~2) for traditional materials (such as metals) with conductivity independent of strain. For other composite materials where the conductivity is strain dependent, the relative resistance change with strain is[5]

$$\frac{\Delta R}{R} = (1+\vartheta)\varepsilon + \frac{\Delta\sigma_{(\varepsilon)}}{\sigma_o} \tag{12.1}$$

where, ϑ is the Poisson's ratio and σ is conductivity. Nanocomposite materials show large resistivity changes since the spacing between the nanomaterials changes significantly with strain.

Graphene is known for its superior electromechanical properties among other nanomaterials. It has been considered a promising material for wearable strain sensor with calculated gauge factors of ~10^3 under 2–6% strain, 10^6 under high strain (>7%), and ~35 under small strain of 0.2% (Figure 12-2).[7a] Graphene can be coated, embedded, or infused in a flexible material. Graphene-woven fabric is formed by coating or embedding a thin film of criss-cross patterned electrical network of graphene on a flexible poly(dimethylsiloxane) (PDMS) substrate (Figure 12-3a).[7] With increasing strain the density of the cracks also increase in the graphene network thereby increasing the resistance and decreasing the current in the pathway (Figure 12-3b). In a similar fashion carbon nanotube (CNT) and nanowire based resistive strain sensors can also be fabricated. However, CNT based strain sensors are limited to sensing limb and joint movements due to their much lower sensitivity.[10] Gold nanowire matrix, though not as sensitive as a graphene-based sensor, has shown great capability for detecting wrist pulses to monitor blood pressure.[6b]

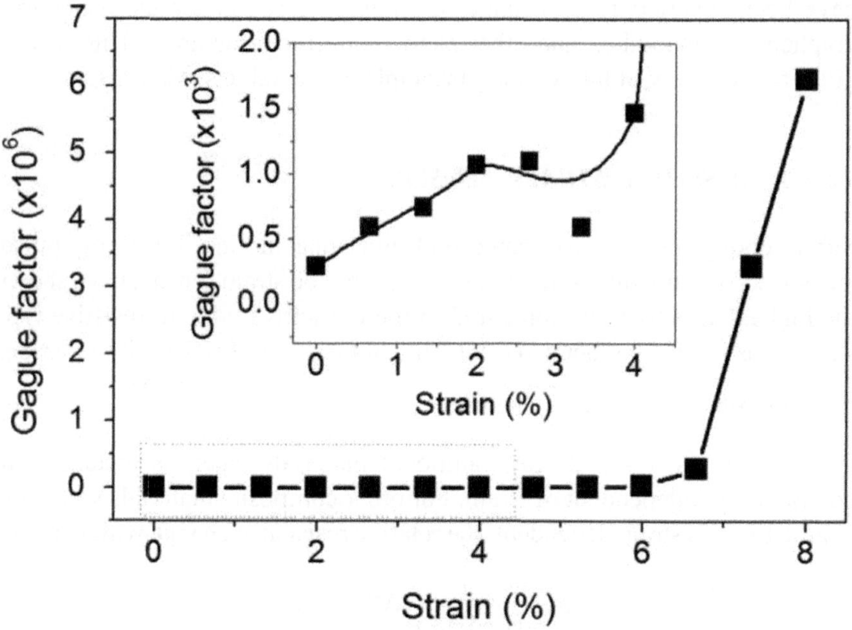

FIGURE 12.2 Graphene gauge gactor vs. strain curve.

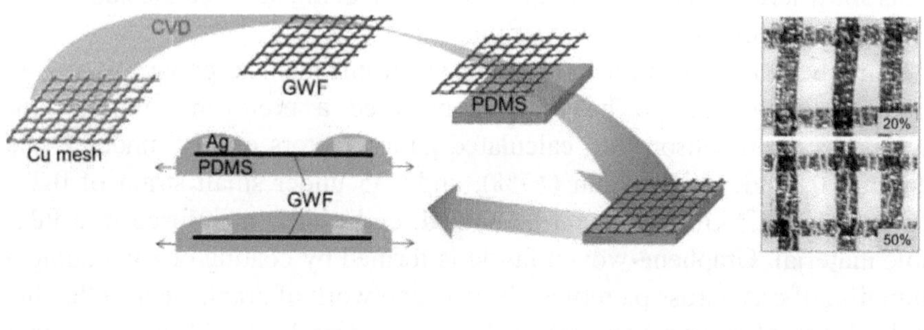

(a) (b)

(b)

FIGURE 12.3 (a) Fabrication steps of graphene woven fabric on a flexible substrate (b) Increase of crack density with strain

On the other hand, graphene-infused elastomers (G-bands) have proven better strain sensors as compared to elastomers coated with graphene. In other words, three-dimensional (3D) nanocomposite matrix is more sensitive than the 2D type. The infusion depth (x) of graphene is related to concentration (c) and soak time (t) as[5]

$$c(x) \propto erfc[\frac{x}{(4Dt)^{\frac{1}{2}}}] \quad (12.2)$$

where $erfc(\cdot)$ is the complementary error function, D is the diffusion coefficient of graphene in the elastomer material. This implies that after short soak times the conduction is dominated by graphene close to surface and bulk graphene dominates after longer soak times. The gauge factor decreases considerably (G~10–15) when graphene is concentrated at the surface compared to that of bulk graphene composite (G~35). The gauge factor also increases with decreasing concentration of bulk graphene. For a minimum bulk concentration of 20%, which is the required concentration to form a continuous conducting path, G is 35 and reduces to 10 for 50% bulk concentration.[5] As mentioned earlier, the lower concentration allows higher resistance changes due to larger changes in the spacing between the nanomatrix.

12.3.2 CAPACITIVE STRAIN SENSOR

In capacitive type strain sensors strain corresponds to changes in capacitance. The gauge factor (G) or sensitivity of a capacitive strain sensor is[8b]

$$G\varepsilon = \frac{\Delta C}{C_o} \quad (12.3)$$

where C_o is the initial capacitance and ΔC is the change in capacitance under strain (ε). The capacitance $C = \epsilon_r \frac{l * w}{t}$, where ϵ_r is the permittivity of the dielectric, t is the thickness of the dielectric, l is the length of the capacitor, and w is the width. Due to the low sensitivity of capacitive type sensor than that of resistive type, these are mostly used in monitoring bodily motions with higher strain such as joint movement or limb movement (Figure 12-4). Capacitive based strain sensor is commonly patterned by screen printing conductive nanomaterials (CNTs, nanowires) on stretchable substrate (PDMS).[6a,8a] A dielectric is coated on top of the substrate and overlapped by another screen-printed substrate aligned orthogonal to the bottom substrate

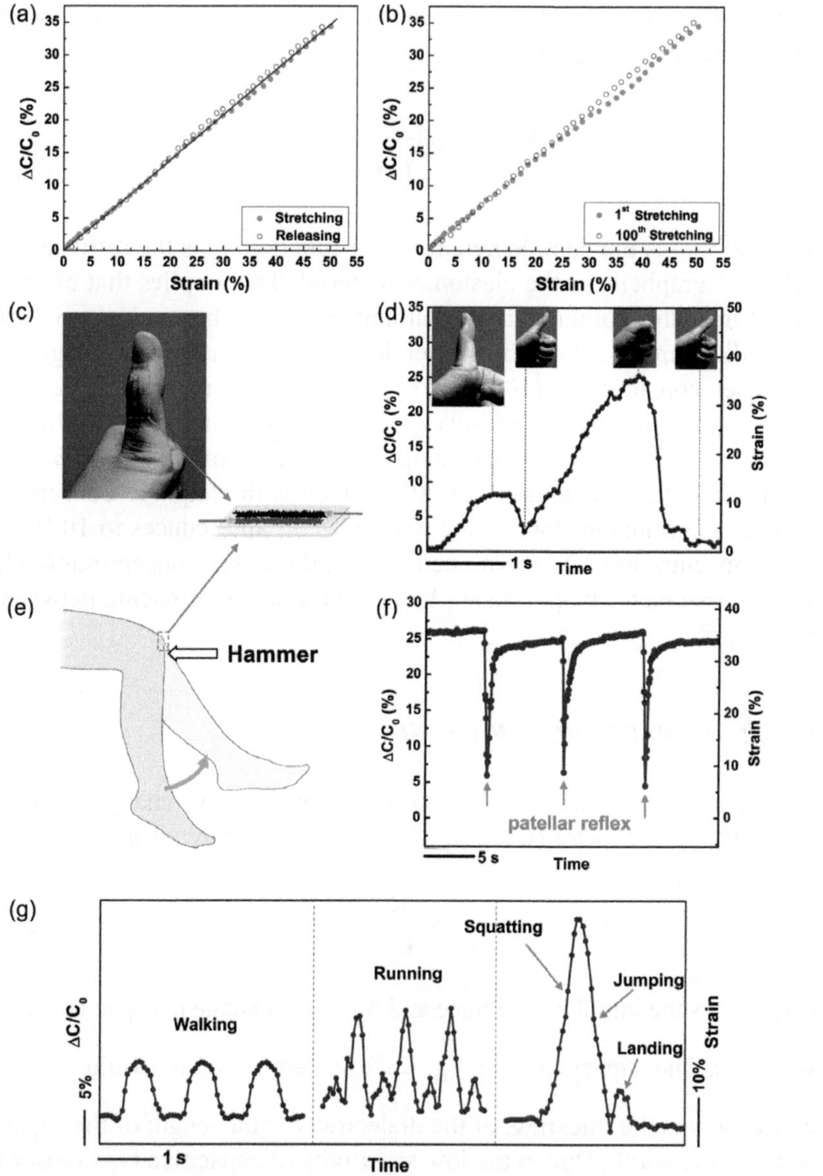

FIGURE 12.4 Body and joint movements that require high strain

(Figure 12-5). This way a pixel of capacitive junctions are formed wherever dielectric is sandwiched between screen-printed electrodes. In this configuration, changes in capacitance due to strain or pressure can be distinguished. Under tensile strain pixels the axis of strain will be affected and pressure

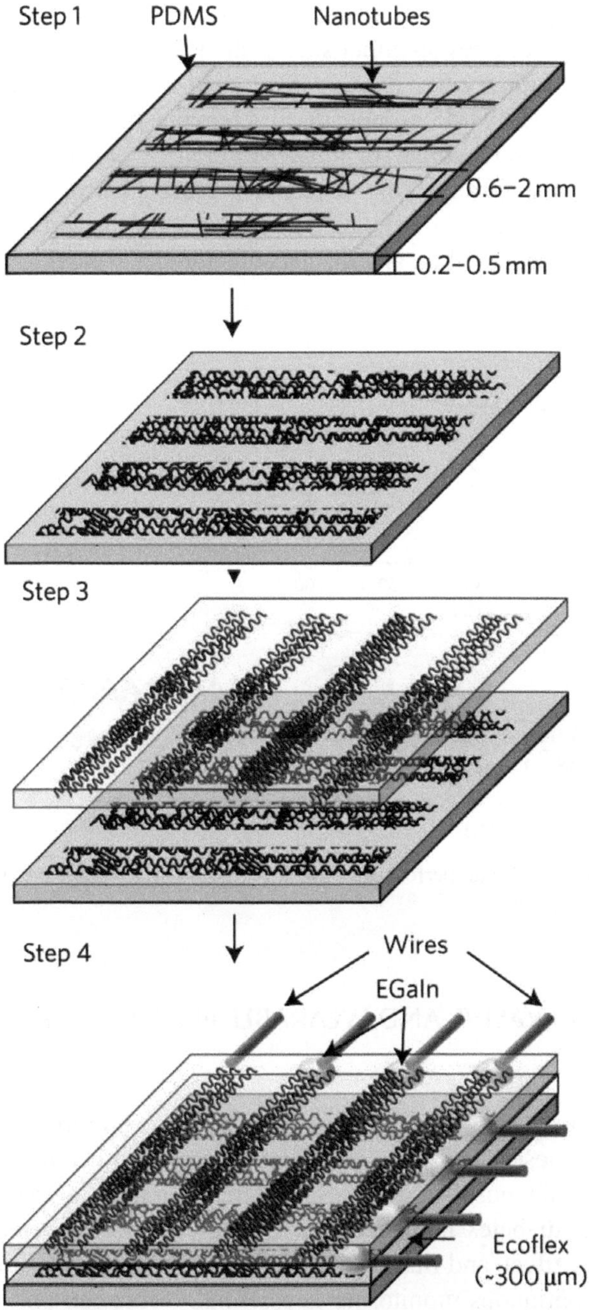

FIGURE 12.5 Fabrication steps of a capacitive strain sensor

will affect the pixels directly under the load.[8a] CNT thin films can also be transferred as electrodes on either side of the PDMS to form a parallel plate capacitor.[8c] In either case, relative resistance change with strain performance of the CNT/PDMS composite film remains the same (Figure 12-6a). Sensitivity and stretchability of the sensor can be enhanced by making a wave-like capacitor rather than a planar (Figure 12-6b). The wavy-structured CNT electrodes show much better electromechanical characteristics such as resistance, stability, and displacement under strain.[8b]

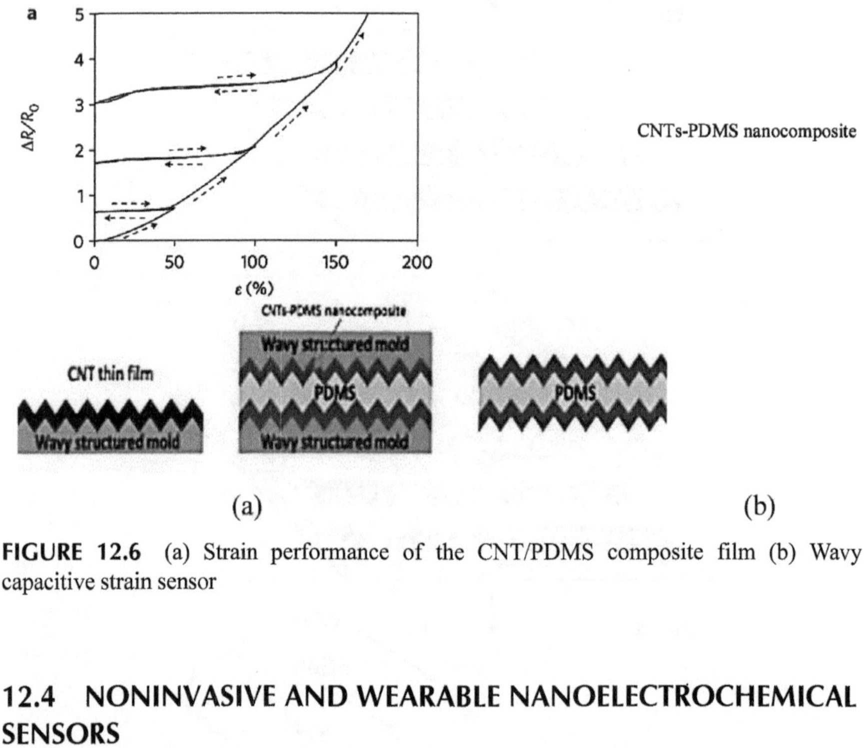

FIGURE 12.6 (a) Strain performance of the CNT/PDMS composite film (b) Wavy capacitive strain sensor

12.4 NONINVASIVE AND WEARABLE NANOELECTROCHEMICAL SENSORS

Continuous monitoring of biomarkers in our body is important as it is highly informative. These biomarkers can alert for possible health risks by early detection and diagnosis of diseases, and help maintain a healthy lifestyle. For example, diabetes management, real-time pathogen detection, monitoring of drug effect, and fitness level during physical activity are few examples where continuous monitoring is essential.[11] This can only be possible if the sensors are not a hurdle in a person's daily activity. Hence, for the past decade the global interest has been in the development of noninvasive wearable electrochemical biosensors. Unlike the wearable motion sensors which are for monitoring vital signs, the noninvasive electrochemical sensing is still in its infancy. The major challenges for these wearable electrochemical sensors are obtaining sensor response in low analyte concentration with small sample volume, biofouling, and biocompatibility. Section 12.4 gives an overview of noninvasive wearable electrochemical sensors based on the type of biofluid they monitor.

12.4.1 SALIVA-BASED SENSOR

Saliva consists of many elements from blood via transcellular or paracellular paths, hence a great alternative to blood.[11] It can indicate neurologic, metabolic, hormonal, and immunologic state of the body.[12] Unlike blood, saliva sampling is simple and has minimal risk of cross-contamination. The challenge in continuous monitoring of saliva is the direct interfacing of nanosensors on biomaterials to reduce health risks. One of the recent developments is a dental tattoo, which is graphene-printed electrode on water-soluble silk, for monitoring bacteria. As shown in Figure 12-7, first the graphene-based sensor and wireless readout coil are printed on water-soluble silk platform and then the sensor is brought in direct contact with the enamel. The sensing element of the biocompatible sensor can be used to detect various target analytes by self-assembly of biofunctional analytes on graphene electrodes. Upon binding of the specific targets to the sensor, the conductivity of the graphene film is modulated and monitored via the graphene readout coil using inductively coupled radio frequency reader device.[13] The sensor has not been implemented for real time on-body applications though.

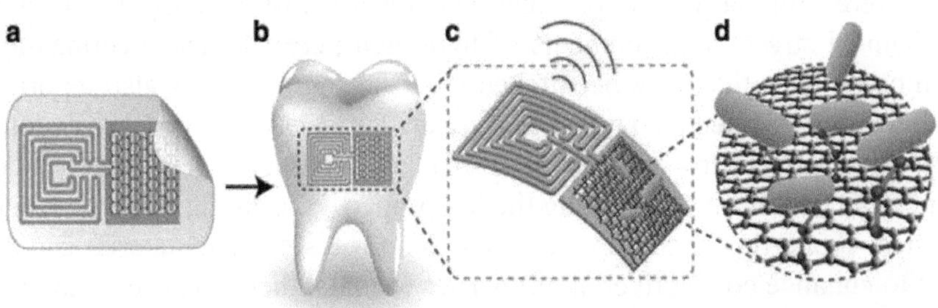

FIGURE 12.7 Schematic depicts (a) Fabrication of graphene printed electrode on water soluble silk (b) adhesion of sensor on tooth enamel (c) wireless transmission of data on (d) target immobilization

Another recent development for monitoring salivary metabolites is a wearable mouth guard which exhibits high selectivity, sensitivity, and stability in human saliva samples.[14] It demonstrated amperometric measurements of lactate via a three-electrode system. The electrodes are screen printed on a flexible polyethylene terephthalate (PET) substrate and attached inside of the mouth guard, where lactate oxidase is immobilized on working electrode by electropolymeric entrapment in poly(o-phynylenediamine) film. However, it is an ongoing effort towards continuous monitoring which will require miniaturization of and integration of amperometric circuit and wireless data transmission and acquisition.

12.4.2 SWEAT-BASED SENSOR

Continuous monitoring of sweat is desirable as it contains electrolytes, metabolites, amino acids, cortisol, proteins, and other numerous biomarkers which can give information on the physical status of the body. Current state-of-the-art wearable sweat analysis devices are either textile based or epidermal tattoo based. Sensors integrated in textile have gain a lot of attention as body sensing networks for remote monitoring physical status of patients, military personnel, and athletes. An ideal fabric should have good tensile strength, adhere well to the sensor, and not impact the electrochemical properties of the analyte. Researchers who conceptualized "smart nanotextiles" have highlighted nanomaterials integrated textile properties as summarized in Table 12-2.[15]

The integrated sensor needs to be flexible and robust to enduring the continuous bending of the fabric, body movement, and strain. Screen-printed amperometric carbon sensors on textile have proved quite robust as there were minimal effects on electrochemical measurements on repeated mechanical stress. Measurements of hydrogen peroxide and nicotinamide adenine dinucleotide have been taken with the sensor. It was also reported that quality of printed ink and physical characteristics of the textile substrate influenced the electrochemical performance.[16] In such a case the performance can be tailored by altering the ink properties, printing technique, and physical properties of the textile. Nanocomposites or nanomaterial can be used to enhance conductive ink properties to make them wear resistant and reduce brittleness.[17] Researchers from the University of Cincinnati demonstrated a completely packaged and functional bandage based radio-frequency identification (RFID) sensor for detecting sweat electrolytes (Figure 12-8),[18] It is a 4-layer device comprising of sweat porous adhesive, microfluidic

TABLE 12-2 Nanomaterials and their application to improve textile performance[15]

Nanomaterial	Textile properties
Conducting polymer	Electrical conductivity
(polypyrrole, polyaniline)	Piezo-sensitive
Carbon nanofibers	Increased tensile strength
	High chemical resistance
	Electrical conductivity
Carbon black nanoparticles	Improved abrasion resistance and toughness
	High chemical resistance
	Electrical conductivity
Carbon nanotubes	100 times the tensile strength of steel at one-sixth the weight
	Electrical conductivity similar to copper
	Good thermal conductivity
Metal–oxide nanoparticles (TiO_2, Al_2O_3, ZnO, MgO)	Photocatalytic ability
	Electrical conductivity
	Ultraviolet (UV) absorption
	Photo-oxidizing capacity against chemical and biological species
	Antimicrobial/self-sterilization
Clay nanoparticles	Electrical, heat, and chemical resistance
	Block UV light
	Flame retardant, anticorrosive
Generation of nanosize porosity in polymer matrix	Lightweight
	Good thermal insulation
	High cracking resistance and mechanical strength

paper pad, RFID/sensor, and vapor-only porous cover. Another bandage based potentiometric sensor for pH monitoring in wound was developed by a research group in the University of Geneva.[19] The sensor exhibited repeatability, reproducibility, and robustness under continuous bending stress. The same research group also developed CNT based yarn electrodes as potential electrochemical sensing textiles instead of printing conductive ink on fabric substrate.[20] The CNT-ink dyed cotton yarns were modified for potassium and ammonium ion selection and pH sensing in sweat. This process simplifies the earlier process as it removes the ink quality issue. Challenges of device integration for wireless transmission and data collection still remain.

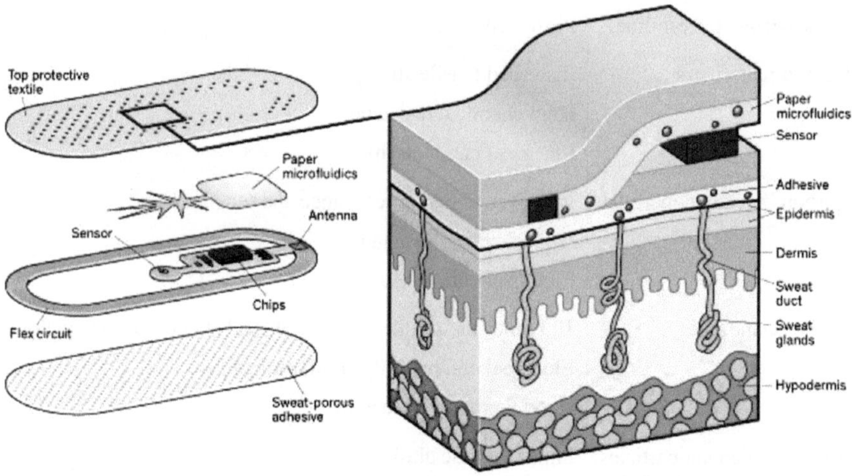

FIGURE12.8 **Bandage** based RFID sensor for detecting sweat electrolytes

The other class of sweat sensors is tattoo based which are in direct contact with the skin, hence, also known as epidermal-based sensors. They overcome the limitation of fabric sensors that cannot be in intimate contact with skin in all the regions on the body. The first tattoo sensor for sweat analysis was fabricated on temporary transfer tattoo base paper.[21] Electrode layers and insulator layer were screen printed on the tattoo paper in sequence, where the tattoo paper was cured after each sequence. The working electrode was then functionalized with CNT, tetrathiafulvalene, LOX enzyme, and chitosan for lactate detection. The completed tattoo sensor performed well when tested for mechanical robustness by applying it to flexible GORE-TEX textile, adherence to skin, and finally tested on neck under different strains. Further, the tattoo was used for lactate concentration in sweat via handheld electrochemical analyzer. The same research group accomplished pH monitoring using the tattoo-based sensor[22] and demonstrated wireless signal transduction for continuous monitoring of sodium in sweat.[23] Ammonium sensing has also been demonstrated in sweat using similar fabrication methods (Figure 12-9) where the sensor could differentiate between aerobic and anaerobic state of the person.[24]

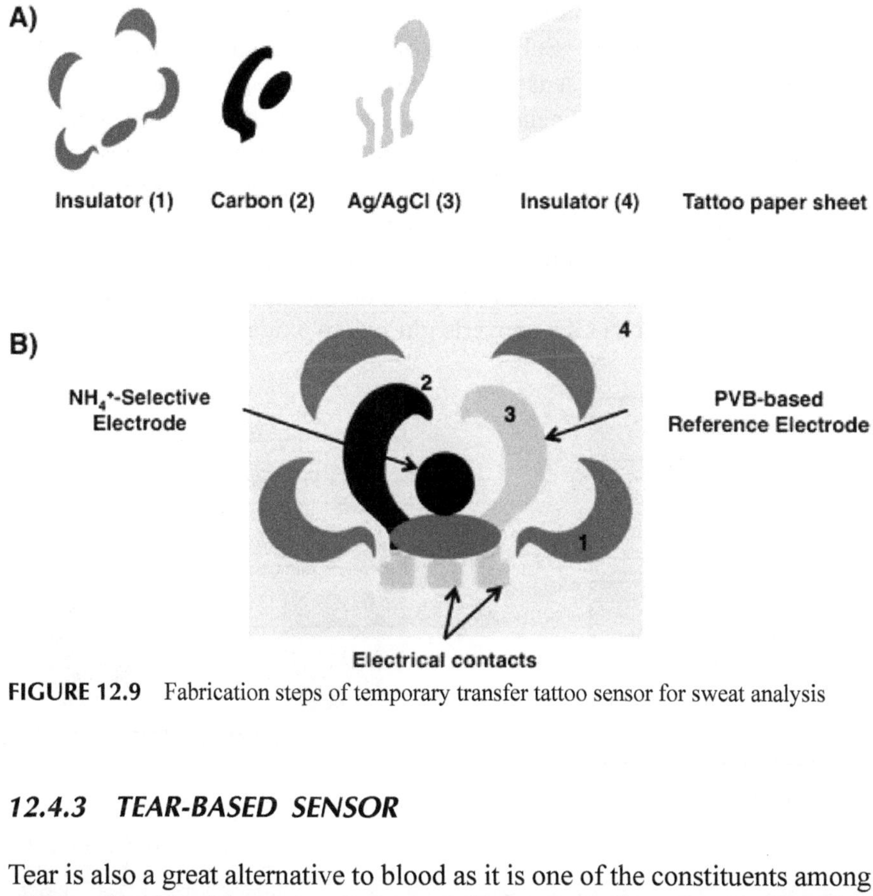

FIGURE 12.9 Fabrication steps of temporary transfer tattoo sensor for sweat analysis

12.4.3 TEAR-BASED SENSOR

Tear is also a great alternative to blood as it is one of the constituents among others. Considerable amount of research has gone into glucose detection in tears and establishing a correlation between glucose in tear and blood.[12] Google is soon to commercialize a contact lens embedded with electrochemical sensor for continuous monitoring of glucose. This amperometric contact lens sensor was developed by researchers in the University of Washington for glucose[25] and lactate detection.[26] The fabrication process involves microfabrication of Ti/Pd/Pt electrodes on flexible PET polymer, cut to 1 cm diameter and heat molded into contact lens (Figure 12-10).[25] The sensor is highly sensitive, repeatable, and responds fast. Further enhancement of the sensor was done by integrating low power sensor interface read out integrated circuit and a loop antenna for power and data transfer using RF power sent from an interrogator.[27] The contact lens sensor was tested for glucose detection on an artificial PDMS eye model. In an effort to control the electrochemical interference from other chemical species, two sensors were fabricated where one had Glucose oxidase (GO_x) and other did not act as the control sensor.[28] Subtraction of control sensor's signal from that of the GO_x-immobilized sensor gives the corrected response. Composite nanostructure of platinum nanoparticle attached on the edges of graphene nanosheets has been reported to enhance the otherwise low glucose signal level in saliva and tear when compared to blood.[29] The nanobiosensor fabrication is simple due to the fact that platinum nanoparticle readily attaches to the dangling, incomplete bonds at the edge of the graphene nanosheet. Then, the GO_x enzyme immobilizes on the platinum nanoparticle and the sensor generates a signal when GO_x converts glucose to oxidase.

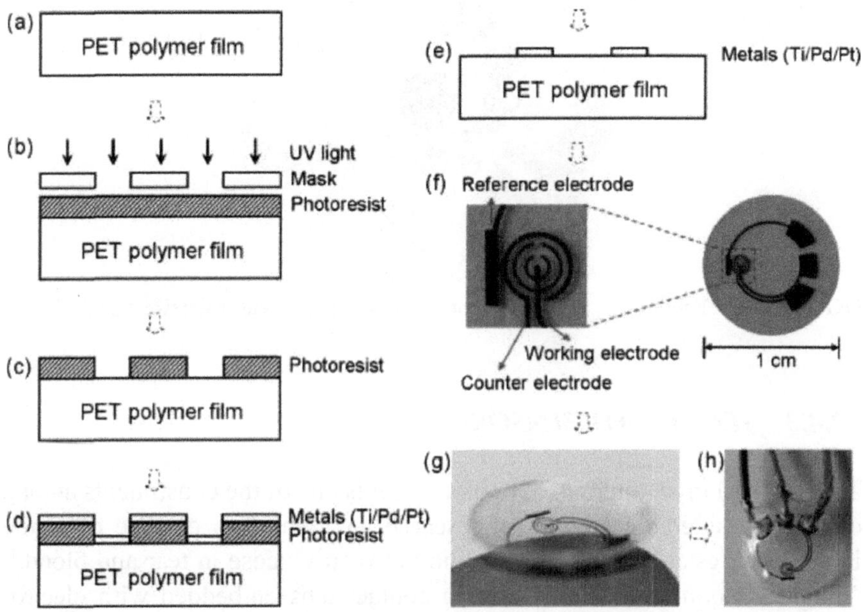

FIGURE 12.10 Fabrication steps of PET contact lens for glucose monitoring in tears

12.5 MAJOR LIMITATION—ENERGY HARVESTING

The existing wearable sensors have typical energy consumption of around 10 μW[30] on an average daily. Powering such consumption with modern batteries becomes a tricky affair in spite of huge leaps in the battery technology. When a wearable application is envisioned, it is best to think of self-powered, self-sustaining system. Batteries are exhaustible and need to be replaced or recharged. They also add bulk to the wearable form that can be uncomfortable to the person wearing the sensor. Apart from looking to reduce the power consumption of the sensor itself, researchers are looking to mitigate the problem of continuously powering a wearable sensor; therefore, various techniques such as wearable solar cells-battery combination, fuel cells, thermoelectric generators, micro/nano kinetic generators, and so on are explored. However, each one of them is beset with its own advantages and problems. In Section 12.5.1, we highlight energy harvesting technique and principles employed for continuously powering wearable sensors.

12.5.1 SOLAR CELLS

The first option that seems to come to mind when one thinks of continuous and free power is a solar cell. It would be a great option to power a wearable device; however, since one does not get continuous supply of light all through the day, its applications to wearable devices are limited to daytime use only or to produce enough power so that it is saved for later use. Also, the technology of stretchable solar cells is not yet developed to allow complete integration with the device. Dye sensitized solar cells and nanomaterials based organic solar cells are good alternatives to crystalline solar cell as they can be fabricated on a flexible substrate. However, they have lower efficiency as compared to crystalline solar cells.

12.5.2 FUEL CELLS

Fuel cells are a very attractive option but miniaturization of fuel cells for wearable options is still in its infancy. Operation of fuel cells requires fuel in liquid or gaseous form which implies the need to carry a miniaturized fuel container. Recently, glucose has been proposed to power fuel cells as it has a good energy and is readily available in living tissues. However, having a fuel cell inside one's body is not a desirabe option unless it is absolutely essential as in case of a pacemaker or other implantable device.[31]

12.5.3 WIRELESS POWER TRANSMITTER

Use of wireless energy transmitters and receivers has dramatically increased with available wireless charging stations for wearable smart watches and smartphones and also the current interest for wearable sensors. Though their footprint has been considerably reduced since the same circuit can be used to transmit and receive data as well as power, the need to be in proximity of a power transmitter is still there.

12.5.4 KINETIC GENERATORS

There has been considerable focus on the research into devices which harvest the energy out of kinetic motion of the body and the heat produced by the body itself. Though the power levels harvested seem tiny, these devices present a great opportunity for wearable power generation. Most of the techniques that involve kinetic generation have an electromagnetic or a piezoelectric approach. A good example of a motion energy generator can be found in the perpetual watches of the yesteryears. The kinetic watch generator had a small flywheel that could produce enough electricity to power the watch.[32] However, the only disadvantage was that the watch should be in continuous motion to charge the batteries. Kinetic generators can be classified into two categories: electromagnetic and electrostatic.

12.5.4.1 ELECTROMAGNETIC GENERATOR

The basic operating principle behind electromagnetic harvesting consists of a vibrating proof mass between thc poles of two permanent magnets. The vibrations of the mass between the poles of the magnet produce electricity that can be used to power a nanoenabled biosensor device. The upper limit of power (P_{max}) from such a device for a human worn application has been shown[33] as

$$P_{max} = 2\frac{Z_l Y_0 m\omega^3}{\pi} \tag{12.4}$$

where Z_1 is the amplitude for the mass movement, Y_0 is the amplitude of the driving motion, m is the mass of the vibrating element, and ω is the angular frequency. It has been reported that a microgenerator using such a method can produce up to 10 μW of electricity[33].

12.5.4.2 ELECTROSTATIC GENERATOR

Electrostatic generators consist of two parallel plates of a capacitor which vibrate with respect to each other. The increase in the separation of the plates of a charged capacitor causes an increase in the voltage and thus generates electrical energy. Such a system has been successfully miniaturized down to the microlevel. However, electrostatic generators fare poorly at low frequencies and large sizes. Hence, it is difficult to work with in order to produce substantial energy.

12.5.5 PIEZOELECTRIC GENERATOR

The perfect nanoscale electricity generators have been achieved by using zinc oxide nanowires. ZnO is an intrinsic piezoelectric material and has been used to make piezoelectric generators that could produce up to 60 pA of currents when stretched.[34] Such generators can be used in a patch form on the skin to produce electricity while the skin stretches due to muscle movement. They can be manufactured in bulk using the conventional micro/nanofabrication techniques, offer high output voltages, and need no external power source to function. However, the need to have high optimal load impedance coupled with fatigue effect limits the use of such generators.

12.5.6 THERMOELECTRIC GENERATOR

Another type of nanogenerator is thermoelectric generator (TEG). These types of devices utilize the difference in temperature between two surfaces to produce electricity (Peltier effect). Since the human body is at a constant temperature which is usually higher than that of the surrounding (ambient) temperature, it is a plausible way of getting a constant supply of heat to power the TEG. A lot of effort has been put to make a truly wearable TEG which can produce continuous electricity. However, the amount of electricity produced is not constant. This is due to the fact that the ambient temperature is not constant and affects the temperature gradient of the TEG device. Also, the placement of the device on the body is a critical issue. The heat flow is more at the head and extremities. However, the skin temperature varies at those points. The skin temperature on the trunk is pretty constant; however, placement of the TEG on the torso is not a practical option due to extremely low heat flows. Local heat flow is variable at different places on the body. Nevertheless, TEGs for wearable applications have shown to have efficiency values that reach ~70%, whereas motion harvesting devices have an efficiency nearby 1%.[35] Several thermoelectric generator designs have been explored which utilize the properties of nanomaterials to efficiently produce electricity from body heat. Some of the nanotechnology-based TEGs are discussed in Table 12-3. Table 12-3 details the latest in nanoenergy harvesting scenario.

TABLE 12-3 Types of wearable nanoenergy harvesters

Type of harvester	Materials used	Power range	Advantages	References
Solar energy harvester	Die-sensitized solar cells incorporating TiO_2 nanoparticles	Energy efficiency is an issue.	Bendable panel for wearable devices; Lifetime	Ref. [36]
Wearable solar sells by stacking textile electrodes	Dye-sensitized solar cells by stacking TiO_2 nanotubes and CNT fabrics—flexible, lightweight, energy efficiency 3.7%	Low energy efficiency 3.7%, limited lifetime	Flexible, lightweight, wearable	Ref. [37]
Biofuel generator/fuel cell	Scalable, good efficiencies, enzymatic redox of organic molecules like glucose, etc. Continuous operation using blood glucose, etc.	4.3–7.9 microwatts maximum output	Better suited for implantable devices	Ref. [38]
Biofuel cell	3D graphene network as substrate	~112 microwatts and 0.96 mAmps	Flexible	Ref. [39]
Nanowire piezoelectric nanogenerators	Flexible, good power efficiencies as compared to solar cells			Ref. [34a]
Nanorods based lead-zirconate-titanate (PZT) electrical triggers	ZnO nanorod array made into an electrical trigger	N/A	–	Ref. [34b]
Hybrid- pyro-piezoelectric	Al, PTFE, Cu composite	146.2 mW m^{-2} with sliding frequency of 4.41 Hz	Self-powered cell and sensor for continuous measurements	Ref. [40]

TABLE 12-3 *(Continued)*

Type of harvester	Materials used	Power range	Advantages	References
Nanomechanical generator	Vertically aligned ZnO nanowires	11 mW/cm^{-3}	Flexible, bendable substrate, high power output	Ref. [41]
PZT based nanomechanical generator	Electrospun PZT nanofibers on interdigitated electrode	0.03 microwatts	Easy fabrication techniques, longer lifetime	Ref. [42]

12.6 FUTURE OUTLOOK

This chapter addressed the recent developments in wearable nanobiosensors for monitoring health and fitness status. In particular, the working principles and challenges of strain based motion sensors and electrochemical sensors were discussed. Energy harvesting being the biggest bottle neck for wearable nanosensors was described. A comprehensive view of nanoenabled energy scavenging alternatives was provided. To summarize, the other important considerations for a wearable sensor which will undergo various deformations such as stretching, bending, twisting, and pressure are resiliency, mechanical robustness, fast response, wear resistance, and body interface. With the dramatically growing interdisciplinary understanding and advances in material science, textile engineering, and nanoengineering there has been a lot of progress in the nanowearables in the past two years. Complete wearable health monitoring is a step closer to reality of being able to monitor fitness, adapt to healthier lifestyle accordingly, and provide accurate statistics to the doctor for faster analysis and treatment. Other challenges which are of serious consideration are data management and setting sensor standards. These challenges will grow with the growing wearable sensor market for continuous monitoring. These topics are beyond the scope of this chapter.

KEYWORDS

- **Wearable technology**
- **Graphene**
- **Carbon nanotube**
- **Energy Harvesting**
- **Solar cells**
- **Kinetic generators**

REFERENCES

1. Atallah, L.; Lo, B.; King, R.; Guang-Zhong, Y. In *Sensor Placement for Activity Detection Using Wearable Accelerometers*, Body Sensor Networks (BSN), 2010 International Conference on, 7–9 June 2010; 2010; pp 24–29.
2. Wearable Technology Database. vandrico.com.
3. Godfrey, A.; Conway, R.; Meagher, D.; ÓLaighin, G. Direct Measurement of Human Movement by Accelerometry. *Med. Eng. Phys.* **2008,** *30* (10), 1364–1386.
4. Future biosensing wearables. rockhealth.com.
5. Boland, C. S.; Khan, U.; Backes, C.; O'Neill, A.; McCauley, J.; Duane, S.; Shanker, R.; Liu, Y.; Jurewicz, I.; Dalton, A. B.; Coleman, J. N. Sensitive, High-Strain, High-Rate Bodily Motion Sensors Based on Graphene–Rubber Composites. *ACS Nano.* **2014,** *8* (9), 8819–8830.
6. (a) Yao, S.; Zhu, Y. Wearable Multifunctional Sensors Using Printed Stretchable Conductors Made of Silver Nanowires. *Nanoscale* **2014,** *6* (4), 2345–2352; (b) Gong, S.; Schwalb, W.; Wang, Y.; Chen, Y.; Tang, Y.; Si, J.; Shirinzadeh, B.; Cheng, W. A Wearable and Highly Sensitive Pressure Sensor with Ultrathin Gold Nanowires. *Nat. Commun.* **2014,** *5*.
7. (a) Li, X.; Zhang, R.; Yu, W.; Wang, K.; Wei, J.; Wu, D.; Cao, A.; Li, Z.; Cheng, Y.; Zheng, Q.; Ruoff, R. S.; Zhu, H. Stretchable and Highly Sensitive Graphene-on-Polymer Strain Sensors. *Sci. Rep.* **2012,** *2*; (b) Wang, Y.; Wang, L.; Yang, T.; Li, X.; Zang, X.; Zhu, M.; Wang, K.; Wu, D.; Zhu, H. Wearable and Highly Sensitive Graphene Strain Sensors for Human Motion Monitoring. *Adv. Funct. Mater.* **2014,** *24* (29), 4666–4670.
8. (a) Lipomi, D. J.; Vosgueritchian, M.; Tee, B. C. K.; Hellstrom, S. L.; Lee, J. A.; Fox, C. H.; Bao, Z. Skin-like Pressure and Strain Sensors Based on Transparent Elastic Films of Carbon Nanotubes. *Nat. Nano.* **2011,** *6* (12), 788–792; (b) Amjadi, M.; Inkyu, P. In *Sensitive and Stable Strain Sensors Based on the Wavy Structured Electrodes*, Nanotechnology (IEEE-NANO), 2014 IEEE 14th International Conference on, 18–21 Aug 2014; 2014; pp 760–763; (c) Cai, L.; Song, L.; Luan, P.; Zhang, Q.; Zhang, N.; Gao, Q.; Zhao, D.; Zhang, X.; Tu, M.; Yang, F.; Zhou, W.; Fan, Q.; Luo, J.; Zhou, W.; Ajayan, P. M.; Xie, S. Super-stretchable, Transparent Carbon Nanotube-Based Capacitive Strain Sensors for Human Motion Detection. *Sci. Rep.* **2013,** *3*.

9. (a) Chun, J.; Kang, N.-R.; Kim, J.-Y.; Noh, M.-S.; Kang, C.-Y.; Choi, D.; Kim, S.-W.; Lin Wang, Z.; Min Baik, J. Highly Anisotropic Power Generation in Piezoelectric Hemispheres Composed Stretchable Composite Film for Self-Powered Motion Sensor. *Nano Energy* **2015,** *11* (0), 1–10; (b) Pang, C.; Lee, G.-Y.; Kim, T.-i.; Kim, S. M.; Kim, H. N.; Ahn, S.-H.; Suh, K.-Y. A Flexible and Highly Sensitive Strain-Gauge Sensor Using Reversible Interlocking of Nanofibres. *Nat. Mater.* **2012,** *11* (9), 795–801.
10. Yamada, T.; Hayamizu, Y.; Yamamoto, Y.; Yomogida, Y.; Izadi-Najafabadi, A.; Futaba, D. N.; Hata, K. A Stretchable Carbon Nanotube Strain Sensor for Human-Motion Detection. *Nat. Nano.* **2011,** *6* (5), 296–301.
11. Bandodkar, A. J.; Wang, J. Non-Invasive Wearable Electrochemical Sensors: A Review. *Trends Biotechnol.* **2014,** *32* (7), 363–371.
12. Makaram, P.; Owens, D.; Aceros, J. Trends in Nanomaterial-Based Non-Invasive Diabetes Sensing Technologies. *Diagnostics* **2014,** *4* (2), 27–46.
13. Mannoor, M. S.; Tao, H.; Clayton, J. D.; Sengupta, A.; Kaplan, D. L.; Naik, R. R.; Verma, N.; Omenetto, F. G.; McAlpine, M. C. Graphene-Based Wireless Bacteria Detection on Tooth Enamel. *Nat. Commun.* **2012,** *3*, 763.
14. Kim, J.; Valdes-Ramirez, G.; Bandodkar, A. J.; Jia, W.; Martinez, A. G.; Ramirez, J.; Mercier, P.; Wang, J. Non-Invasive Mouthguard Biosensor for Continuous Salivary Monitoring Of Metabolites. *Analyst* **2014,** *139* (7), 1632–1636.
15. (a) Coyle, S.; Wu, Y.; Lau, K.-T.; De Rossi, D.; Wallace, G.; Diamond, D. Smart Nanotextiles: A Review of Materials and Applications. *MRS Bull.* **2007,** *32* (05), 434–442; (b) Qian, L., Hinestroza, Juan P. Application of Nanotechnology for High Performance Textiles. *J. Text. Apparel, Technol. Manage.* **2004,** *4* (1), 5.
16. Yang, C. C.; Hsu, Y. L. A Review of Accelerometry-Based Wearable Motion Detectors for Physical Activity Monitoring. *Sensors* **2010,** *10* (8), 7772–7788.
17. (a) Stoppa, M.; Chiolerio, A. Wearable Electronics and Smart Textiles: A Critical Review. *Sensors* **2014,** *14* (7), 11957–11992; (b) Shyamkumar, P.; Rai, P.; Oh, S.; Ramasamy, M.; Harbaugh, R.; Varadan, V. Wearable Wireless Cardiovascular Monitoring Using Textile-Based Nanosensor and Nanomaterial Systems. *Electronics* **2014,** *3* (3), 504–520.
18. Rose, D. P.; Ratterman, M.; Griffin, D. K.; Hou, L.; Kelley-Loughnane, N.; Naik, R. R.; Hagen, J. A.; Papautsky, I.; Heikenfeld, J. Adhesive RFID Sensor Patch for Monitoring of Sweat Electrolytes. *Biomed. Eng., IEEE Trans. on* **2014,** *PP* (99), 1–1.
19. Guinovart, T.; Valdés-Ramírez, G.; Windmiller, J. R.; Andrade, F. J.; Wang, J. Bandage-Based Wearable Potentiometric Sensor for Monitoring Wound pH. *Electroanalysis* **2014,** *26* (6), 1345–1353.
20. Guinovart, T.; Parrilla, M.; Crespo, G. A.; Rius, F. X.; Andrade, F. J. Potentiometric Sensors Using Cotton Yarns, Carbon Nanotubes and Polymeric Membranes. *Analyst* **2013,** *138* (18), 5208–5215.
21. Jia, W.; Bandodkar, A. J.; Valdés-Ramírez, G.; Windmiller, J. R.; Yang, Z.; Ramírez, J.; Chan, G.; Wang, J. Electrochemical Tattoo Biosensors for Real-Time Noninvasive Lactate Monitoring in Human Perspiration. *Anal. Chem.* **2013,** *85* (14), 6553–6560.
22. Bandodkar, A. J.; Hung, V. W. S.; Jia, W.; Valdes-Ramirez, G.; Windmiller, J. R.; Martinez, A. G.; Ramirez, J.; Chan, G.; Kerman, K.; Wang, J. Tattoo-Based Potentiometric Ion-Selective Sensors for Epidermal pH Monitoring. *Analyst* **2013,** *138* (1), 123–128.
23. Bandodkar, A. J.; Molinnus, D.; Mirza, O.; Guinovart, T.; Windmiller, J. R.; Valdés-Ramírez, G.; Andrade, F. J.; Schöning, M. J.; Wang, J. Epidermal Tattoo Potentiometric

Sodium Sensors with Wireless Signal Transduction for Continuous Non-Invasive Sweat Monitoring. *Biosens. Bioelectron.* **2014,** *54* (0), 603–609.
24. Guinovart, T.; Bandodkar, A. J.; Windmiller, J. R.; Andrade, F. J.; Wang, J. A Potentiometric Tattoo Sensor for Monitoring Ammonium in Sweat. *Analyst* **2013,** *138* (22), 7031–7038.
25. Yao, H.; Shum, A. J.; Cowan, M.; Lähdesmäki, I.; Parviz, B. A. A Contact Lens with Embedded Sensor for Monitoring Tear Glucose Level. *Biosens. Bioelectron.* **2011,** *26* (7), 3290–3296.
26. Thomas, N.; Lähdesmäki, I.; Parviz, B. A. A Contact Lens with an Integrated Lactate Sensor. *Sens. Actuators, B: Chem.* **2012,** *162* (1), 128–134.
27. Yu-Te, L., Huanfen, Yao, Lingley, A., Parviz, B., Otis, B. P. A 3 microWatt CMOS Glucose Sensor for Wireless Contact-Lens Tear Glucose Monitoring. *EEE J. Solid-State Circuits* **2012,** *47* (1), 335–344.
28. Yao, H.; Liao, Y.; Lingley, A. R.; Afanasiev, A.; Lähdesmäki, I.; Otis, B. P.; Parviz, B. A. A Contact Lens with Integrated Telecommunication Circuit and Sensors for Wireless and Continuous Tear Glucose Monitoring. *J. Micromech. Microeng.* **2012,** *22* (7), 075007.
29. Claussen, J. C.; Kumar, A.; Jaroch, D. B.; Khawaja, M. H.; Hibbard, A. B.; Porterfield, D. M.; Fisher, T. S. Nanostructuring Platinum Nanoparticles on Multilayered Graphene Petal Nanosheets for Electrochemical Biosensing. *Adv. Funct. Mater.* **2012,** *22* (16), 3399–3405.
30. Mitcheson, P. D. Energy Harvesting for Human Wearable and Implantable Bio-Sensors. *Conf. Proc. IEEE Eng. Med. Biol. Soc.* **2010,** *6* (10), 5627952.
31. MacVittie, K.; Halamek, J.; Halamkova, L.; Southcott, M.; Jemison, W. D.; Lobel, R.; Katz, E. From "Cyborg" Lobsters to a Pacemaker Powered by Implantable Biofuel Cells. *Energy Environ. Sci.* **2013,** *6* (1), 81–86.
32. Kinetron. Micro Generator Technology. http://www.kinetron.eu/micro-generator-technology.
33. Mitcheson, P. D.; Yeatman, E. M.; Rao, G. K.; Holmes, A. S.; Green, T. C. Energy Harvcsting From Human and Machine Motion for Wireless Electronic Devices. *Proc. IEEE* **2008,** *96* (9), 1457–1486.
34. (a) Gao, P. X.; Song, J.; Liu, J.; Wang, Z. L. Nanowire Piezoelectric Nanogenerators on Plastic Substrates as Flexible Power Sources for Nanodevices. *Adv. Mater.* **2007,** *19* (1), 67–72; (b) Zhou, J.; Fei, P.; Gao, Y.; Gu, Y.; Liu, J.; Bao, G.; Wang, Z. L. Mechanical–Electrical Triggers and Sensors Using Piezoelectric Micowires/Nanowires. *Nano Lett.* **2008,** *8* (9), 2725–2730; (c) Qi, Y.; Kim, J.; Nguyen, T. D.; Lisko, B.; Purohit, P. K.; McAlpine, M. C. Enhanced Piezoelectricity and Stretchability in Energy Harvesting Devices Fabricated from Buckled PZT Ribbons. *Nano Lett.* **2011,** *11* (3), 1331–1336.
35. Bonfiglio, A., De Rossi, D. *Wearable Monitoring Systems*. 1st edn; Springer: New York, 2011; p. 296.
36. (a) Chen, X.; Mao, S. S. Titanium Dioxide Nanomaterials: Synthesis, Properties, Modifications, and Applications. *Chem. Rev.* **2007,** *107* (7), 2891–2959; (b) Benkstein, K. D.; Kopidakis, N.; van de Lagemaat, J.; Frank, A. J. Influence of the Percolation Network Geometry on Electron Transport in Dye-Sensitized Titanium Dioxide Solar Cells. *The J. Phys. Chem. B* **2003,** *107* (31), 7759–7767; (c) O'Regan, B.; Gratzel, M. A Low-Cost, High-Efficiency Solar Cell Based on Dye-Sensitized Colloidal TiO2 Films. *Nature* **1991,** *353* (6346), 737–740; (d) Grätzel, M. Dye-Sensitized Solar Cells. *J. Photochem. Photobiol., C: Photochem. Rev.* **2003,** *4* (2), 145–153; (e) Law, M.;

Greene, L. E.; Johnson, J. C.; Saykally, R.; Yang, P. Nanowire Dye-Sensitized Solar Cells. *Nat. Mater.* **2005,** *4* (6), 455–459; (f) Li, L.; Zhai, T.; Bando, Y.; Golberg, D. Recent Progress of One-Dimensional ZnO Nanostructured Solar Cells. *Nano Energy* **2012,** *1* (1), 91–106.

37. Pan, S.; Yang, Z.; Chen, P.; Deng, J.; Li, H.; Peng, H. Wearable Solar Cells by Stacking Textile Electrodes. *Angew. Chem. Int. Ed.* **2014,** *53* (24), 6110–6114.
38. Wu, X. E.; Guo, Y. Z.; Chen, M. Y.; Chen, X. D. Fabrication of Flexible and Disposable Enzymatic Biofuel Cells. *Electrochim. Acta* **2013,** *98* (0), 20–24.
39. Zhang, Y.; Chu, M.; Yang, L.; Tan, Y.; Deng, W.; Ma, M.; Su, X.; Xie, Q. Three-Dimensional Graphene Networks as a New Substrate for Immobilization of Laccase and Dopamine and Its Application in Glucose/O2 Biofuel Cell. *ACS Appl. Mater. Interfaces* **2014,** *6* (15), 12808–12814.
40. Zi, Y.; Lin, L.; Wang, J.; Wang, S.; Chen, J.; Fan, X.; Yang, P.-K.; Yi, F.; Wang, Z. L. Triboelectric–Pyroelectric–Piezoelectric Hybrid Cell for High-Efficiency Energy-Harvesting and Self-Powered Sensing. *Adv. Mater.* **2015,** *27* (14), 2340–2347.
41. Zhu, G.; Yang, R.; Wang, S.; Wang, Z. L. Flexible High-Output Nanogenerator Based on Lateral ZnO Nanowire Array. *Nano Lett.* **2010,** *10* (8), 3151–3155.
42. Chen, X.; Xu, S.; Yao, N.; Shi, Y. 1.6 V Nanogenerator for Mechanical Energy Harvesting Using PZT Nanofibers. *Nano Lett.* **2010,** *10* (6), 2133–2137.

CHAPTER 13

CHALLENGES AND FUTURE PROSPECTS OF NANOENABLING SENSING TECHNOLOGY

CHANDRA K. DIXIT[1,*] and AJEET KAUSHIK[2]

[1]*Department of Chemistry, University of Connecticut, Storrs, CT, USA*

[2]*Centre for Personalized Medicine, Institute of NeuroImmuno Pharmacology, Department of Immunology, Herbert Wertheim College of Medicine, Florida International University, Miami, FL, USA*

**E-mail: chandra.kumar_dixit@uconn.edu*

CONTENTS

Nanosensors are a potential candidate for being inducted as a technology of future; it is visible with the "nano" tagging of products by all major players in the field of analytics and diagnostics. However, in our opinion there is a long road ahead to achieve this goal. There are several challenges that hinder the utility of nanotechnology and nanosensing platforms in our routine. In this chapter, we summarize all the potential drawbacks associated with the use of this technology and how future will be if we find a potential solution to these bottlenecks.

13.1 CHALLENGES

Miniaturization of analytical tools and diagnostics systems in health-care industry comes at the price of background capital invested in research and the amount of it that reaches to the end users. Research involving miniaturized nanosensing platforms, such as wearable technology (described in chapter 8), nanobiosensors for disease diagnosis (Chapter 10 and 12), and in-vivo applications (Chapter 11), involve tremendous amount of time and labor.

In principle, money, manpower, market, and mindset of the general public towards nanobased alternatives govern the future of nanobiosensors. According to a 2014 report released by Research And Markets (the largest market research institution) the market of medical devices based on nanotechnology was around5 billion US $ in 2014 and is bound to increase to $8.5 billion by 2019.[1] Rise in the aging population, and huge capital involved in research and development currently are the main driving factors for this market. Attributing to these factors, the overall cost of nanoenabled platforms in health care has increased drastically. Further, foreign direct investment (FDI) approvals add tremendous financial burden on the already suffering field. Therefore, besides commercial availability of the nanoparticles for research purposes, there are only few FDI-approved products for diagnostics and therapeutic use. In this regard, nanobased implants hold the largest share of the market in health-care sector.

In addition, there is an ongoing debate regarding the stability of nanoenabled sensors and their toxicity.[2,3] For example, most of the polymer based drug delivery vehicles have monomeric units with high cytotoxicity.[3] This restricts the free use of such delivery tools for diseases and disorders that have safer alternatives in terms of controlled release of pill or capsule formats. There are more than 2000 publications reporting the development of novel nanodelivery vehicles but realization of this technology suffers severely.

We must address these drawbacks as soon as possible before nanobiosensors lose their share in the market to the conventional diagnostics tools and upcoming but very promising field of microfluidics.[1]

13.2 FUTURE PROSPECTS

Apart from the bottleneck mentioned in the Section 13.2, this technology has the potential to change the ways of conventional diagnostics and therapeutics.

In our opinion, merger of nanotechnology with biotechnology is revolutionizing several domains with explicit real-world applications, such as bio-based supercapacitors, bioelectronics, and so on[4–6]; however, in the field of diagnostics and therapeutics, involvement of human subjects complicates the whole scenario, which causes significant challenge in delivering this technology to the end users.[7,8] In future, this must be addressed either by developing novel material with high biocompatibility[9] or by incorporating additional features that can outsmart the current state-of-the-art.

Addressing scientific challenges associated with the development and stability processes should also be sufficed with an outreach program where clinicians and diagnostic laboratories must be made aware of the benefits of this technology.[10] For example, gold nanoparticle-mediated tumor therapy[11] has demonstrated the potential to eradicate cancer from a patient's body and clinical trials are ongoing but a big population of doctors and patients are least informed about the existence of such a technology.

On a concluding remark, nanobiosensing technology, if not suitable as a stand-alone solution, can be incorporated with other existing tools and techniques, such as microfluidic systems, to address the drawbacks that a specific nanobiosensor has.

KEYWORDS

- **Nanosensors**
- **Nanoenabled**
- **Nanotechnology**
- **Biotechnology**
- **Microfluidic systems**

REFERENCES

1. Research & Market report: Nanotechnology in medical devices market by product, applications – Global forecast to 2019. http://www.researchandmarkets.com/research/kjgx54/nanotechnology_in.
2. Xue, H. Y.; Liu S.; Wong H. L. Nanotoxicity: A Key Obstacle to Clinical Translation of siRNA-Based Nanomedicine. *Nanomedicine(Lond)* **2014**, *9* (2), 295–312.
3. Marin, E.; Briceno, M. I.; Caballero-George, C. Critical Evaluation of Biodegradable Polymers Used in Nanodrugs. *Int. J. Nanomed.* **2013**, *8*, 3071–3091.
4. Chen, T.; Dai, L. Carbon Nanomaterials for High-Performance Supercapacitors. *Materialstoday* **2013**, *7–8*, 272–280.
5. Deutscher, G. Superconductivity: The Imaginary is Real. *Nat. Nanotechnol.* **2013**, *8*, 10–11.
6. Kim, D-H.; Lee, Y. Bioelectronics: Injection and Unfolding. *Nat. Nanotechnol.* **2013**. (Advance article: doi:10.1038/nnano.2015.129.)
7. Davenport, M. Closing the Gap for Generic Nanomedicines. *Chem. Eng. News.* **2014**, *92* (45), 10–13.
8. Pillai, G. Nanomedicines for Cancer Therapy: An Update of FDA Approved and Those Under Various Stages of Development. *SOJ Pharm. Pharm. Sci.* **2014**, *1* (2), 13.
9. Kapoor, D. N.; Dhawan, S. *Biocompatible Nanomaterials for Targeted and Controlled Delivery of Biomacromolecules*; ASME Press: Park Avenue, New York, 2013.
10. Accelerating commercialization of nanobiotechnology. Centre of Innovation for Nanobiotechnology. https://www.ncbiotech.org/sites/default/files/COIN%20Flyer%20-%20Revised%20December%202010.pdf.
11. Kodiha, M.; Wang, Y. M.; Hutter, E.; Maysinger, D.; Stochaj, U. Off the Organelles—Killing Cancer Cells with Targeted Gold Nanoparticles. *Theranostics* **2015**, *5* (4), 357–370.

INDEX

D

N

Y

Z

* 9 7 8 1 7 7 1 8 8 3 2 8 3 *
#0002 - 030816 - C0 - 234/156/21 [23] - CB - 9781771883283